Cyclodextrins

Cyclodextrins

Properties and Industrial Applications

Sahar Amiri
Department of Polymer Engineering
Islamic Azad University, Tehran, Iran

Sanam Amiri
Department of Textile Engineering
Amirkabir University of Technology
Tehran, Iran

This edition first published 2017
© 2017 John Wiley & Sons Ltd

All rights reserved. No part of this publication may be reproduced, stored in a retrieval system, or transmitted, in any form or by any means, electronic, mechanical, photocopying, recording or otherwise, except as permitted by law. Advice on how to obtain permission to reuse material from this title is available at http://www.wiley.com/go/permissions.

The right of Sahar Amiri and Sanam Amiri to be identified as the authors of this work has been asserted in accordance with law.

Registered Office(s)
John Wiley & Sons, Inc., 111 River Street, Hoboken, NJ 07030, USA
John Wiley & Sons Ltd, The Atrium, Southern Gate, Chichester, West Sussex, PO19 8SQ, UK

Editorial Office
The Atrium, Southern Gate, Chichester, West Sussex, PO19 8SQ, UK

For details of our global editorial offices, customer services, and more information about Wiley products visit us at www.wiley.com.

Wiley also publishes its books in a variety of electronic formats and by print-on-demand. Some content that appears in standard print versions of this book may not be available in other formats.

Limit of Liability/Disclaimer of Warranty
In view of ongoing research, equipment modifications, changes in governmental regulations, and the constant flow of information relating to the use of experimental reagents, equipment, and devices, the reader is urged to review and evaluate the information provided in the package insert or instructions for each chemical, piece of equipment, reagent, or device for, among other things, any changes in the instructions or indication of usage and for added warnings and precautions. While the publisher and authors have used their best efforts in preparing this work, they make no representations or warranties with respect to the accuracy or completeness of the contents of this work and specifically disclaim all warranties, including without limitation any implied warranties of merchantability or fitness for a particular purpose. No warranty may be created or extended by sales representatives, written sales materials or promotional statements for this work. The fact that an organization, website, or product is referred to in this work as a citation and/or potential source of further information does not mean that the publisher and authors endorse the information or services the organization, website, or product may provide or recommendations it may make. This work is sold with the understanding that the publisher is not engaged in rendering professional services. The advice and strategies contained herein may not be suitable for your situation. You should consult with a specialist where appropriate. Further, readers should be aware that websites listed in this work may have changed or disappeared between when this work was written and when it is read. Neither the publisher nor authors shall be liable for any loss of profit or any other commercial damages, including but not limited to special, incidental, consequential, or other damages.

Library of Congress Cataloging-in-Publication Data applied for
ISBN: 9781119247524 (cloth) | 9781119247548 (epdf) | 9781119247920 (epub)

Cover design by Wiley
Cover images: (Background) © Mandy Disher Photography/Gettyimages; (Small molecule) © Molekuul/Gettyimages; (Big molecule) © theasis/Gettyimages

Set in 10/12pt WarnockPro by SPi Global, Chennai, India
Printed and bound in Malaysia by Vivar Printing Sdn Bhd

10 9 8 7 6 5 4 3 2 1

The basis of a life as a scientist is a warm and supportive home life. It would have been impossible without my mother and father who supported me throughout my life.

Contents

Preface *xv*

1   **Introduction** *1*
1.1   History of Cyclodextrins *2*
1.2   Cyclodextrin Properties *3*
1.2.1   Toxicity Considerations *5*
1.2.2   Inclusion Complex Formation *6*
1.3   Inclusion Complex Formation Mechanism *8*
1.3.1   Hydrophobic Interaction *10*
1.3.2   van der Waals Interaction *11*
1.3.3   Hydrogen-Bonding Interaction *11*
1.3.4   Release of Enthalpy-Rich Water *11*
1.3.5   Release of Conformational Strain *12*
1.3.6   Inclusion Complex Formation with Various Environments *12*
1.4   Important Parameters in Inclusion Complex Formation *15*
1.4.1   Effects of Temperature *15*
1.4.2   Use of Solvents *16*
1.4.3   Effects of Water *16*
1.4.4   Solution Dynamics *16*
1.4.5   Volatile Guests *17*
1.5   Inclusion Complex Formation Methods *17*
1.5.1   Coprecipitation *17*
1.5.2   Slurry Complex Formation *18*
1.5.3   Paste Complexation *18*
1.5.4   Wet Mixing and Heating *18*
1.5.5   Extrusion *18*
1.5.6   Dry Mixing *18*
1.6   Methods for Drying of Complexes *19*
1.6.1   Highly Volatile Guests *19*
1.6.2   Spray Drying *19*
1.6.3   Low-Temperature Drying *19*

1.7	Release of the Complex	19
1.8	Application of Inclusion Compounds	20
1.8.1	Characterization of Inclusion Complexes	20
1.9	Applications of Cyclodextrins	21
1.9.1	Cosmetics, Personal Care, and Make Up	22
1.9.2	Foods and Flavors	22
1.9.3	Pharmaceuticals	23
1.9.3.1	Drug Delivery	24
1.9.3.2	Gene Delivery	26
1.9.4	Cyclodextrin Applications in Agricultural and Chemical Industries	27
1.9.5	Encapsulation of Various Guest Molecules	28
1.9.6	Supramolecular Polymer Base on Host–Guest Complexes	29
1.10	Characterization and Experimental Techniques	29
1.10.1	NMR Spectroscopy	30
1.10.2	FTIR Spectroscopy	30
1.10.3	X-ray Diffraction	30
1.10.4	Thermogravimetric Analysis	30
1.10.5	UV–Vis Spectral Changes	31
1.10.6	Phase Solubility Technique	31
	References	31
2	**Supramolecular Chemistry and Rotaxane**	**41**
2.1	What Is Supramolecular Chemistry	41
2.1.1	History of Supramolecular Chemistry	42
2.1.2	Concept of Supramolecular Chemistry	43
2.1.2.1	Noncovalent Interactions	43
2.1.2.2	Electrostatic Interactions	44
2.1.2.3	Hydrogen Bonding	44
2.1.2.4	Van der Waals Interactions	46
2.1.2.5	Hydrophobic Effect	47
2.2	Host–Guest Chemistry	47
2.2.1	Cyclodextrins as Supramolecular Hosts	48
2.3	Cyclodextrin-Containing Supramolecular Structures	48
2.3.1	Cyclodextrins	49
2.3.2	Cyclodextrin Shape and Inclusion Complex Formation	49
2.4	Supramolecular Chemistry	49
2.4.1	Cyclodextrin Rotaxanes	50
2.4.2	Studies on Responsive CD-Based Polymers	52
2.4.2.1	Cyclodextrin Dimers	52
2.4.2.2	Catenanes	52
2.4.2.3	Rotaxanes	53

2.5	Cyclodextrin-based Rotaxanes and Pseudorotaxanes	54
2.5.1	Pseudopolyrotaxanes	57
2.5.1.1	Synthetic Route	58
2.5.2	Polyrotaxanes and Pseudopolyrotaxanes Based on Cyclodextrins	58
2.5.3	Cyclodextrin-based Polyrotaxanes	60
2.5.4	Cyclodextrin Molecular Tubes	62
2.5.4.1	Cyclodextrin-Based Nanotube Structure	63
	References	65
3	**Smart Polymers**	**75**
3.1	Introduction	75
3.2	Supramolecular Self-Assembly	76
3.3	Synthesis of Block Copolymers	76
3.3.1	Free and Living Radical Polymerization	76
3.3.2	Block Copolymers	79
3.4	Self-Assembly of Amphiphilic Block Copolymers	79
3.4.1	Smart Polymers Synthesized Based on Living Controlled Radical Polymerization	81
3.4.2	Definition of Self-Assembly	82
3.4.3	Self-Assembled Structures Based on Block Copolymers	83
3.4.3.1	Micelles	84
3.4.3.2	Vesicles	84
3.4.3.3	Dendrons and Dendrimers	86
3.5	Stimuli-Sensitive Supramolecular Structures	87
3.5.1	Stimuli-Responsive Polymers Based on Cyclodextrins	89
3.5.2	pH-Responsive Systems	90
3.5.3	Temperature-Responsive Systems	93
3.5.4	Redox-Responsive Systems	96
3.5.5	Other Stimuli-Responsive Hydrogels	96
3.5.5.1	Light-Sensitive Materials	96
3.5.5.2	Photoresponsive Polymers	97
3.5.5.3	Photoresponsive Liposomes	97
3.5.5.4	Photoresponsive Micelles	97
3.5.5.5	Photoresponsive Vesicles	99
3.5.5.6	Electroresponsive Polymers	100
3.5.5.7	Magnetic-Responsive Polymers	102
3.6	Polymers with Dual-Stimuli Responsiveness	103
3.6.1	Cyclodextrins for Synthesis of Responsive Supramolecules	104
3.6.2	pH-Responsive Inclusion Complexes	104
3.6.3	Temperature-Responsive Inclusion Complexes	107
3.6.4	Photoresponsive Inclusion Complexes	109

3.6.5	pH-Sensitive Polyrotaxane	*110*
3.7	Stimuli-Sensitive Polyrotaxane for Drug Delivery	*111*
3.7.1	Photoresponsive Inclusion Complex Application	*113*
3.7.2	Redox-Responsive Inclusion Complexes Applications	*117*
3.8	Multi-Stimuli-Responsive Inclusion Complexes	*120*
3.8.1	pH- and Temperature-Responsive Inclusion Complexes Applications	*120*
3.8.2	pH- and Redox-Responsive Inclusion Complexes Applications	*121*
3.8.3	Temperature- and Photoresponsive Inclusion Complexes Applications	*121*
3.8.4	Temperature- and Redox-Responsive Inclusion Complexes Applications	*123*
	References	*124*
4	**Basics of Corrosion**	*141*
4.1	Introduction to Corrosion and Its Types	*142*
4.1.1	Corrosion	*142*
4.1.2	Forms of Corrosion	*143*
4.1.2.1	Uniform Corrosion	*143*
4.1.2.2	Galvanic Corrosion	*144*
4.1.2.3	Pitting Corrosion	*145*
4.1.2.4	Crevice Corrosion	*146*
4.1.2.5	Intergranular Corrosion (IGC)	*147*
4.1.2.6	Dealloying Corrosion	*147*
4.1.2.7	Stress Corrosion Cracking (SCC)	*148*
4.1.2.8	Erosion Corrosion	*149*
4.2	Corrosion Protection	*149*
4.2.1	Anticorrosion Methods	*149*
4.2.2	Anticorrosion Coating	*150*
4.3	An Introduction to Self-healing Coatings	*151*
4.3.1	Classification of Self-healing Approaches	*152*
4.3.1.1	Materials with Intrinsic Self-healing	*152*
4.3.1.2	Capsule-based Sealing Approach	*154*
4.3.1.3	Vascular Self-healing Materials	*157*
4.3.1.4	Active Anticorrosion Coatings	*158*
4.4	Protective Coatings Containing Corrosion Inhibitors	*160*
4.5	An Introduction to Sol–Gel	*160*
4.5.1	Sol–Gel Chemistry	*161*
4.5.2	General Procedures Involved in the Preparation of Sol–Gel Coatings	*162*
4.5.3	Applications of Sol–Gel-Derived Coating	*163*
4.5.3.1	Corrosion Protective Sol–Gel Coatings	*163*

4.5.3.2	Organic–Inorganic Hybrid (OIH) Sol–Gel Coatings	*164*
4.6	Addition of Corrosion Inhibitors to Sol–Gel Coating Micro-/Nanoparticles	*166*
4.6.1	Direct Addition of Inhibitor	*166*
4.6.1.1	Inorganic Inhibitors	*166*
4.7	Self-healing Coating Containing Corrosion Inhibitor Capsules	*168*
4.7.1	Self-healing Anticorrosion Coatings Based on Nano-/Microcontainers Loaded with Corrosion Inhibitors	*168*
4.7.2	Preparation of Supramolecular Corrosion Inhibitor Nanocontainers for Self-protective Hybrid Nanocomposite Coatings	*169*
4.7.2.1	Formation of Cyclodextrin–Inhibitor Inclusion Complexes	*171*
4.7.2.2	Characterization of Encapsulated Organic Corrosion Inhibitors	*172*
4.8	Morphology of the Smart Corrosion Inhibitor Nanocontainers	*178*
4.8.1	Microstructural Characterization	*180*
4.8.2	Self-healing Mechanism of Corrosion Inhibitor Nanocontainers	*181*
4.8.3	EIS Measurement of Coating Containing Inhibitor Nanocontainers	*182*
4.8.4	Salt Spray Test for Investigation of Anticorrosive Performance of the Nanocapsules Incorporated Coating	*185*
4.8.5	Controlled Release of Inhibitors from Nanocontainers Obtained by Encapsulation of Inhibitor Corrosion in CDs	*187*
4.9	Concluding Remarks	*188*
	References	*189*
5	**Phytochemicals**	*201*
5.1	Phenolic Acids	*202*
5.1.1	Polyphenolic Antioxidant Property	*204*
5.1.2	Phenolic Compounds and Free Radicals	*204*
5.1.3	Extraction of Plant Polyphenols	*205*
5.2	Flavanoids	*206*
5.2.1	Flavones	*207*
5.2.2	Catechins	*207*
5.2.3	Isoflavonoids	*208*
5.2.4	Tannins	*208*
5.2.5	Anthocyanidins	*208*
5.2.6	Lignans and Stilbenes	*208*

5.2.7	Alkaloids and Other Nitrogen-containing Metabolites *209*	
5.3	Phytochemical Importance *209*	
5.3.1	Oxidative Stress and Phenolic Compounds in Foods *209*	
5.3.2	Important Parameters for Phenolic Efficiency *210*	
5.4	Encapsulation *211*	
5.4.1	Polyphenol Encapsulation *212*	
5.4.2	Physical Methods *213*	
5.4.2.1	Spray-drying *214*	
5.4.2.2	Fluid Bed Coating *214*	
5.4.2.3	Extrusion–Spheronization Technique *215*	
5.4.2.4	Centrifugal Extrusion *215*	
5.4.2.5	Supercritical Fluids *215*	
5.4.3	Physicochemical Methods *216*	
5.4.3.1	Spray-cooling/Chilling *216*	
5.4.3.2	Encapsulation by Emulsions *216*	
5.4.3.3	Coacervation *217*	
5.4.3.4	Ultrasonication *217*	
5.4.4	Chemical Methods *218*	
5.4.4.1	Micelles *218*	
5.4.4.2	Liposomes *219*	
5.4.4.3	In Situ Polymerization *220*	
5.4.4.4	Interfacial Polymerization *221*	
5.4.4.5	Freeze-drying *221*	
5.5	Encapsulation of Phenolic Compounds Via Cyclodextrin *222*	
5.5.1	Cyclodextrin Inclusion Complexes Formation *223*	
5.5.2	Polyphenol Encapsulation in Cyclodextrins *223*	
5.5.3	Solubilization and Stabilization of Polyphenols *224*	
5.6	Why Encapsulation by Cyclodextrin? *224*	
5.6.1	Cyclodextrins and Flavonoids *225*	
5.7	Concluding Remarks *227*	
	References *227*	

6 Cyclodextrins Application as Macroinitiator *239*

6.1 Cyclodextrins Application as Macroinitiator in Polyrotaxane Synthesis Via ATRP *239*
6.2 Inclusion Complexes of PDMS and γ-CD Without Utilizing Sonic Energy *241*
6.3 Supramolecular Pentablock Copolymer Containing Via ATRP of Styrene and Vinyl Acetate Based on PDMS/CD Inclusion Complexes as Macroinitiator *246*
6.3.1 Complex Formation of γ-CD with Br–PDMS–Br *248*
6.3.2 Characterization of Polyrotaxane-based Pentablock Copolymers *249*

6.4	Synthesis and Characterization of Poly(vinylacetate)-*b*-Polystyrene-*b*-(Polydimethyl siloxane/cyclodextrin)-*b*-Polystyrene-*b*-Poly(vinyl acetate) Pentablock Copolymers	256
6.4.1	Preparation of PDMS Macroinitiator (Br–PDMS–Br)	258
6.4.2	Polymerization of St and PVAc Initiated by PDMS–CDs Macroinitiator	259
6.4.3	Microstructural Studies of the Pentablock Copolymers	263
6.5	Conclusion	265
	References	265
7	**Cyclodextrin Applications**	**269**
7.1	Cyclodextrin Industrial Applications	269
7.1.1	Pharmaceutical Applications of Cyclodextrins	270
7.1.2	Inclusion Complex Formation Advantages with Drugs	271
7.1.2.1	Water Solubility	271
7.1.2.2	Drug Bioavailability	271
7.1.2.3	Drug Safety	271
7.1.2.4	Drug Stability	272
7.1.2.5	Mask Unpleasant Odor, Taste, and Side Effects of Drugs	272
7.1.2.6	Drug Stability	272
7.1.2.7	Drug Solubility and Dissolution	273
7.1.2.8	Reduction in Volatility	273
7.1.3	CD-based Carriers	273
7.2	Drug Delivery Systems Based on Cyclodextrins	273
7.2.1	Oral Drug Delivery System	274
7.2.2	Rectal Drug Delivery System	274
7.2.3	Nasal Drug Delivery System	275
7.2.4	Transdermal Drug Delivery System	275
7.2.5	Ocular Drug Delivery System	275
7.2.6	Liposomal Drug Delivery	276
7.2.7	Osmotic Pump Tablet	277
7.2.8	Peptide and Protein Delivery	277
7.3	Cyclodextrin-based Targeting Systems	278
7.3.1	Nanoparticles	279
7.3.2	Nanocapsules	279
7.3.3	Microsphere	280
7.3.4	Nanosponges	280
7.4	CDs in the Food Industry	281
7.5	Cyclodextrins in Skin Delivery and Cosmetic	283
7.6	Agricultural Applications	284
7.7	Self-healing Coating	285
	References	287

Index *301*

Preface

Cyclodextrins (CDs) are a family of cyclic oligosaccharides consisting of (α-1,4)-linked α-D-glucopyranose units. CDs are obtained from the enzymatic digestion of the most essential polysaccharides, starch, and cellulose. The specific donate shape (truncated cone) of CDs is due to the chair conformation of the glucopyranose units and leads to a hydrophobic cavity and a hydrophilic surface. CDs can form host–guest interaction with a wide range of hydrophobic and hydrophile segments and encapsulate molecules in their hydrophobic cavity. Encapsulation in a CD cavity may alter or improve the physical, chemical, biological characteristics, and stability of the guest molecule. The formation of inclusion complexes between CDs as host and guest molecules is based on noncovalent interaction such as hydrogen bonding or van der Waals interactions and leads to the formation of supramolecular structures. These structures can be used as macroinitiators to initiate various types of reactions. CDs are widely used in many industrial products such as pharmacy, food and flavors, chemistry, chromatography, catalysis, biotechnology, agriculture, cosmetics, hygiene, medicine, textiles, drug delivery, packing, separation processes, environment protection, fermentation, and catalysis.

One of the most attractive applications of CDs is their role as molecular encapsulants in food and drug industries. Encapsulation of phytochemicals and flavors in food industry allows the quality and quantity of the flavor to be preserved to a greater extent for longer periods compared to other encapsulants; it provides longevity to the food item, and also masks its unpleasant odor. CDs are potential candidates for increasing the solubility of hydrophobic drugs, delivering the required amount of drug to the targeted site for the necessary period of time, both efficiently and precisely, and limiting undesirable properties of drug molecules. These characteristics have lowered drug production costs.

Inclusion complex formation between CDs and guest molecules is based on reversible and noncovalent bonding strategies and can be used as self-healing agents for various purposes. CDs can encapsulate corrosion inhibitors, become active in corrosive electrolytes, slowly diffuse out of the host material to ensure both continuous and controlled delivery of the inhibitors to corrosion sites and long-term corrosion protection.

Each year, many patents, research articles, books, and scientific abstracts are published about CDs and their applications in various fields, but many aspects of CDs and their derivatives are still unknown and attractive. That is why we focused on CDs and their applications in various fields.

This book reflects cyclodextrins structure, their properties, formation of inclusion complex with various compounds, and their applications. The purpose of this book is to cover both basic and applied science in chemistry, biology, and physics of CDs. We hope that this book will arouse the interest of scientists and engineers who wish to diversify their research fields.

Tehran, June 2017

Sahar Amiri
Sanam Amiri

1

Introduction

Cyclodextrins, also known as cycloamyloses, cyclomaltoses, or Schardinger dextrins, are cyclic oligosaccharides consisting of six, seven, eight, or more glucopyranose units composed of α-(1,4)-linked glucopyranose subunits synthesized from the enzymatic degradation of starch [1].

Cyclodextrins are chemically and physically stable macromolecules produced by intramolecular transglycosylation reaction from enzymatic degradation of starch with glucanotransferase (CGTase) enzyme [2]. Due to steric factors, cyclodextrins built by less than six glucopyranose units do not exist; however, cyclodextrins with more than eight glucopyranose units have been synthesized [3]. Because of chair conformation of glucopyranose unit, their molecular structure, and the lack of free rotation about the bonds connecting the glucopyranose units, CDs have a unique toroid or truncated cone shape with hydrophilic outer surface and hydrophobic cavity [4]. The most common ones are α, β, and γ cyclodextrins consisting of 6, 7, and 8 glucopyranose units that are crystalline, homogeneous, and nonhygroscopic substances produced by the enzymatic degradation of starch [5]. Glucosyltransferases of starch caused degradation of amylose fraction: one or several turns of the amylose helix are hydrolyzed off and their ends are joined together, thereby producing cyclic oligosaccharides. Per year using environmentally friendly technologies, thousands of tons of CDs are produced, with prices acceptable for most of the industrial purposes [6]. Absorption of CDs is negligible, so they are harmless; they have been widely used because of low toxicity both orally and intravenously. Unmodified CDs are completely resistant to β-amylase. α-Amylase is capable of hydrolyzing CDs only at a slow rate. After intravenous injections, CDs are mainly excreted in their intact form by renal filtration as they are minimally susceptible to hydrolytic cleavage or degradation by human enzymes [5, 7].

Chemical reactions of cyclodextrins led to intramolecular interactions based on noncovalent bonding such as hydrogen or van der Waals bonding and formed supramolecular structures. Specific structure of cyclodextrins with truncated shape causes the formation of complex between cyclodextrins and a wide range of molecules, which is called host–guest or inclusion complex. Formation of inclusion complex modifies or improves the physical, chemical, and/or biological characteristics of the guest molecule [8]. Because of negligible toxicity and also the formation of inclusion complex with various compounds, cyclodextrins can be used in various industrial products such as carriers, stabilizing agent, food and flavors, cosmetics, packing, textiles, separation processes, fermentation, catalysis, and drug delivery systems [9].

1.1 History of Cyclodextrins

CDs were first discovered in 1891, when in addition to reducing dextrins, a small amount of crystalline material was obtained from starch digestion of *Bacillus amylobacter*. Antoine Villiers worked on the action of enzymes on various carbohydrates, particularly using the butyric ferment *Bacillus amylobacter* on potato starch. He called this crystalline product "cellulosine." After this period, Schardinger isolated two crystalline products in 1903 and isolated a new organism that was able to produce acetone and ethyl alcohol from sugar and starch-containing plant material [1]. By inoculating the amylaceous paste with the bacillus, a slightly acidic liquid with butyric-acid smell was formed. After purification of fractional precipitation, the dextrins (called so by Schardinger) presented very different optical rotation properties. It was difficult to hydrolyze them any further [10]. Crystalline structures of α- and β-cyclodextrin were determined by X-ray crystallography in 1942 [11]. In 1948–1950, the X-ray structure of γ-cyclodextrin was discovered and it was found that CDs can form inclusion complexes [12].

CDs are fractionalized to pure components by enzymic production. They were characterized physically and chemically by French [11] and Cramer in the 1950s [13]. Their ability to form inclusion complex was discovered by Cramer's group [14]. Various patents were published about application of CDs in drug formulations and protection of easily oxidizable substances against atmospheric oxidation, the enhancement of solubility of poorly soluble drugs, and reduction of the loss of highly volatile substances. In 1970, numerous industrial applications of cyclodextrins were discovered and industrial-scale production of CDs was started. Traditionally, three factors stood on the way of their industrial development:

(i) high production costs; (ii) incomplete toxicological studies; and (iii) lack of sufficient scientific knowledge of native CDs and their derivatives [8].

From the 1980s, with a more accurate picture of their toxicity and better understanding of molecular encapsulation, several inclusion complexes appeared in market, especially in drug preparations, food industry, macromolecular chemistry [15–17], supramolecular chemistry [18, 19], catalysis [20, 21], membranes [22], foods [23], biotechnology [24], enzyme technology [25], cosmetics [26–28], pharmacy and medicine [29–32], textiles [28, 33, 34], chromatography [35, 36], agrochemistry [37], microencapsulation [38], nanotechnologies [39, 40], and analytical chemistry [41].

The most important and amazing property of CDs is their ability to form inclusion complexes with several hydrophobic and hydrophilic compounds [5, 42–44]. Cyclodextrins are truncated cone or torus rather than perfect cylinder because of the chair conformation of glucopyranose units. Secondary hydroxyl groups (C_2 and C_3) are located on the wider edge of the ring and the primary hydroxyl groups (C_6) on the other edge and the apolar C_3 and C_5 hydrogens and ether-like oxygens are at the inside of the torus-like molecules. Therefore, the outside of cyclodextrins is hydrophilic and inside is hydrophobic. CDs are water soluble, biocompatible in nature with hydrophilic outer surface and lipophilic cavity [4]. As a result of this cavity, cyclodextrins are able to form inclusion complexes with a wide variety of hydrophobic guest molecules. One or two guest molecules can be entrapped by one, two, or three cyclodextrins.

1.2 Cyclodextrin Properties

Cyclodextrins are crystalline, homogeneous, nonhygroscopic, nontoxic with truncated shape and are made up of glucopyranose units. They are classified into three common types: α-cyclodextrin (Schardinger's α-dextrin: cyclomaltohexaose, cyclohexaglucan, and cyclohexaamylose), β-cyclodextrin (Schardinger's β-dextrin: cyclomaltoheptaose, cycloheptaglucan, and cycloheptaamylose), and γ-cyclodextrin (Schardinger's γ-dextrin: cyclomaltooctaose, cyclooctaglucan, and cyclooctaamylose) and are referred to as first generation or parent cyclodextrins. α-, β-, and γ-CD are composed of six, seven, and eight α-(1,4)-linked glycosyl units, respectively (Figure 1.1) [4]. β-Cyclodextrin is the most accessible, priced the lowest, and generally the most useful. Their main properties are given in Table 1.1. On the side where the secondary hydroxyl groups

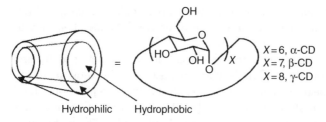

Figure 1.1 Schematic diagram of cyclodextrins.

Table 1.1 Cyclodextrin properties.

Property	α-Cyclodextrin	β-Cyclodextrin	γ-Cyclodextrin
Number of glucopyranose units	6	7	8
Molecular weight (g/mol)	972	1135	1297
Solubility in water at 25°C (%w/v)	14.5	1.85	23.2
Outer diameter (Å)	14.7	15.3	17.5
Cavity diameter (Å)	5.1	6.2	8.1
Height of torus (Å)	7.8	7.8	7.8
Cavity volume (Å^3)	174	262	427
Surface tension (MN/m)	71	71	71
Melting temperature range (°C)	255–260	255–265	240–245
Crystal water content (wt%)	10.2	13–15	8–18
Water molecules in cavity	6	11	17

are situated, the cavity is wider than on the other side where free rotation of the primary hydroxyls reduces the effective diameter of the cavity [45, 46].

All secondary hydroxyl groups are situated on one of the two edges of the ring, whereas all the primary hydroxyl groups are placed on the other edge, so CDs have a doughnut- or wreath-shaped truncated cone. CDs have high electron density and Lewis-base character because of non-bonding electron pairs of the glycosidic-oxygen bridges that are directed toward the inside of the cavity. H-bonds determined rigidity of CDs. In α-CD, one glucopyranose unit is in distorted position and H-bond belt is incomplete, but in β-CD, a complete secondary intramolecular H-bond is formed and causes rigid structure and lowest water solubility of β-CD among all CDs. The γ-CD is noncoplanar and more flexible; therefore, it is the most soluble of the three CDs [43, 47].

Depending on the type of cyclodextrin and the guest compound, cyclodextrins' inclusion complex has two main types of crystal packing:

channel structures and cage structures. Cyclodextrin derivatives have been synthesized by aminations, esterifications, or etherifications of primary and secondary hydroxyl groups of the cyclodextrins; and their solubility is usually different from that of their parent cyclodextrins. The volume of hydrophobic cavity of cyclodextrins has been changed. This can improve solubility, stability against light or oxygen and help control the chemical activity of guest molecules [1, 2, 4].

Various conditions, such as regioselective reagents, optimization of reaction conditions, and a good separation of products, are needed for the synthesis of uniform cyclodextrin derivatives. Various reactions can be substituted the OH-groups cyclodextrins with azide ions, halide ions, thiols, inorganic acid derivatives as sulfonic acid chloride, thiourea, and amines, which requires activation of the oxygen atom by an electron-withdrawing group and formed ethers and esters, epoxides, acyl derivatives, isocyanates [2].

One of the most important properties of cyclodextrins is their ability to form supramolecular complexes with various hydrophobic and hydrophilic compounds as guest, such as catenanes, rotaxanes, polyrotaxanes, and tubes [4, 48]. Various studies describe the various applications of cyclodextrins (over 1000 patents or patent applications in the past 5 years) [2–4].

1.2.1 Toxicity Considerations

The most important application of cyclodextrins is in drug-based compounds, so toxicity and safety profiles of cyclodextrins are very important for the researchers [49, 50]. In general, CDs and their hydrophilic derivatives are only able to permeate lipophilic biological membranes, such as the cornea, with considerable difficulty. Even the somewhat lipophilic randomly methylated cyclodextrins do not readily permeate lipophilic membranes and interact more readily with membranes than the hydrophilic cyclodextrin derivatives [51, 52].

Orally administered cyclodextrins are practically nontoxic. It is due to the lack of absorption from the gastrointestinal tract, but at very high concentrations, CDs can extract cholesterol and other lipid membrane components from cells, leading to the disruption of cell membranes and may show toxic properties [51–53].

Because they are natural, relatively nontoxic, have a low price, are commercially available, and possess the ability to form inclusion complexes with a wide range of guest molecules, CDs have been used in many areas including but not limited to pharmaceutical [29–32], food [23], cosmetic [26–28, 52], and textile industries [5, 28, 33, 34].

Using CDs and their derivatives in various applications is well evidenced by the increasing number of marketed or approved medicinal, skin, and cosmetic products containing CDs. This potentially broadens the application areas of both cyclodextrins and functionalized compound by CDs [4]. α-Cyclodextrins have side effects, such as ability of binding some lipids, irritating after intramuscular injection, absorption after oral administration to rats, and cleavage only by the intestinal flora of caecum and colon. Absorption of β-cyclodextrin and irritation after its intramuscular injection are less than those of α-cyclodextrin and are altered by bacteria in caecum and colon. Therefore, it is the most common cyclodextrin in pharmaceutical formulations and, thus, probably the best studied cyclodextrin in humans [49, 50].

Of all the CD derivatives available, HP-β-CD is the safest, as it does not permeate the membranes. HP-β-CD has been shown to have a reduced hemolytic potential, making it suitable for parenteral as well as oral applications. There are several references in the literature about the parenteral safety profile of HP-β-CD, including the parenteral infusion of HP-β-CD in human volunteers. HP-β-CD is well tolerated in most species, particularly, if dosed orally. It shows limited toxicity, depending upon the dose and route of administration. Many pharmaceutical and cosmetic products with HP-β-CD are already on the market, such as sporanox itraconazole formulation as injectable and oral dosage forms, indomethacin eyedrop solution. HP-β-CDs are also used in skin-care and hair-care topical products [54, 55].

γ-CD is used widely in industries because it causes low irritation after intramuscular injection, faces rapid and complete degradation by intestinal enzymes to glucose, and is the least toxic cyclodextrin, at least of the three natural cyclodextrins. For these reasons, γ-CD is promoted as food additive by its main manufactures; complexing abilities, in general, less than those of α-cyclodextrin and the water soluble β-cyclodextrin derivatives, but its complexes frequently have limited solubility in aqueous solutions and tend to aggregate in aqueous solutions, which makes the solution unclear [51–55].

1.2.2 Inclusion Complex Formation

The ability to form solid inclusion complexes (host–guest complexes) with different guest compounds, such as solid, liquid, and gaseous compounds, is the most amazing property of the cyclodextrins.

Cyclodextrin's structural features allow for the selective formation of inclusion complexes with a range of other molecules. This ability is also known as molecular recognition, or chiral recognition when dealing with enantiomeric compounds. The guest compounds are partially or fully located inside the hydrophobic cavity of cyclodextrins (host), which involve noncovalent bonding in the process of complex formation [56–58].

As the cavity of cyclodextrins is hydrophobic, the inclusion of a molecule in the cyclodextrin cavity is basically a substitution of the water inside the cavity with a less polar substance. The substitution of water from the cavity with a more nonpolar guest is energetically favorable for both the cyclodextrin and the guest. Different molecular interactions such as hydrophobic interaction has been considered as being responsible for the formation of cyclodextrin inclusion complexes in an aqueous solution and recovering high-energy water from the cyclodextrin cavity upon inclusion of substrate [59].

Hydrophobicity of cyclodextrin cavity provides a suitable microenvironment for interaction between host and guest molecule that leads to the formation of inclusion complex. Hydrophilicity of outer sphere of cyclodextrins allows hydrogen-bonding cohesive interactions. Therefore, cyclodextrins can form inclusion complexes with a wide variety of hydrophobic organic compounds and induce physicochemical and biological property changes in the guest molecules, such as enhancing the therapeutic potential, solubility, diffusion and decreasing the decomposition of drugs before they enter tissues [59, 60].

The hydrophilic outer surface and the hydrophobic interior surface of the cone structure enable the complexation of various amphiphilic and hydrophobic guest molecules in water which have an appropriate molecular structure equivalent to the CD ring size [2]. The complexes exist of noncovalent interactions such as hydrogen bonds, van der Waals forces, and hydrophobic interactions between the host and the guest molecule [61].

The driving force of the complexation of guest molecule in the hydrophobic cavity of the CDs is controlled by the release of the displaced water molecules from the torus. During the release and the increased mobility of water molecules the entropy increases. Furthermore, the formation of new H-bonds between the water molecules and the increase of the cohesion forces lead to decrease of enthalpy [59, 60]. van der Waals forces have only a very short range, so that inclusion compounds are more stable in general, when the cavity of the CD is filled out perfectly by the guest molecule. Dipole–dipole interactions stabilize only complexes of guests with strong dipole moments because of the axial dipole moment of the CDs [62, 63].

The formation of cyclodextrin inclusion complexes directly depends on the dimensions of the cyclodextrin cavity and the guest molecule. If the guest molecule is too large or bulky, it will not fit completely into the cyclodextrin cavity and likewise very small size guest molecules will not form stable complexes with cyclodextrins as they will slip out of the cavity [55].

Most frequently, complexes are formed at a 1:1 CD:guest ratio, which is related to size of the guest molecule. When a too long guest molecule is reacted with cyclodextrin, it will be completely fitted into one CD cavity. Multiple CDs can be threaded onto the guest, thereby creating 2:1, 3:1, and so on (CD:guest) ratios [64]. Various ratios are possible in inclusion complex if low-molecular-weight molecules are used as guest, because more than one guest may fit into the cyclodextrin cavity [65]. Due to the steric requirement of complexation, the different cyclodextrins show different capabilities to form inclusion complexes with the same guest molecules. Cavity depth in all cyclodextrins is same (~7.8 Å). However, the determining factor for internal diameter of the cavity and its volume is the number of glucose units that have diameter of approximately 6, 7, and 9 Å in α-CD, β-CD, and γ-CD, respectively [66, 67].

1.3 Inclusion Complex Formation Mechanism

A guest molecule is trapped within the cavity of the cyclodextrin as a host molecule and inclusion complex is formed, which is directly dependent on the dimensional fit between host cavity and guest molecule. One of the important parameters in the formation of inclusion complex is the lipophilic cavity of cyclodextrin molecules that provides a microenvironment and leads to entrance of appropriately sized nonpolar moieties [59, 60]. Formation of complex dependents on hydrogen bonding, no covalent bonds are broken or formed, various thermodynamic factors also affect. Removal of water molecule from hydrophobic cavity and formation of van der Waals bonding, hydrophobic, and hydrogen-bond interactions are driving force for formation of inclusion complex [47].

The method to synthesize host–guest complex depends on the properties and nature of the guest molecule. When a hydrophobic guest molecule is added to cyclodextrin solution, water molecules in the cyclodextrin cavity are substituted by the guest molecules. Inclusion complex formation induces structural changes in the cyclodextrin and changes guest properties [47].

If the guest molecule is larger than the cyclodextrin cavity, it cannot be fully included in cavity. Only partially included in the host cavity, the guest molecules are in contact with inner surface of the macrocyclic ring, adjacent cyclodextrin molecules, and solvent molecules. Hydrogen bonds, van der Waals interaction, and electrostatic interactions are the driving forces to stabilize the structure of the inclusion complexes [68, 69].

When the guest molecule is hydrophobic, water molecules can be displaced by guest molecules present in the solution. This leads to an apolar–apolar association, decreases cyclodextrin ring strain, and causes more stable structure with lower energy state [2]. The binding of guest molecules within cyclodextrin as host is not fixed or permanent but is a dynamic equilibrium. Inclusion strength depends on dimensional fit of guest molecule and cyclodextrin cavity and on the specific local interactions between surface atoms. Inclusion complex formation can happen under various systems such as solution, cosolvent system, presence of any nonaqueous solvent or in the crystalline state where water is typically the solvent of choice [2, 47].

The special shape of cyclodextrin with hydrophobic cavity and hydrophilic surface led to formation of inclusion complex with various molecules with a wide range of chemical properties that are different from those of noncomplexed guest molecules. Cyclodextrins can form inclusion complex with a wide range of guest. They can be linear or branched chain aliphatics, aldehydes, ketones, alcohols, organic acids, fatty acids, aromatics, gases, and polar compounds such as halogens, oxyacids, and amines [70].

Inclusion complex formation of cyclodextrins with guest molecules, demonstrate a significant effect on the physical and chemical properties of guest molecules and induce appropriate modifications of guest molecules, such as increase solubility of insoluble or volatile guests, stabilization of volatile and unstable guests against the degradative effects of oxidation, visible or UV light and heat, control of volatility and sublimation, physical isolation of incompatible compounds, chromatographic separations, taste modification by masking off flavors, unpleasant odors and controlled release of drugs and flavors [5, 7].

Due to the presence of multiple reactive hydroxyl groups in inner and outer surfaces of CDs, various chemical modifications are possible. They cause various functionalities of cyclodextrins by substituting various functional groups on the primary and/or secondary face in molecular recognition, which are useful as enzyme mimics, targeted drug delivery and analytical chemistry [1, 2].

Scheme 1.1 presents the process of inclusion complex formation; it shows water molecules as small circles and drug molecules. Hydrophobic drug molecules and the hydrophobic cavity of the truncated CD cylinder repelled water molecules which is the main driving force for inclusion complex formation and is mainly the substitution of the polar–apolar interactions (between the apolar CD cavity and polar water) for apolar–apolar interactions (between the drug and the CD cavity). The main driving force for complex formation is thought to be due to the

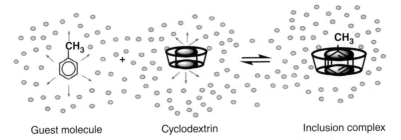

Scheme 1.1 Schematic presentation of inclusion complex formation.

release of enthalpy-rich water from the cavity after the entrapping of guest molecules [62, 71].

No covalent bonds are formed or broken during drug–CD complex formation. Weak van der Waals forces, hydrogen bonds, and hydrophobic interactions keep the complex together. Due to the limitation in size and apolar character of the CD cavity, solubilization strategy using cyclodextrin complexation is not suitable for very small compounds, or compounds that are too large such as peptides, proteins, enzymes, sugars, and polysaccharides; however, the side chain in macromolecules may contain suitable groups that can react with CDs in aqueous solutions and form a partial complex with CDs [47, 61].

1.3.1 Hydrophobic Interaction

Hydrophobic interaction occurs when nonpolar molecules tend to cluster together in an aqueous environment due to the removal of apolar surfaces from contact with water. The structure of the surrounding water is a critical factor in classical hydrophobic interaction. The interaction results in a slightly positive $H°$ and a large positive $S°$ at low temperature, and its thereby said to be entropy driven [71, 72]. The fact that the entropy change is positive, even though the molecules are clustering together, shows that there must be a contribution to the entropy from the solvent and that solvent molecules must be more free to move once the solute molecules have herded into small aggregates [73].

Since the majority of the cyclodextrin complexation is enthalpy driven, it seems obvious that hydrophobic interactions are of minor contribution compared to the other driving forces and therefore several authors have reported that hydrophobic interactions do not need to be taken into consideration. However, cyclodextrin is a semipolar molecule where semipolar means a cavity more hydrophobic than water but less hydrophobic than n-octanol base on the dielectric constant of toluidinyl groups after inclusion, which provides an environment suitable for interaction with hydrophobic guests. If hydrophobic interactions occur, there should be no expectation that the classical system is applicable in the cyclodextrin system [71–73].

1.3.2 van der Waals Interaction

When two molecules are brought close together, they both attract or repel each other depending on the distance that separates them. The attraction force of the molecules is caused by the instantaneous and short-lived imbalance in the electron distribution of an atom that generates a temporary dipole. These short-living induced dipoles result in an induction electron distribution of the neighboring atom that generates a temporary polarization. This polarization minimizes the electron–electron repulsion between the atoms, also known as induced dipole–induced dipole interaction or London dispersion forces. Other forces involved are dipole-induced dipole and permanent dipole. Common for all these repulsive and attractive forces, known as van der Waals forces, are that they are neither noncovalent nor nonionic. These forces are usually weak for all kinds of interactions but are likely to be numerous in the cyclodextrin cavity and thereby have to be taken into consideration [74, 75].

1.3.3 Hydrogen-Bonding Interaction

If hydrogen is close to an atom that is very good at attracting electrons (such as N, O, or F), the hydrogen end of the bond becomes very positively charged and the other atom becomes negatively charged (i.e., polar). Hydrogen is the smallest atom in the periodic table, which makes it possible for hydrogen atom and the other atom to get very close together. The combination of high polarity and close approach result in the interaction being particularly strong due to the force of attraction between two opposite charges. This is proportional to the magnitude of their charges divided by the square of the distance between them. In fact, the interaction is so strong that it dwarfs all other dipole–dipole attractions [76].

The hydrogen bonding is considered to play an important role in the stability of the cyclodextrin complexes in aqueous solution. It may, furthermore, contribute to a conformational change either in the cyclodextrin, the guest, or both, which results in a more stable complex [72, 73].

1.3.4 Release of Enthalpy-Rich Water

When water is substituted from the cavity of the cyclodextrin, a decrease in energy occurs. This is caused due to an increase in solvent solvent interaction, since the surface contact between solvent and cyclodextrin cavity, as well as between solvent and guest molecule, is reduced. Furthermore, water inside the cyclodextrin cavity cannot possess its tetrahedral hydrogen-bonding capacity compared to those in the surrounding solvent, and it is therefore often reported as high-energy water or enthalpy-rich water. One of the main driving forces for complexation could, therefore, be the release of this high-energy water from the cyclodextrin cavity, which, allowing them to form their full complement of hydrogen, bonds with the surrounding water [71, 72].

1.3.5 Release of Conformational Strain

α-Cyclodextrin has lowest number of glucose units. Torsion of the cyclodextrin molecules would be affected upon penetration of the guest molecule into the cavity; thus, release energy and conformational strain become important [71, 72].

1.3.6 Inclusion Complex Formation with Various Environments

Properties of the guest molecule, active material, the equilibrium kinetics, thermodynamic parameters, formulation of ingredients and processes and the use dosage form affected on the method choice for inclusion complex formation. Methods that are used for inclusion complex formation are dry mixing, mixing in solutions and suspensions followed by a suitable separation, the preparation of pastes, and several thermomechanical techniques. By increasing the number of water molecules in the surrounding environment, rate of inclusion complex breaking becomes faster. In highly dilute and dynamic system such as the body, the guest molecule has difficulty finding another cyclodextrin to reform the complex and is left free in solution [62, 77].

By the formation of inclusion complex between cyclodextrins and guest molecule, a crystalline structure is formed and is divided into cage-type, channel-type, and layer-type arrangement, which is dependent on guest molecule. Small guest molecules can be enclosed in the host cavity, so the arrangement is cage type, where both the ends of the host cavity are closed by the adjacent molecules to create an isolated cage and the arrangement of cyclodextrin molecules are in a zigzag mode. If guest molecule is a large molecule, such as an alkyl chain or a linear polymer, the arrangement is channel type or column structure. The guest molecule can be incorporated in cyclodextrin cavity and make an infinite cylindrical channel [6, 78]. The other types of cyclodextrin arrangements are head-to-head and head-to-tail arrangements.

In head-to-head arrangement, secondary hydroxyl groups of two cyclodextrins face each other and are connected by hydrogen bonds to create a barrel-like cavity, but in the head-to-tail arrangement, the primary hydroxyl side faces the secondary hydroxyl side of the next molecules exposing hydrogen bonds and cyclodextrin rings are linearly stacked. When the guest molecule is large, it cannot be fully incorporated in cyclodextrin cavity and layer arrangement is observed [79]. The relation between complex formation, cyclodextrin, and guest molecule is described by Equation 1.1, and dissociation constant K_D can be quantitatively described by Equation 1.2.

$$\text{cyclodextrin} + \text{guest} \xrightleftharpoons{K_D} \text{complex} \tag{1.1}$$

$$K_D = \frac{[\text{complex}][\text{guest}]}{[\text{complex}]} \tag{1.2}$$

Formation of inclusion complex in solution state involved entire the guest molecule inside the CD cavity which led to rupture water molecules inside the CD ring and around the guest, and interactions between the guest molecules and hydroxyl groups on CD cavity are formed, crystalline complex formed and water molecules is reconstructed around the complex [46, 47]. These interactions between the components of the system are favorable driving force for complex formation and they replaced the guest into the cyclodextrin cavity, which led to decrease in the repulsive forces when polar water molecules are displaced from the apolar cyclodextrin cavity to join the larger pool, so that the number of hydrogen bonds formed increases. In the inclusion formation process, cyclodextrin ring strain decreases and high-energy conformation of the CD–water complex shifts to the lower energy conformation, and an apolar–apolar association, resulting in a more stable lower energy state overall [43, 47, 56].

The formation of inclusion complex entrapped guest molecules and increased their shelf life under various conditions and against environmental parameters such as oxidation or chemical reaction. By dissolving a dried complex in water and increasing water content in the surrounding environment, complex separation is increased, so the concentration gradient shifts the equilibrium in Equation 2.7 to the left [56, 57].

Body is a highly dilute and dynamic system. The rates of formation/dissociation of complexes are close to diffusion-controlled limits because low guest availability in the body and increasing temperature weaken the complex and contribute to the dissociation of guest [43, 47]. The temperature plays an important role when the guest is lipophilic and has access to tissue but not cyclodextrin; so the tissue acts as a sink and causes dissociation of the complex following the simple mass action principles. The inclusion complex fully or partially covered the guest molecule and let to a stable product. It changed physical and chemical properties of the guest molecule. Cyclodextrins may enhance the aqueous solubility of highly insoluble drugs or insoluble compounds because of their ability to form hydrogen bonds with insoluble drugs and form a partially water-soluble inclusion complex, which is the most important applications of cyclodextrin inclusion complexes along with controlled release of the drug and inhibitor corrosion compounds. Various techniques are used for the formation of inclusion complexes. They are dependent on the natural properties of selected guest molecule, final application of guest molecule, and equipments available [43, 47, 56, 57].

One of the most commonly used techniques is adding the guest molecule to an aqueous solution of cyclodextrin with stirring under different conditions. The complex either precipitates out immediately or upon slow cooling and/or evaporation (based on stirring condition) and the precipitate is collected. This method is called coprecipitation. It is used for a wide range of guest molecules, but scaling it up is difficult because a large quantity of water is required. The main advantage of this technique is that complex formation is visible by the disappearance of the guest [80, 81].

Another method has been investigated. It needs less water. This method is called slurry complexation. In this method, cyclodextrin at a 50–60% solids concentration suspended, the aqueous phase becomes saturated with cyclodextrin in solution and, as guest molecules complex with the dissolved cyclodextrin, the complex precipitates out of the aqueous phase [80, 81]. Another method is paste complexation that involves adding a small amount of water to the cyclodextrin to form a paste (most suitable for industrial synthesis). The resulting complex can be dried directly and milled to obtain a powder if a hard mass is formed [80–82].

Heat is a double-effect parameter. First it increases the amount of CD dissolved and increases the probability of complexation. Second, it destabilizes the complex, with most complexes beginning to decompose at 50–60 °C. Though the heat stability varies with different guest molecules, some can be highly thermally stable. For extremely fine and unstable complexes, a desiccator or freeze dryer may be used to dry the complexes in order to minimize the decomposition of the volatile guest [81, 82].

Due to the solubility of cyclodextrin and easily displacement from the cavity, water is usually the preferred solvent for complexation techniques. If guests have low solubility in water, complexation can be very slow or impossible. An organic solvent can be used to dissolve the guest, but the solvent must be removed by evaporation or other methods. Furthermore, the use of too much solvent can cause the guest to be so dilute that it does not come in contact with the cyclodextrin in a sufficient amount to facilitate complexation [56, 57].

For high molecular weight oils, interaction with cyclodextrins is rather than themselves, for this reason the amount of solvent is increased, accompanied by sufficient mixing in order to disperse and separate the guest molecules from each other, so different complexation techniques, such as paste or dry mixing could be more efficient for these types of guest molecules [58].

Inclusion complex of cyclodextrin can exist in both liquid and solid state, with their structures differing significantly in each state. In solution, the formation of inclusion complexes is not a fixed state but rather a dynamic equilibrium between complexed and noncomplexed guest molecules where the guest molecule continuously associates and dissociates from the host [81, 82].

The guest is trapped within the cavity, and the entire complex is solvated by water molecules. In the crystal state, the guest may be included in a void space of a lattice or merely be aggregated to the outside of the cyclodextrin, forming a microcrystalline or amorphous powder and this may lead to the formation of nonstoichiometric inclusion complexes. One of the benefits of preparing inclusion complexes in solution is that more cyclodextrin molecules become available for complexation. In the crystalline form, only the surface molecules of the cyclodextrin crystal are available for complexation [56, 57, 82].

1.4 Important Parameters in Inclusion Complex Formation

Inclusion complex of cyclodextrin and guest molecule is a stoichiometric molecular phenomenon in which cyclodextrin cavity traps guest molecules and interacts with them. The interaction is due to noncovalent forces such as van der Waals forces, hydrophobic interactions, and other forces and is responsible for the formation of a stable complex [46, 47].

Depending upon the weight of a guest molecule, number of trapped guests in cyclodextrin cavity is determined. Although generally one guest molecule is included in one cyclodextrin molecule. For guests with low molecular weight, more than one guest molecule may fit into the cavity; and for guests with low molecular weights it is possible that more than one cyclodextrin molecule bind to the guest. In principle, only a portion of the molecule must fit into the cavity to form a complex. As a result, 1:1 molar ratios are not always achieved, especially with high- or low-molecular-weight guests [59, 60].

1.4.1 Effects of Temperature

Temperature has double effect upon cyclodextrin complex formation. Increasing temperature may increase solubility of the complex but may destabilize it. Therefore, controlling temperature is very critical. It needs to be balanced. With change in the guest molecule, the heat stability of the complex varies, but generally it starts to decompose at 50–60 °C; if the guest is strongly bound or the complex is highly insoluble, complexes are stable at higher temperatures [47, 56, 59, 60].

1.4.2 Use of Solvents

Solvent has an important role in the formation of inclusion complex between cyclodextrin and guest molecule. Water is the most commonly used solvent. By increasing solubility of cyclodextrin in the solvent, more number of molecules are available for the formation of inclusion complex and yield of reaction is increased. The guest molecule must be able to displace the solvent from the cavity if the solvent forms a complex with the cyclodextrin and at the end of reaction solvent must be easily removed if solvent-free complexes are desired. Inclusion complex formation is affected by solubility of guest molecules which low solubilized guest in water slow or stopped complexation and organic solvent must be used to dissolve the guest as desirable. Used solvent can be easily removed by evaporation and did not complex well with cyclodextrin such as ethanol, acetone, and diethyl ether [47, 56, 59, 60].

1.4.3 Effects of Water

The solubility of both cyclodextrin and guest increases straightly with increase in the amount of water as the complexation occurs more readily. However, the amount of water should not be increased considerably, because as the amount of water is further increased, the cyclodextrin and the guest may be so dilute that they do not get in contact as easily as they do in a more concentrated solution. An optimum level of water is desirable to ensure that complex formation is obtained at a sufficiently fast rate. When oils with high molecular weight are used, more water is needed for dispersion and separation of the oil molecules by good mixing to separate or isolate the oil molecules from each other, which increases the solubility of the oil [47, 56, 59, 60].

1.4.4 Solution Dynamics

In solution state, more cyclodextrin molecules become available than in crystalline state, because in crystalline form, only the surface molecules of the cyclodextrin crystal are available for complexation. By heating the system, the solubility of the cyclodextrin and guest increases, so the probability of complex formation increases [47, 56].

1.4.5 Volatile Guests

Fine and unstable guests can evaporate during complex formation, especially if heat is used. In such cases, this problem can be prevented by using a sealed reactor or by refluxing the volatile guests back to the mixing vessel [47, 56].

1.5 Inclusion Complex Formation Methods

Because of both hydrophobic and hydrophilic properties, cyclodextrins can form inclusion complex with various organic and inorganic compounds, especially with various guest molecules with suitable properties, polarity, and dimensions. Driving force in formation of inclusion complex between host and guest is noncovalent bonding such as hydrogen bonding, van der Waals interaction, and hydrophobic interactions. Several methods are used to form cyclodextrin complexes. They are described in the following sections [47, 56, 59, 60].

1.5.1 Coprecipitation

Coprecipitation is the most common method in formation of inclusion complex of various compounds with cyclodextrin, in which cyclodextrin is dissolved in water and the guest molecule is added while stirring the cyclodextrin solution. By choosing the appropriate concentration of guest and host, the solubility of the cyclodextrin–guest complex will be increased as the complex formation reaction proceeds or as cooling is applied. In many cases, for formation of inclusion complex, the solution of cyclodextrin and guest must be cooled while stirring before a precipitate is formed [47, 56].

For collection of the precipitate, various methods can be used such as decanting, centrifugation, or filtration. For purification of the precipitate, it may be washed with a small amount of water or other water-miscible solvent such as ethyl alcohol, methanol, or acetone. After formation of inclusion complex, crystalline complex may be washed with solvent to remove residual impurities or unreacted monomers [47, 56, 59, 60].

In this method, high volume of water is needed because of the limited solubility of the cyclodextrin and guest molecule. Therefore, time and energy for heating and cooling may become important cost factors, which are uneconomical. Also, at the end of reaction, removing the residual water is difficult and may be a concern, which can be reduced in many cases by recycling the mother liquor. In this method, nonionic surfactants reduce complex formation; in such cases additives, such as ethanol, can promote complex formation in the solid or semisolid state [47, 56].

1.5.2 Slurry Complex Formation

In slurry method, cyclodextrin is not completely dissolved in the aqueous solution and inclusion complex between cyclodextrin and guest is formed in a saturated aqueous phase. After saturation of aqueous phase, the complex will crystallize or precipitate out of the aqueous phase and the complex can be collected with the coprecipitation method in which the complexation time depends on the guest. In this method, less amount of water is needed; size of the reactor is small; and ambient temperature is maintained (most important factor), but rate of reaction is slow. To increase the rate of reaction, temperature can be increased, but it may destabilize the complex and the complexation reaction may not be able to take place completely [61–63].

1.5.3 Paste Complexation

This method is same as slurry method, but in this method a small amount of water is mixed with the cyclodextrin using a mortar and pestle, or on a large scale using a kneader. Required time for complex formation is dependent on the guest and the amount of water used in the paste. The resulting complex can be dried directly or washed with a small amount of water and collected by filtration or centrifugation [61–63].

1.5.4 Wet Mixing and Heating

In this method, the amount of water needed for reaction is low which may be found in a filter cake from the coprecipitation or slurry methods. The guest and cyclodextrin are thoroughly mixed and placed in a sealed container, heated to about 100 °C, and then the content is removed and dried. Efficiency of this method is strongly dependent to amount of water; the degree of mixing and the heating time have to be optimized for each guest [61–63].

1.5.5 Extrusion

In this method, the inclusion complex is formed after heating and mixing of cyclodextrin, guest, and water, premixed or mixed when they are added to the extruder. Degree of mixing, amount of heating, and time can be controlled in the barrel of the extruder. Resulted complex may be dried by heating or cooling. However, the heat-labile guests decompose in this method [61–63].

1.5.6 Dry Mixing

This method is the simplest method for formation of inclusion complex in which no water is needed. It is performed at ambient temperature.

In large-scale mixing, the complex formation is hard and the process is time consuming. Oils or liquid guests can be complexed by simply adding guest to the cyclodextrin and mixing them together. After inclusion, a washing step is needed [61–63].

1.6 Methods for Drying of Complexes

After the formation of inclusion complex, the complexes can be dried in an oven, fluid-bed dryer, or any other dryer, but it is important that the complex does not get destroyed during the drying process. The heating rate and degree are dependent on the type of guest [61–63].

1.6.1 Highly Volatile Guests

For volatile guests with boiling temperatures under 100 °C, drying must be done at a lower temperature otherwise they will decompose. The drying temperature must be a few degrees below the boiling temperature to protect the guest [61–63].

1.6.2 Spray Drying

To avoid too large and blocking particles, precipitation must be controlled. Therefore, spray drying can be used. When the guests are volatile, some optimization of drying conditions is required in order to reduce the losses. In this method, highly volatile and heat-sensitive guests may decompose [61–63].

1.6.3 Low-Temperature Drying

When low temperature is used for drying the complex, loss of extremely volatile guests is minimum. Desiccator or freeze dryer may be used to dry these complexes. Freeze drying is especially useful for heat-unstable guests and soluble complexes such as hydroxypropylated cyclodextrin complexes [61–63].

1.7 Release of the Complex

After the formation, an inclusion complex and dry processing, the crystalline complex is very stable at ambient temperatures and dry conditions, but water and heat can displace the complexed guest by another guest. When complex is in water, it dissolves in water, and then it is displaced by water molecules and released. Each guest complex has special solubility and specific rate of release. There is an equilibrium between free, complexed cyclodextrin and the guest. In stimuli-responsive copolymers which multiple guest components are formed inclusion complex with CD, released rate of guest molecules is not same as in the original guest mixture [83–85].

1.8 Application of Inclusion Compounds

Until 1970, use of cyclodextrins was at small scale. In the last decade, biotechnological advancements have imparted improvements in cyclodextrin production and their usage. The wide-range application of cyclodextrin led to commercial availability of purified cyclodextrins and its derivatives, which has lowered their production costs. Cyclodextrins are actually nontoxic and their ability to change guest properties via formation of inclusion complex leads to a large number of applications related to pharmaceutical chemistry, food technology, analytical chemistry, chemical synthesis, and catalysis. As already mentioned earlier, increased solubility of guest molecule, stabilization of unstable compounds, and long-term protection of color, odor, flavor, and guest molecule against oxidation, chemical reactions, and environmental parameters are promising ways with various applications [61–63, 83–85].

1.8.1 Characterization of Inclusion Complexes

Inclusion complexes of guest molecules and cyclodextrins can be characterized by a variety of spectrometric methods, such as ^{13}C NMR (13-carbon nuclear magnetic resonance), ^1H NMR (hydrogen nuclear magnetic resonance), XRD (X-ray diffraction), DSC (differential scattering calorimetry), FTIR (Fourier transform infrared spectrometry), and SEM (scanning electron microscope). The most precise method to investigate the formation of inclusion complex between cyclodextrin and guest molecules in solution is proton nuclear magnetic resonance (^1H NMR) spectroscopy. Insertion of a guest molecule into the hydrophobic cavity of the cyclodextrin will result in the chemical shift of guest and host molecules in the NMR spectra. The magnitude of the shifts increases slightly on going from α-CD to β-CD and dramatically on going to γ-CD. The large chemical shifts will be observed for the protons attached to C_3 and C_5, which are located in the inner cavity of cyclodextrin due to the inclusion phenomena and move significantly up-field in the ^1H NMR spectrum in the presence of a guest. The chemical shift change ($\Delta\delta$) is defined as the difference in chemical shift change, positive sign means a down-field shift and negative sign means an up-field shift [83–85].

Due to the rigid conical shape of the cyclodextrins, the shift for the C_5 protons is larger than that for the C_3 protons. C_1, C_2, and C_4 protons located at the exterior of the cavity therefore show only a marginal up-field shift. In the ^{1}H NMR spectra of inclusion compounds, the signals of the guests are significantly broadened. By incorporation of guest molecule in cyclodextrin cavity, physical and chemical properties of guest molecule, such as electronic absorption, fluorescence, phosphorescence, and optical rotation properties electronic absorption, fluorescence, phosphorescence, chemical shift, and optical rotation properties, may be changed and confirmed formation of inclusion complex and entrapment of guest molecule in the cavity of host molecule.

Crystalline structure of inclusion compound can be detected with X-ray diffraction, especially when guests are organometallic compounds [83–87].

FTIR is useful to approve the existence of guest and host molecules in their inclusion complexes. Thermal analysis is another method which indicated formation of inclusion complex between cyclodextrin and guest molecule which melting peaks of the guests is destroy after complex formation. Thermal analysis investigates decomposition behavior of guests present inside cyclodextrins and the increase of the stability of the guest when included above the decomposition temperature of the uncomplexed molecule, which is used in a variety of applications [83–87].

1.9 Applications of Cyclodextrins

Cyclodextrins and their derivatives have been used widely in various industries because of the formation of inclusion complexes and entrapped hydrophilic and hydrophobic guest molecules in their capacity. This property of led to their applications in pharmaceutical, chemical, and food industries. In pharmaceutical industry, formation of inclusion complex between drug and cyclodextrin increased water solubility and bioavailability of drug. They can act as carrier and deliver drug to target cell and respond to some parameters such as pH, temperature, and light by releasing the drug molecules from the CD cavities at controlled rate and slowly. Encapsulation may lead to beneficial changes in the chemical and physical properties of the guest molecules and increase shelf life of guest molecules:

- stabilizes light-sensitive, oxygen-sensitive, and highly volatile components;

- modifies chemical and physical properties;
- increases solubility of guest molecules with low solubility;
- protects against degradation by microorganisms, UV light, oxidation, and chemical reactions;
- masks unpleasant smell or taste, pigments, or the color of guest molecules.

Because of these properties, cyclodextrins or their derivatives are suitable for applications in analytical chemistry, agriculture, chemical catalysis, the pharmaceutical field, food, cosmetic, skin, and toiletries [49].

1.9.1 Cosmetics, Personal Care, and Make Up

Cyclodextrins act as a new way to achieve new and effective products endowed with a satisfactory biological activity and an efficient delivery on the skin, hair, and nails.

One of the most important uses of cyclodextrins or their derivatives is in cosmetic preparation, mainly in volatile suppression of perfumes, room fresheners, and detergents by controlled release of fragrances from inclusion complex. They have major benefits in this area. They solubilize and stabilize sensitive components; improve solubility of oil guest molecule, such as vitamins; improve the absorption of active compounds; stabilize emulsions; reduce or eliminate bad smells of some cosmetic ingredients; protect flavor and deliver flavor in lipsticks and creams; are water solubility; enhance thermal stability of oils; and reduce the loss of active component through photo-destruction, oxidation, volatilization [26–28, 88].

Cyclodextrins are used in toothpaste, skin creams and lotions, liquid and solid fabric softeners, paper towels, tissues and underarm shields. Their major benefits in this section are stabilization of volatile segments, fragrance control, converting liquid ingredient to a solid form. Skin lotions and creams based on cyclodextrins have stable perfumes against evaporation and oxidation; and antimicrobial efficacy of these products is also improved because of encapsulation [29].

Cyclodextrins mask odors in dishwashing and laundry detergent compositions, increase the availability of drug with low solubility, increase the availability of antimicrobial triclosan in toothpastes, limit the interaction between the UV filter and the skin in sunscreen lotions, stabilize skin lotions against UV and light, increase performance and shelf life of self-tanning emulsions or creams [26–28, 88].

1.9.2 Foods and Flavors

In food industry, the formation of inclusion complex between cyclodextrins and volatile oils or liquids, flavors, and colors provides a promising method for encapsulating taste and odor of food products and stabilizing them during use. They are used in food industry to remove cholesterol from products such as milk, butter and eggs, have a texture-improving effect on pastry and on meat products, reduce bitterness, ill smell, and taste, stabilize flavors, emulsions, such as mayonnaise, margarine, or butter creams, encapsulate flavor throughout many rigid food-processing methods of freezing, thawing, and microwaving, stabilize fish oils, enhance flavor for alcoholic beverages, such as whisky and beer, and to remove virulence of citrus fruit juices [15–17, 23].

Another promising alternative of cyclodextrins is encapsulation of natural and artificial flavors in the form of volatile oils or liquids, which is important in food industry [51, 88]. The most usual use of CD in food industry is to remove unsafe oils and cholesterol from animal products, such as eggs, dairy products; thus, they improve the frying property of fat, reduce smoke formation, foaming, browning, and deposition of oil residues on surfaces. Some additives such as flavonoids and terpenoids are antioxidative and antimicrobial, so they are good for human health because of their properties, but they cannot be utilized in food stuffs owing to their very low aqueous solubility and bitter taste [15–17, 23, 89].

1.9.3 Pharmaceuticals

Many drugs are hydrophobic and show low solubility in water, which restrict their application in drug delivery systems, by formation of inclusion complexes provides numerous advantages in pharmaceutical formulation development for oral, intravenous, ocular, and subcutaneous administration are microencapsulated to achieve controlled, targeted, or triggered release of active ingredients and increase solubility, safety, and bioavailability of poorly soluble drugs by increasing the drug solubility. A drug substance has to have a certain level of water solubility to be readily delivered to the cellular membrane, but it needs to be hydrophobic enough to cross the membrane. Formation of inclusion complex between drug and cyclodextrin improves drug solubility, stability, and delivery through biological membranes [29–32].

Hydrophobic drugs entrapped in cyclodextrins cavity and cyclodextrins act as host for drug molecules and carry the drug molecule by maintaining the hydrophobicity in solution and delivering them to the surface of the biological membrane, for example, skin, mucosa, or the cornea, where they partition into the membrane. Cyclodextrins increase drug solubility in water, so availability of drug at the surface of the biological barrier increases and permeation of the drug is enhanced. Therefore, they can be used in various drug delivery systems [3, 29–32, 40].

Drugs usually did not have sufficient solubility in water and combination of organic solvents, surfactants, and extreme pH conditions must be used, which often causes irritation or other adverse reactions in organisms. In general, cyclodextrins have numerous applications in the pharmaceutical industry. They increase water solubility of hydrophobic substances, improve bioavailability, increase the pharmacological applicability and effect, reduce the drug dosage, simplify the handling and storage of volatile products, drug delivery to target cell, improve the stability of substances, increase resistance of guest molecule to hydrolysis, oxidation, heat, light, and metal salts, reduce skin damage for the dermal route, and reduce the effects of bitter or irritant tasting and bad smelling drugs [3, 29–32, 49].

When cells and enzymes are encapsulated, bioreactors' and biocatalysts' efficiency and thermal and operational stabilities have been improved; their recovery also becomes easy. Cell encapsulation is used to achieve high-throughput screening in directed evolution experiments in molecular biology. In recent years, cyclodextrins have significant effect in anticancer drugs because of their limited aqueous solubility (hydrophobicity), ability to degrade in gastrointestinal fluids, insufficient in vitro stability, low bioavailability, short in vivo stability, affinity for intestinal and liver cytochrome, poor intestinal permeability. Strong dose-dependent side effects of promising anticancer drugs have long been obstacles in the treatment of cancer [39]. Cyclodextrins are competent enough to overcome certain forms of these associated drawbacks of anticancer drugs and also other hydrophobic drugs [32, 40].

1.9.3.1 Drug Delivery

Cyclodextrin-based nanocarriers are good candidate for delivering anticancer drugs to target cell when other cells are not damaged. In this manner, anticancer drug is incorporated into cyclodextrin cavity as a carrier. Encapsulation of drug in cyclodextrin may increase the drug-loading capacity and solubility, entrapment efficiency, prolong the existence of the drug in systemic circulation, reduce toxicity, control drug release, protect drug, release in a slow and targeted manner. Some drugs such as anticancer drugs must be released at target cell and must not be absorbed from gastrointestinal tract, so encapsulation of these drugs is the most important way because of their bulky and hydrophilic nature; if any absorption occurs, it is by passive diffusion [3, 32, 40].

In addition to encapsulation, polymeric hydrogels based on cyclodextrins are good candidates for biomedical applications because of their biocompatibility and led to designing a delivery formulation that promotes gelation and drug loading simultaneously. Supramolecular hydrogels are formed based on inclusion complex formation between

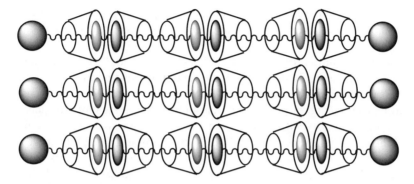

Figure 1.2 Inclusion complex between high-molecular-weight PEO and α-CD [90–93].

CDs and polymers in which intermolecular conjugation between drugs and hydrogels is due to a cross-linking reaction and is used as drug delivery systems [32, 90–93].

Cyclodextrin and PEO are biocompatible and bioabsorbable; supramolecular hydrogels containing these polymers are used as injectable drug delivery systems and would become effective for implementation. Li *et al.* reported inclusion complex-based self-assemble supramolecular hydrogels via formation of inclusion complex formation between α-CD and poly(ethylene oxide) (PEO) based on hydrogen bonding, which are reversible and modified solubility of PEO (Figure 1.2) [90–93].

Formation of inclusion complex of bioactive agents such as drugs, proteins, vaccines, or plasmid DNAs with cyclodextrins changes viscosity of the system. Hydrogel networks formed at room temperature without any contact with organic solvents that can be injected into the tissue under pressure because of the thixotropic property, and served as a depot for controlled release. By increasing the molecular weight of PEO, release rate can be decreased sharply. This is because of the chain entanglement effect and different complex stability [94].

Supramolecular hydrogels based on PEO/α-CD are soluble in aqueous environments due to the hydrophilic nature of PEO, which is a drastic problem for in vivo use of high-molecular-weight PEO of more than 10,000 Da; another problem is that high-molecular-weight PEO is not biodegradable and cannot be filtered through human kidney membranes due to the large hydrodynamic radius [94]. Hydrogel network based on α-CD and Pluronic (PEO–PPO–PEO) triblock copolymers via intermolecular hydrophobic interaction of the middle blocks with cyclodextrin was synthesized and characterized by

Ni et al. This can further strengthen the hydrogel network for formation of more stable hydrogels for long-term controlled release of drugs [95].

Inclusion complexes are formed between α-CD and PEO blocks. They formed a hydrogel network in which gelation regions became larger by increasing the concentration of α-CD. Bioactive agents such as drugs, proteins, vaccines, or plasmid DNAs can be incorporated into the gel at room temperature without any contact with organic solvents [95]. Due to the hydrophilic nature of PEO, dissociation of α-CD/PEO hydrogels in aqueous environments is rapid and release of bioactive agents from α-CD/PEO hydrogels is fast [94]. Li et al. reported another hydrogel based on PEO, α-CD, and L-phenylalanine in which peptide linkages were enzymatically degraded by papain (Figure 1.3) [93–95].

1.9.3.2 Gene Delivery

Cationic polymers can be modified and functionalized via formation of inclusion complex by cyclodextrins and can be used as gene-delivery carriers, which formed linkage between the polymers and DNA. Li and coworkers reported that inclusion complex formation between α-CD and PEO repeating units from PEO-ran-PPO led to grafted oligoethylenimine (OEI) chains on the CD moieties [94]. Inclusion complex formation improved mobility and flexibility of the polyrotaxanes and caused interaction between cationic α-CD rings with DNA or the cellular membranes [94–96].

Inclusion complex formation between dimethylaminoethyl-modified α-CDs and PEO was synthesized which chain capped by cleavable end groups where the α-CD possesses tertiary amines which is a vital process in the intracellular trafficking of the DNA after being delivered into the cells by the networks [97–103].

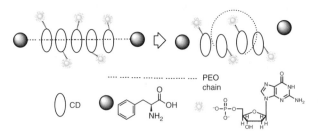

Figure 1.3 Schematic illustration of drug-conjugated polyrotaxane and the concept of drug release [93–95].

1.9.4 Cyclodextrin Applications in Agricultural and Chemical Industries

Cyclodextrins form inclusion complexes with a wide variety of hydrophobic or hydrophilic guest molecules such as agricultural chemical agents including insecticides, fungicides, herbicides, rodenticides, attractants, repellents, pheromones, and growth regulators. In agriculture, cyclodextrins are used for pesticide complexation. Formation of inclusion complex between cyclodextrin and agricultural chemical agents increased their stability against UV light, oxidation, and environmental reactions, standardized their compositions, improved wettability, controlled dosing, simplified handling of dry powders, reduced packing and storage costs, made the manufacturing processes more economical, and reduced labor costs [20, 27].

Cyclodextrins delay germination of seed and inhibit degradation of the starch with amylases, so initially the plant growth is more slow. Because of cyclodextrins' specific adsorption character, they have been frequently applied for the improvement of the separation parameters (mainly chiral-separation capacity) of various chromatographic methods. The efficiency of chromatographic separation technologies can be increased by the modification of both the stationary and the mobile phase of the system. This can be obtained with cyclodextrins as additives of stationary phase or modifier of the mobile phase [20–22].

The stationary phases of these columns contain immobilized cyclodextrins or derived supramolecular architectures. Cyclodextrins and their derivatives show specific adsorption property and chiral-separation capacity. Therefore, they can be used for the separation of enantiomers with markedly different biological activity. Cyclodextrins can form complexes with a wide variety of organic and inorganic compounds. They are useful for separate complex mixtures of organic and inorganic compounds at trace level in complicated accompanying matrices such as liquid chromatography, thin layer, gas–liquid, high-performance liquid, gas chromatography, size exclusion chromatography, gel permeation chromatography, and electrically driven separation methods, capillary electrophoresis, supercritical fluid chromatography, and ultra-performance liquid chromatography [22–25].

Chiral activity or mask-contaminating compounds are indicated by nuclear magnetic resonance, allowing more accurate determinations [31, 37]. Cyclodextrin led to catalytic ability of guest molecule bind with hydrophobic cavity of cyclodextrin and can be used as enzyme mimics and also as biocatalytic processes by increasing the enantioselectivity. Cyclodextrins can enhance solubility of low-solubility guest molecules, improve bioavailability, biodegradation, and bioremediation in microbial and plant growth, accelerate the degradation rate of hydrocarbons, produce higher biomass yield and better utilization of hydrocarbon as a carbon and energy source, and decrease the toxicity of soil. Low cost, biocompatibility, and effective degradation make cyclodextrins a useful tool for bioremediation process [37].

Cyclodextrin causes higher enantioselectivity and increases preferential attack by the reagent by the formation of inclusion complex with the guest molecule and accelerates the rate of hydrolysis [31, 37]. Inclusion complex formation of cyclodextrin with guest molecules modified guest properties such as solubility of organic contaminants, masked of toxic or organic pollutants, water and atmosphere, remove toxic aromatic compounds from wastewaters, improve stabilizing action, encapsulation, adsorption and masked pollutant substances [20, 31, 37].

1.9.5 Encapsulation of Various Guest Molecules

Encapsulation is a process to entrap one substance (active agent or core) within another substance (carrier material, membrane, shell, capsule, or matrix), which is useful in various applications such as to improve delivery corrosion inhibitors, drugs especially anticancer drugs, bioactive molecules and living cells into foods, proteins and lipids, polysaccharides, chemical agents, and so on. Microencapsulation process can be used in various industries such as pharmaceuticals products, food industry, cosmetics and fragrances, textiles, paints and coatings, adhesives, toner applications [83–87].

Encapsulation shell must be able to form a barrier between the guest molecule and its surrounding environment to save internal phase from degradation and control their release rate. Encapsulation of compounds can be done via various techniques such as spray drying, spray bed drying, fluid-bed coating, spray chilling, spray cooling, or melt injection. Spray drying is one of the oldest and the most widely used encapsulation techniques, which is a flexible, continuous, and an economical operation, but it has several disadvantages such as complex equipments, nonuniform conditions in the drying chamber, and it is not always easy to control particle size. One of the most important and useful method for encapsulation of organic and inorganic compounds is inclusion complex formation between guest molecule and cyclodextrins, which can remove bad taste or smell, separate taste and aroma of guest, stabilize food ingredients, increase guest bioavailability, and also provide barriers between sensitive bioactive materials and the environment [83–87].

Encapsulation of active ingredients show main advantages such as improve guest stability in final products and during processing, stabilized volatile actives such as aroma, led to controlled/triggered/targeted release of guest molecule, remove color, taste, and odor of the actives or unpleasant feelings during eating, protect the guest molecule against oxidation or deactivation due to reactions with reactive species from the environment and handling of toxic materials became easier and more safe [28, 32, 33].

Encapsulation has been used in textile industry and it enhances the properties of finished textile and shows absorption or releases heat in response to changes in environmental temperatures. Increasing temperature causes change in materials' melt phase, absorbs excess heat, and feels cool. Encapsulation may be done for various purposes such as to entrap color in paint industries; mask taste, odor, or color in food industry; protect ingredients from oxidation or degradation reactions caused by exposure to light, heat, moisture, or oxygen; stabilize lactic acid bacteria; and increase the shelf life of guest molecule and final products [28, 32, 33, 83–87].

1.9.6 Supramolecular Polymer Base on Host–Guest Complexes

Cyclodextrins have hydrophobic cavity and hydrophilic surface, so they are responsible for the adjustment of various kinds of molecules and formed inclusion compounds with wide range molecules. Self-assembly reactions have been used in formation of large molecular, supramolecular, and polymolecular structures. They illustrate supramolecular chemistry that contains catenanes and rotaxanes in nanometer-scale structures with potential applications via incorporating the aspect of entanglement into chemical systems [17, 57, 83–87].

Supramolecular structures based on inclusion complex formation is by the formation of hydrophobic interactions, van der Waals interaction, hydrogen bonding, donor–acceptor interactions, and transition metal coordination and they are able formed in both organic solvents and aqueous media. In host–guest inclusion complex formation, hydrophobic guest molecules can incorporate in host cavity, where host is a large molecule or aggregate such as a synthetic cyclic compound [83–87]. Cyclodextrins' internal cavity can preserve organic host molecules and is used to thread more than one cyclodextrin rings in solution resulting in certain supramolecular structures [17, 57].

1.10 Characterization and Experimental Techniques

After synthesis of inclusion, complex cyclodextrin with various guest molecules, these complexes must be characterized. Some of the techniques are discussed here.

1.10.1 NMR Spectroscopy

The simplest and informative methods for investigation of the formation of inclusion complex of cyclodextrins are ^{1}H NMR and ^{13}C NMR spectroscopies. By comparing NMR spectroscopy of pure cyclodextrin, guest and inclusion complex, formation of complex can be proven [83–87].

The recent advances of NMR techniques have allowed a much more detailed structural elucidation of CDs and their complexes by measuring chemical shift changes as a function of concentration ^{1}H NMR relies on changes in chemical shifts caused by the guest and the host on each other. Hydrogens H_3 and H_5 are located inside the cavity of cyclodextrin and other hydrogens (H_1, H_2, and H_4) are located on the exterior of the cavity. H_3 is located near the wider rim of the cyclodextrin cavity, while the H_5 hydrogens form a ring near the narrower rim of the methylene (H_6) bearing the primary hydroxyl groups. By formation inclusion complex and incorporation of guest molecule in cyclodextrin cavity, resonance positions are the average chemical shifts in the free and complexed cyclodextrins weighted by the fractional population in each state and H_3 and H_5 showed up-field shift which this change indicates that the guest molecule is inserted into the cyclodextrin cavity [83–87].

1.10.2 FTIR Spectroscopy

FTIR is an effective analytical tool for identification of unknowns, sample screening and profiling samples based on vibrate and bonds and groups of bonds at characteristic frequencies which allows detection of complex formations in solid phase and implication of the different functional groups of guest and host molecules in the inclusion. FTIR provides information about the chemical bonding or molecular structure of materials, whether organic or inorganic. FTIR analyzes the significant changes in the shape and position of the absorbance bands of guest, cyclodextrin, physical mixture, and inclusion complexes [83–87].

1.10.3 X-ray Diffraction

Powder XRD is one of the most precise methods for characterizing inclusion complex formation between cyclodextrin and guest molecules. By changing the position of peak and also intensities of the XRD peaks, formation of inclusion complex between cyclodextrin and guest molecules can be identified [83–87].

1.10.4 Thermogravimetric Analysis

Thermogravimetric analysis shows mass changes continuously as a function of temperature or time as the temperature by temperature and a plot of mass or mass percentage as a function of time is called a thermogram or a thermal decomposition curve. Inclusion complex formation stabilizes guest molecule and delays mass decomposition of guest molecule. [83–87].

1.10.5 UV–Vis Spectral Changes

Equilibrium constants, binding constant, and formation of inclusion complex between guest molecule and cyclodextrin are determined by UV–Visible absorption spectroscopy. In inclusion complex process, cyclodextrin acts as active site analogs and causes absorption of drugs in the body, separates aromatic hydrocarbons in petroleum industry and reduces volatility of insecticides in agriculture. Pure cyclodextrin did not show any UV–vis absorption, so conventional UV–vis standard calibration plot for CD could not be prepared and an indirect UV–vis spectroscopy method for the determination of the conjugation degree of CD and guest molecule is used. Binding constant is related to incorporation the guest molecule into hydrophobic cavity of CD and by formation of inclusion complex, a significant change in the UV–vis spectrum of guest molecule and the absorbance change seen and can be used for the preparation of the reciprocal plot from the Hildebrand–Benesi equation [83–87].

1.10.6 Phase Solubility Technique

Association constants and stoichiometry of inclusion complex formation also can be determined by phase solubility technique. In this method, an equal weight of the guest molecule is added into each of several vials or ampoules to be complexed and a constant volume of solvent is added to each container. After that, portions of the complexing agent are added to the vessels and closed and the contents brought to solubility equilibrium by prolonged agitation at constant temperature. The solution phases are then analyzed for total solute content. A phase diagram is constructed by plotting the molar concentration of dissolved solute, found on the vertical axis, against the concentration of complexing agent added on the horizontal axis [104].

References

1 Villiers, A. (1891) Sur la fermentation de la feculepar l'action du ferment butyrique. *C. R. Hebd. Seances Acad. Sci.*, **112**, 536–538.
2 Szetjli, J. (1998) Introduction and general overview of cyclodextrin chemistry. *Chem. Rev.*, **98**, 1743–1753.

3 Stella, V.J. and Rajewski, R.A. (1997) Cyclodextrins: their future in drug formulation and delivery. *Pharm. Res.*, **14**, 556–567.
4 Szetjli, J. (1989) Downstream processing using cyclodextrins. *TIB-TRCH*, **7**, 171–174.
5 Hedges, R.A. (1998) Industrial applications of cyclodextrins. *Chem. Rev.*, **98**, 2035–2044.
6 Dass, C.R. and Jessup, W. (2000) Cyclodextrins and liposomes as potential drugs for the reversal of atherosclerosis. *J. Pharm. Pharmacol.*, **52**, 731–761.
7 Eastburn, S.D. and Tao, B.Y. (1994) Applications of modified cyclodextrins. *Biotechnol. Adv.*, **12**, 325–339.
8 Szejtli, J. (1988) *Cyclodextrin Technology*, Kluwer Academic Publisher, Dordrecht.
9 Jansook, P., Ritthidej, G.C., Ueda, H. et al. (2010) γ-CD/HP γ-CD mixtures as solubilizer: solid-state characterization and sample dexamethasone eye drop suspension. *J. Pharm. Pharm. Sci.*, **13**, 336–350.
10 Schardinger, F. (1903) Uber die Zulassigkeit des Warmhaltens von zum Gebus bestimmten Nahrungsmittel mittelst Warme speichernder Apparate, sog. Thermophore. Wien. *Klin. Wochenschr.*, 468–474.
11 Schardinger, F. (1903) Uber Thermophile Bakterien aus verschiedenen Speisen und Milch, sowie uber einige Umsetzungsprodukte derselben in kohlenhydrathaltigen Nahrlosungen, darunter krystallisierte Polysaccharide (Dextrine) aus Starke. *Z. Untersuch. Nahr. Genussm.*, **6**, 865–880.
12 Schardinger, F. (1911) Bildung kristallisierter Polysaccharide (Dextrine) aus Starkekleister durch Microben. *Zentralbl. Bakteriol. Parasitenk. Abt. II*, **29**, 188–197.
13 French, D., Knapp, D.W. and Pazur, H. (1950) Studies of the Schardinger dextrins. VI. The molecular size and structure of β-dextrin. *J. Am. Chem. Soc.*, **72**, 5120–5152.
14 Freudenberg, K.,Cramer, F., and Plieninger, H. (1953) Ger. Patent 895,769.
15 Tonelli, A.E.J. (2008) Cyclodextrins as a means to nanostructure and functionalize polymers. *J. Incl. Phenom. Macrocyl. Chem.*, **60**, 197–202.
16 Chen, Q.R., Liu, C. and Liu, F.Q. (2010) Application of cyclodextrins in polymerization. *Prog. Chem.*, **22**, 927–937.
17 Steed, J.W. and Atwood, J.L. (2009) Molecular guests in solution, in *Supramolecular Chemistry*, 2nd edn, John Wiley & Sons, Ltd., UK Chapter 6, p. 307.

18 Li, J. and Loh, X.J. (2008) Cyclodextrin-based supramolecular architectures: syntheses, structures, and applications for drug and gene delivery. *Adv. Drug Deliv. Rev.*, **60**, 1000–1017.
19 Harada, A., Takashima, Y. and Yamaguchi, H. (2009) Cyclodextrin-based supramolecular polymers. *Chem. Soc. Rev.*, **38**, 875–882.
20 Komiyama, M. and Monflier, E. (2006) Cyclodextrin catalysis, in *Cyclodextrins and Their Complexes. Chemistry, Analytical Methods, Applications* (ed. H. Dodziuk), Wiley-VCH, Verlag GmbH & Co.KGaA, Weinheim, Germany Chapter 4, p. 93.
21 Karakhanov, E.A. and Maximov, A.L. (2010) Molecular imprinting technique for the design of cyclodextrin based materials and their application in catalysis. *Curr. Org. Chem.*, **14**, 1284–1295.
22 Cabral Marques, H.M. (2010) A review on cyclodextrin encapsulation of essential oils and volatiles. *Flavour Frag. J.*, **25**, 313–326.
23 Singh, J., Dartois, A. and Kaur, L. (2010) Starch digestibility in food matrix: a review. *Trends Food Sci. Technol.*, **21**, 168–180.
24 Li, J.J., Zhao, F. and Li, J. (2011) Polyrotaxanes for applications in life science and biotechnology. *Appl. Microbiol. Biotechnol.*, **90**, 427–443.
25 Villalonga, R., Cao, R. and Fragoso, A. (2007) Supramolecular chemistry of cyclodextrins in enzyme technology. *Chem. Rev.*, **107**, 3088–3116.
26 Buschmann, H.J. and Schollmeyer, E. (2002) Applications of cyclodextrins in cosmetic products: a review. *J. Cosmet. Sci.*, **53**, 185–191.
27 Hashimoto, H. (2006) Cyclodextrin applications in food, cosmetic, toiletry, textile and wrapping materiel fields, in *Cyclodextrins and Their Complexes. Chemistry, Analytical Methods, Applications* (ed. H. Dodziuk), Wiley-VCH, Verlag GmbH & Co. KGaA, Weinheim, Germany Chapter 16, p. 452.
28 Tarimci, N. (2011) Cyclodextrins in the cosmetic field, in *Cyclodextrins in Pharmaceutics, Cosmetics and Biomedicine* (ed. E. Bilensoy), John Wiley & Sons, Inc., London Chapter 7, p. 131.
29 Macaev, F., Boldescu, V., Geronikaki, A. and Sucman, N. (2013) Recent advances in the use of cyclodextrins in antifungal formulations. *Curr. Top. Med. Chem.*, **21**, 2677–2683.
30 Van de Manakker, F., Vermonden, T., van Nostrum, C.F. and Hennink, W.E. (2009) Cyclodextrin-based polymeric materials: synthesis, properties, and pharmaceutical/biomedical applications. *Biomacromolecules*, **10**, 3157–3175.

31 Hincal, A.A., Eroglu, H. and Bilensoy, E. (2011) Regulatory status of cyclodextrins in pharmaceutical products, in *Cyclodextrins in Pharmaceutics, Cosmetic, and Biomedicine: Current and Future Industrial Applications* (ed. E. Bilensoy), John Wiley & Sons, Hoboken, NJ.

32 Brewster, M.E. and Loftsson, T. (2007) Cyclodextrins as pharmaceutical solubilizers. *Adv. Drug Deliv. Rev.*, **59**, 645–666.

33 Voncina, B. and Vivod, V. (2013) Cyclodextrins in textile finishing, in *Textile Dyeing* (ed. M. Günay), InTech, Tijeka, Croatia Chapter 3, p. 53.

34 Voncina, B. (2011) Application of cyclodextrins in textile dyeing, in *Textile Dyeing* (ed. P. Hauser), InTech, Tijeka, Croatia Chapter 17, p. 373.

35 Xiao, Y., Ng, S.C., Tan, T.T.Y. and Wang, Y. (2012) Recent development of cyclodextrin chiral stationary phases and their applications in chromatography. *J. Chromatogr. A*, **1269**, 52–68.

36 West, C. (2014) Enantioselective separations with supercritical fluids. *Curr. Anal. Chem*, **10**, 99–120.

37 Morillo, E. (2006Chapter 16) Application of cyclodextrins in agrochemistry, in *Cyclodextrins and Their Complexes. Chemistry, Analytical Methods, Applications* (ed. H. Dodziuk), Wiley-VCH, Verlag GmbH & Co. KGaA, Weinheim, Germany, p. 459.

38 Guo, R. and Wilson, L.D. (2013) Cyclodextrin-based microcapsule materials – their preparation and physiochemical properties. *Curr. Org. Chem.*, **17**, 14–21.

39 Davis, F. and Higson, S. (2011) Cyclodextrins, in *Macrocycles. Construction, Chemistry and Nanotechnology Applications* (eds F. Davis and S. Higson), John Wiley & Sons, Ltd., UK Chapter 6, pp. 190–254.

40 Chilajwar, S.V., Pednekar, P.P., Jadhav, K.R. et al. (2014) Cyclodextrin-based nanosponges: a propitious platform for enhancing drug delivery. *J. Expert Opin. Drug Deliv.*, **11**, 111–120.

41 Tong, J. and Chen, L.G. (2013) Review: preparation and application of magnetic chitosan derivatives in separation processes. *Anal. Lett.*, **46**, 2635–2656.

42 Lu, X. and Chen, Y. (2002) Chiral separation of amino acids derivatized with fluoresceine-5-isothiocyanate by capillary electrophoresis and laser induced fluorescence detection using mixed selectors of beta-cyclodextrin and sodium taurocholate. *J. Chromatogr. A*, **955**, 133–240.

43 Baudin, C., Pean, C., Perly, B. and Goselin, P. (2000) Inclusion of organic pollutants in cyclodextrin and derivatives. *Int. J. Environ. Anal. Chem.*, **77**, 233–242.

44 Koukiekolo, R., Desseaux, V., Moreau, Y. et al. (2001) Mechanism of porcine pancreatic alpha-amylase inhibition of amylose and maltopentaose hydrolysis by alpha-, beta- and gammacyclodextrins. *Eur. J. Biochem.*, **268**, 841–848.
45 Freudenberg, K., Plankenhorn, E. and Knauber, H.A. (1947) Über Schardingers Dextrine aus Stärke. *Chem. Justus Liebigs*, **558**, 1–10.
46 Thoma, J.A. and Stewart, L. (1965) Cycloamyloses, in *Starch, Chemistry and Technology*, vol. **1** Fundamental Aspects, Chapter IX (eds R.L. Whistler and E.F. Paschall), Academic Press, New York, p. 209.
47 Szejtli, J. (1982) *Cyclodextrins and Their Inclusion Complexes*, Akadémiai Kiadó, Budapest, Hungary.
48 Buschmann, H.J. and Schollmeyer, E. (2002) Applications of cyclodextrins in cosmetic products: a review. *J. Cosmet. Sci.*, **53**, 575–592.
49 Irie, T. and Uekama, K. (1977) Pharmaceutical applications of cyclodextrins. III.Toxicological issues and safety evaluation. *J. Pharm. Sci.*, **86**, 147–162.
50 Thompson, D.O. (1997) Cyclodextrins-enabling excipients: their present and future use in pharmaceuticals. *Crit. Rev. Ther. Drug Carrier Syst.*, **14**, 1–104.
51 Antlsperger, G. and Schmid, G. (1996) Toxicological comparison of cyclodextrins, in *Proceedings of the Eighth International Symposium on Cyclodextrins* (eds J. Szejtli and L. Szente), Kluwer Academic Publishers, Dordrecht, p. 149.
52 Arima, H., Motoyama, K. and Irie, T. (2011) Recent findings on safety profiles of cyclodextrins, cyclodextrin conjugates, and polypseudorotaxanes, in *Cyclodextrins in Pharmaceutics, Cosmetics, and Biomedicine: Current and Future Industrial Applications* (ed. E. Bilensoy), Wiley, Hoboken, pp. 91–122.
53 Totterman, A.M., Schipper, N.G., Thompson, D.O. and Mannermaa, J.P. (1997) Intestinal safety of water-soluble cyclodextrins in paediatric oral solutions of spironolactone: effects on human intestinal epithelial $CaCO_2$ cells. *J. Pharm. Pharmacol.*, **49**, 43–48.
54 Frömming, K.H. and Szejtli, J. (1994) *Cyclodextrins in Pharmacy. Topics in Inclusion Science*, Kluwer Academic Publishers, Dordrecht.
55 Loftsson, T. (2002) Cyclodextrins and the biopharmaceutical classification system of drugs. *J. Incl. Phenom. Macrocycl. Chem.*, **44**, 63–67.
56 Hirayama, F. and Uekama, K. (1987) Methods of investigating and preparing inclusion compounds, in *Cyclodextrins and Their Industrial Uses* (ed. D. Duchêne), Editions de Santé, Paris, pp. 131–172.

57 Chen, G. and Jiang, M. (2011) Cyclodextrin-based inclusion complexation bridging supramolecular chemistry and macromolecular self-assembly. *Chem. Soc. Rev.*, **40**, 2254–2266.
58 Nimse, S.B. and Kim, T. (2013) Biological applications of functionalized calixarenes. *Chem. Soc. Rev.*, **42**, 366–386.
59 Semsarzadeh, M.A. and Amiri, S. (2012) Preparation and characterization of inclusion complexes of poly(dimethylsiloxane)s with gamma-cyclodextrin without sonic energy. *Silicon*, **4**, 151–156.
60 Semsarzadeh, M.A. and Amiri, S. (2013) Preparation and properties of polyrotaxane from α-cyclodextrin and poly (ethylene glycol) with poly (vinyl alcohol). *Bull. Mater. Sci.*, **36**, 989–996.
61 Buchwald, P. (2002) Complexation thermodynamics of cyclodextrins in the framework hydrogels of β-cyclodextrin crosslinked by acylated poly(ethylene glycol): synthesis and properties of a molecular size-based model for nonassociative organic liquids that includes a modified hydration-shell hydrogen-bond model for water. *J. Phys. Chem. B*, **106**, 6864–6870.
62 Liu, L. and Guo, Q.X. (2002) The driving forces in the inclusion complexation of cyclodextrins. *J. Incl. Phenom. Macrocycl. Chem.*, **42**, 1–14.
63 Rekharsky, M.V. and Inoue, Y. (1998) Complexation thermodynamics of cyclodextrins. *Chem. Rev.*, **98**, 1875–1918.
64 Moya-Ortega, M.D., Messner, M., Phatsawee, J. *et al.* (2011) Cross-drug loading in cyclodextrin polymers: dexamethasone model drug. *J. Incl. Phenom. Macrocycl. Chem.*, **69**, 377–382.
65 Glisoni, R.J., García-Fernández, M.J., Pino, M. *et al.* (2013) β-Cyclodextrin hydrogels for the ocular release of antibacterial thiosemicarbazones. *Carbohydr. Polym.*, **93**, 449–457.
66 Lui, Y.Y., Fan, X.D., Kang, T. and Sun, L.A. (2004) Cyclodextrin microgel for controlled release driven by inclusion effect. *Macromol. Rapid Commun.*, **25**, 1912–1916.
67 Gil, E.S., Li, J., Xiao, H. and Lowe, T.L. (2009) Quaternary ammonium beta-cyclodextrin nanoparticles for enhancing doxorubicin permeability across the in vitro blood–brain barrier. *Biomacromolecules*, **10**, 505–516.
68 Muñoz-Botella, S., Castillo, B. and Martyn, M.A. (1995) Cyclodextrin properties and applications of inclusion complex formation. *Ars Pharm.*, **1995** (36), 187–198.
69 Schneiderman, E. and Stalcup, A.M. (2000) Cyclodextrins: a versatile tool in separation science. *J. Chromatogr. B*, **745**, 83–102.
70 Schmid, G. (1989) Cyclodextrin glucanotransferse production: yield enhancement by overexpression of cloned genes. *Trends Biotechnol.*, **7**, 244–248.

71 Loftsson, T. and Brewster, M.E. (2010) Pharmaceutical applications of cyclodextrins: basic science and product development. *J. Pharm. Pharmacol.*, **62**, 1607–1621.
72 Connors, K.A. (1997) The stability of cyclodextrin complex in solution. *Chem. Rev.*, **97**, 1325–1357.
73 Easton, C. and Lincoln, S. (1999) *Modified Cyclodextrin*, Imperial College Press, London, England.
74 Van Holde, K., Johnson, W.C. and Ho, P.S. (1998) *Principles of Physical Biochemistry*, Prentice-Hall Inc., Upper Saddle River, NJ, USA.
75 Martin, A.N. (1993) *Physical Pharmacy*, 4th edn, Lippincott Williams & Wilkins, Baltimore, MD, USA.
76 Chang, R. (1990) *Physical Chemistry with Application to Biological Systems*, Macmillan Publishing Company, New York, USA.
77 Hirose, K.A. (2001) Practical guide for the determination of binding constants. *J. Incl. Phenom. Macrocycl. Chem.*, **39**, 193–209.
78 Kawaguchi, Y., Nishiyama, T., Okada, M. *et al.* (2000) *Macromolecules*, **33**, 4472–4477.
79 Harada, A., Li, J., Kamachi, M. *et al.* (1998) Structures of polyrotaxane models. *Carbohydr. Res.*, **305**, 127–129.
80 Saenger, W.R., Jacob, J. and Gessler, K. (1998) Structures of the common cyclodextrins and their larger analogues beyond the doughnut. *Chem. Rev.*, **98**, 1787–1802.
81 Denadai, A.M.L., Teixeira, K.I., Santoro, M.M. *et al.* (2007) Supramolecular self-assembly of beta-cyclodextrin: an effective carrier of the antimicrobial agent chlorhexidine. *Carbohydr. Res.*, **342**, 2286–2296.
82 Nascimento, J.C.S., Anconi, C.P.A., Lopes, J.F. *et al.* (2007) An efficient methodology to study cyclodextrin clusters: application to alpha-CD hydrated monomer, dimer, trimer and tetramer. *J. Incl. Phenom. Macrocycl. Chem.*, **59**, 265–277.
83 Amiri, S. and Rahimi, A. (2014) Preparation of supramolecular corrosion inhibitor nanocontainers for self-protective hybrid nanocomposite coatings. *J. Polym. Res.*, **21**, 566, DOI: 10.1007/s10965-014-0566-5, 2014.
84 Amiri, S. and Rahimi, A. (2014) Self-healing hybrid nanocomposite coatings containing encapsulated organic corrosion inhibitors nanocontainers. *J. Polym. Res.*, **22**, 624, DOI: 10.1007/s10965-014-0624-z, 2014.
85 Amiri, S. and Rahimi, A. (2015) Synthesis and characterization of supramolecular corrosion inhibitor nanocontainers for anticorrosion hybrid nanocomposite coatings. *J. Polym. Res.*, **22**, 66, DOI: 10.1007/s10965-015-0699-1, 2015.

86 Amiri, S. and Rahimi, A. (2015) Self-healing anticorrosion coating containing 2-mercaptobenzothiazole and 2-mercaptobenzimidazole nanocapsules". *J. Coat. Polym. Res.*, **23**, 6. doi: 10.1007/s11998-014-9652-1.

87 Amiri, S. and Rahimi, A. (2015) Anti-corrosion hybrid nanocomposite coatings with encapsulated organic corrosion inhibitors. *J. Coat. Technol. Res.*, **12**, 587–593.

88 Del Valle, M.E.M. (2004) Cyclodextrins and their uses: a review. *Process Biochem.*, **39**, 1033–1046.

89 Roux, M., Perly, B. and Djedaini, P.F. (2007) Self-assemblies of amphiphilic cyclodextrins. *Eur. Biophys. J.*, **36**, 861–867.

90 Duchene, D., Ponchel, G. and Wouessidjewe, D. (1999) Cyclodextrins in targeting: application to nanoparticles. *Adv. Drug Deliv. Rev.*, **36**, 29–40.

91 Emara, L.H., Badr, R.M. and Elbary, A.A. (2002) Improving the dissolution and bioavailability of nifedipine using solid dispersions and solubilizers. *Drug Dev. Ind. Pharm.*, **28**, 795–807.

92 Loftsson, T., Brewster, M.E. and Masson, M. (2004) Role of cyclodextrins in improving oral drug delivery. *Am. J. Drug Deliv.*, **2**, 261–275.

93 Carrier, R.L., Miller, L.A. and Ahmed, I. (2007) The utility of cyclodextrins for enhancing oral bioavailability. *J. Control Release*, **123**, 78–99.

94 Li, J., Ni, X. and Leong, K.W. (2003) Injectable drug-delivery systems based on supramolecular hydrogels formed by poly(ethylene oxide)s and alpha-cyclodextrin. *J. Biomed. Mater. Res. A*, **65A** (2), 196–202.

95 Ni, X., Cheng, A. and Li, J. (2009) Supramolecular hydrogels based on self-assembly between PEO–PPO–PEO triblock copolymers and alpha-cyclodextrin. *J. Biomed. Mater. Res. A*, **88A** (4), 1031–1036.

96 Pun, S.H. and Davis, M.E. (2002) Development of a nonviral gene delivery vehicle for systemic application. *Bioconjugate Chem.*, **13** (3), 630–639.

97 Ooya, T., Choi, H.S., Yamashita, A. et al. (2006) Biocleavable polyrotaxane plasmid DNA polyplex for enhanced gene delivery. *J. Am. Chem. Soc.*, **128** (12), 3852–3853.

98 Ulya, N., Tangül, S., Nilüfer, T. et al. (2007) Use of cyclodextrins as a cosmetic delivery system for fragrance materials: Linalool and benzyl acetate. *AAPS Pharm. Sci. Tech.*, **8**, DOI: 10.1208/pt0804085.

99 Loftsson, T., Leeves, N., Bjornsdottir, B. et al. (1999) Effect of cyclodextrins and polymers on triclosan availability and substantivity in toothpastes in vivo. *J. Pharm. Sci.*, **88** (12), 1254–1258.

100 Lim A.H. (2008) Interactions of Amphiphilic Molecules with Dendrimers. Masters of Engineering thesis, Nanyang Technological University.
101 Esfand, R. and Tomalia, D.A. (2001) Poly(amidoamine) (PAMAM) dendrimers: from biomimicry to drug delivery and biomedical applications. *Drug Discov. Today*, **6** (8), 427–436.
102 Tomalia, D.A. and Dvornic, P.R. (1994) What promise for dendrimers? *Nature*, **372** (6507), 617–618.
103 Klajnert, B. and Bryszewska, M. (2001) Dendrimers: properties and applications. *Acta Biochim. Pol.*, **48** (1), 199–208.
104 Higuchi, T. and Connors, K.A. (1965) Phase-solubility techniques. *Adv. Anal. Chem. Instrum.*, **4**, 117–212.

2

Supramolecular Chemistry and Rotaxane

Cyclodextrins (CDs) are cyclic oligomers based on glucopyranose subunits and an important class of host–guest systems in supramolecular chemistry. Their unique hydrophobic cavities enable the sequestration of organic molecules by inclusion complexation due to noncovalent host–guest interactions [1, 2]. Supramolecular host–guest systems are an interaction of two or more organic molecules, ions, or radicals in a unique structural relationship using forces other than full covalent bonds. The most common types of noncovalent host–guest interactions such as hydrogen bonds, ionic bonds, van der Waals forces, and hydrophobic interactions are observed in supramolecular structures [3, 4].

Guest molecules incorporated in CD cavities via perfect binding interactions provide an interesting combination of supra- and macromolecular chemistry with nanotechnology. The combination of CDs and nano- and microgels provides some advantages and may improve the properties and application possibilities for both. Hence, nano- and microgels as carrier may promote self-bonding and fixation of CDs and may form inclusion complexes for various purposes with target applications for long-time activation of the chemical agents. Formation of inclusion complexes between guest molecules and CD may moderate reaction condition, controlled guest uptake and release from inclusion or microgels, and also increased stability of guest molecule against degradation via various methods [5, 6].

2.1 What Is Supramolecular Chemistry

Supramolecular chemistry has been defined by Jean-Marie Lehn, Donald J. Cram, and Charles J. Pedersen, who were awarded the Nobel Prize for their work in this field in 1987 [7].

It can also be expressed as the chemistry of noncovalent bonds. From the definition of supramolecular chemistry, the obvious differences between supramolecular and molecular chemistry can be derived. Molecular chemistry is based on the formation of covalent bonding between various molecules, while supramolecular chemistry is based on the formation of noncovalent interactions (NCIs) between molecules that are associated, combined, or aggregated through hydrogen bonding, electrostatic forces, donor–acceptor interactions, van der Waals interactions. One of the most important aims of this discipline is to create molecular devices that can mimic biomolecular functions [7–9].

2.1.1 History of Supramolecular Chemistry

The first definition of supramolecular chemistry dates back to the end of the nineteenth century. CD inclusion compounds were discovered by Villiers and Hebd in 1891. Coordination chemistry and lock-and-key concept were introduced by Werner (1893) and Emil Fischer (1984), respectively [8–10]. Supramolecular chemistry led to the formation of a wide range of structures such as crown ethers, which were discovered by Pedersen in the 1960s, and development of macrocyclic ligands designed for metal cations and used as ligands in supramolecular complexes with cationic alkali metals [9].

The other type of supramolecular chemistry as ligand type is cryptands, which were reported by Jean-Marie Lehn and led to a great contribution in the synthesis of new materials to the development of modern supramolecular chemistry. Spherand is the other type of macrocyclic systems featuring fully preorganized binding sites, developed by Cram and coworkers in the 1970s [10] (Figure 2.1).

Innovations in supramolecular chemistry based on noncovalent and reversible interactions became a challenge after the 1990s, which were used to develop sensors and methods for electronic and biological interfacing by Willner and coworkers and led to evolution of fast-growing fields in the chemical sciences [11, 12]. By formation of supramolecular

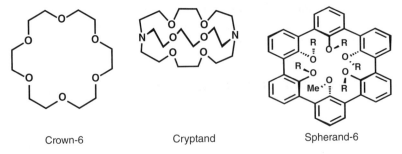

Figure 2.1 Schematic structure of Crown-6, Cryptand, and Spherand-6.

molecules, electrochemical and photochemical motifs are gathered into supramolecular systems and functionality is increased [13]. Supramolecular chemistry can be considered a multipurpose method for a wide range of applications via incorporation of inorganic chemistry and organic chemistry [14].

2.1.2 Concept of Supramolecular Chemistry

Formation of supramolecular structures is possible via intermolecular noncovalent binding interactions of certain chemical, physical, and biological species of greater complexity than molecules themselves that are held them together to organized and features various properties with interdisciplinary field usage [15–17]. Degrees of strength, physical, and chemical properties of supramolecular structures is related to interaction between species that are used [18–20].

As defined by one of its leading proponents, Jean-Marie Lehn, a Nobel laureate, it is the "chemistry of molecular assemblies and of intermolecular bond" or more colloquially as the "chemistry beyond the molecule." Supramolecular species are characterized by the spatial arrangement of their components, their architecture or supramolecular structure, and the nature of the intermolecular bonds that holds these components together [14, 21, 22].

Formation of supramolecular self-assembled compounds is a fast and spontaneous interaction between species, which caused polymer chain extension, but it is important that noncovalent binding is formed between two entities – a host and a guest [23–25].

2.1.2.1 Noncovalent Interactions

Supramolecular chemistry refers to the domain of chemistry beyond that of molecules and focuses on chemical systems consisting of a discrete number of assembled molecular subunits or components. While traditional chemistry focuses on the covalent bond, supramolecular chemistry examines the weak and reversible NCIs between molecules. Therefore, the study of NCIs is crucial to understand supramolecular chemistry. Many biological processes depend on these noncovalent forces for structure and function [23–26].

For instance, in biological processes, storage and replication of genetic information depends on hydrogen bonding in the DNA double helix. There are four types of noncovalent forces in supramolecular chemistry: electrostatic interactions, hydrogen bonding, van der Waals (VdW) interactions, and hydrophobic effects. A brief overview of these NCIs is presented in the following sections [26].

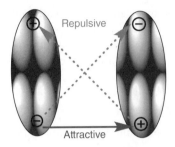

Scheme 2.1 Schematic presentation of electrostatic interaction.

2.1.2.2 Electrostatic Interactions

These interactions are based on attractive or repulsive forces between charges including dipoles and induced dipoles and they may be charge–charge, charge–dipole, and dipole–dipole interactions (Scheme 2.1).

The electrostatic interaction energy is given by Coulomb's law (Equation 2.1):

$$E = \frac{q_1 q_2}{4\pi\varepsilon\varepsilon_0 r^2} \tag{2.1}$$

where E is the interaction energy, q_1 and q_2 are the charges on the two atoms, ε_0 is the permittivity of vacuum, ε is the dielectric constant of the medium, which plays an important role in the strength of the electrostatic interaction, and r is the distance between the two atoms. In principle, with increasing dielectric constant, the free ion-pair interaction energy decreases significantly. Thus, the strongest electrostatic interaction energy for any free ion pair will be observed in vacuum because it has the smallest dielectric constant ($\varepsilon = 1$) among all media. Salts can be assumed to form a tight contact ion pair in apolar solvents such as dioxane or benzene [26–28]. The interaction between two point charges can be attractive or repulsive, depending on the sign of the charges [27].

Similarly, the force between a charge and a dipole or between two dipoles can become attractive or repulsive, depending on the geometric arrangement. However, the interactions between charges or dipoles and apolar molecules are always attractive because the presence of a charge in proximity to the molecule induces a dipole that is oriented such that the stabilization is maximized [28].

2.1.2.3 Hydrogen Bonding

Hydrogen bonds formed between two different molecules are called intermolecular hydrogen bonds, whereas those formed within one molecule are called intramolecular hydrogen bonds. It is the intermolecular hydrogen bonds that are responsible for the existence of liquid water

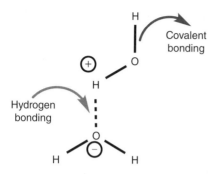

Scheme 2.2 Schematic presentation of hydrogen bonding.

under ambient conditions and its high boiling point of 100 °C, compared to, for example, hydrogen sulfide, which builds no hydrogen bonding (Scheme 2.2). Intramolecular hydrogen bonds are partly responsible for the secondary, tertiary, and quaternary structures of proteins and nucleic acids [24, 29–31].

The hydrogen atom in a hydrogen bond is partly shared by two relatively electronegative atoms such as nitrogen (N), oxygen (O), or fluorine (F). It is in proximity to one of the electronegative atoms, the so-called hydrogen bond donor (D), than to the hydrogen bond acceptor. The hydrogen bond donor atom pulls electron density away from the covalently bound hydrogen atom to create a partial positive charge (δ^+). Subsequently, the hydrogen atom is attracted through electrostatic interaction to the hydrogen bond acceptor, having a partial negative charge (δ^-) [32–35].

If the hydrogen bond acceptor is an anionic compound, the formed hydrogen bond is much stronger than for a neutral acceptor and is called ionic hydrogen bond. The hydrogen bond thus is not a truly covalent chemical bond, but it has some features of covalent bonds: it is strong and directional, produces interatomic distances shorter than the sum of the VdW radii, and usually involves a limited number of interaction partners, which can be interpreted as a type of valence [35, 36].

By comparison, the bond energy for a common C—H covalent bond is about 98.0 kcal/mol, and for a common C—C covalent bond it is about 84.0 kcal/mol. Thus, even the strongest hydrogen bond cannot compete with covalent bonds. The total binding energy in complexes supported by multiple hydrogen bonds may sometimes be bigger than that of a single covalent bond. Nevertheless, one must note that the strength of multiple hydrogen bonds would be modulated by the so-called secondary interactions, which can be attractive or repulsive [37–39]. The presence of relatively strong hydrogen bonds can be characterized by typical shifts of the proton NMR signals and of the IR bands for the D—H⋯A stretching vibrations. Hydrogen bonds are ubiquitous in supramolecular chemistry. In particular, they are responsible for the overall shape of many proteins, for the recognition of substrates by numerous enzymes, and for the double-helix structure of DNA [40–42].

2.1.2.4 Van der Waals Interactions

This kind of interaction is based on the sum of the attractive and repulsive forces between molecules (or between parts of the same molecule) other than those due to covalent bonds, electrostatic interaction of ions, or hydrogen bonds (Scheme 2.3). van der Waals interactions arise from the polarization of an electron cloud by the proximity of an adjacent nucleus, resulting in a weak electrostatic attraction, for instance, between two permanent dipoles, between a permanent dipole and a corresponding induced dipole, or between two instantaneously induced dipoles [42–45].

Even in the absence of a net dipole moment in an unpolarized atom or molecule, a dipole is always induced in this atom or molecule as a result of local fluctuations of the electron density distribution in a second, nearby atom or molecule. Strictly speaking, van der Waals interactions may be divided into dispersion and exchange-repulsion types. The dispersion interaction is the attractive component that results from the interactions between fluctuating multipoles in adjacent molecules, while the exchange repulsion balances dispersion at short range and defines molecular shape. Although they are very weak, they are additive over the entire association surface and may become the most important attractive force between apolar molecules [24, 28–30].

van der Waals interactions play an important role in formation of inclusion complexes, which is energetically most favorable distance between atoms or groups. Smaller distance between atoms than their van der Waals radii lead to strongly repulsive interaction [23–25].

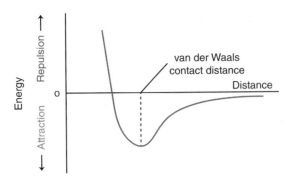

Scheme 2.3 Schematic presentation of van der Waals interactions.

2.1.2.5 Hydrophobic Effect

The hydrophobic effect is a key property driving apolar molecules to form intermolecular aggregates in aqueous solution. The name literally describes the apparent repulsion between water and apolar substances. The hydrophobic effect explains the separation of a mixture of oil and water into two layers or the beadings of water on hydrophobic surfaces such as lotus leafs. At the molecular level, the hydrophobic effect is a very important driving force for biological structures and is responsible for protein folding, formation of lipid bilayers, and micelles [46–49].

The water molecules in pure water always form as many strong hydrogen bonds as thermodynamically feasible. When an apolar molecule is dissolved in water, the water molecules directly surrounding the apolar molecule cannot form as many such hydrogen bonds as in bulk water and cannot also favorably interact with this apolar surface [31, 32]. Upon association of apolar molecules, the overall apolar surface is reduced. The total number of water molecules directly surrounding these aggregates decreases compared to the nonaggregate situation, and some of them are released into bulk water [50–53].

These released water molecules form additional hydrogen bonds with other water molecules, which results in a decrease in enthalpy for the whole system. Furthermore, the water molecules around the apolar surfaces are at a higher level of order than in the bulk solution. The release of many water molecules upon aggregation of apolar surfaces thus also increases the entropy of the system. The larger the apolar surface of the associated molecules, the more favorable is the entropic driving force [33]. Therefore, the aggregation of apolar molecules in water is favorable both enthalpically and entropically. It is controlled by surface desolvation and is called the classical hydrophobic effect [51, 54].

2.2 Host–Guest Chemistry

Host–guest is obtained by binding of guest molecules (a wide range of organic and inorganic compounds, monatomic cation, inorganic anion or hormones, and pheromones) with host molecules (large molecules or aggregates such as an enzyme, synthetic cyclic compound with a sizeable central hole or cavity) [35]. Formally, a host is defined as the molecular entity possessing convergent binding sites and the guest possesses divergent binding sites. Host–guest supramolecules are composed via electrostatic forces other than full covalent bonds and stabilize guest molecule and host–guest complex in solution and to the topological relationships between the guest and host [55–57]. Cavitands are hosts possessing intramolecular cavities, which are strictly

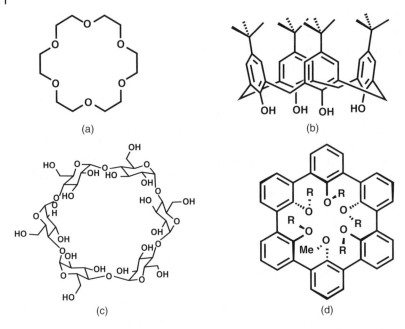

Figure 2.2 Host or guest molecules which can be used for formation of supramolecular structures. (a) Crown ether, (b) calixarene, (c) cyclodextrin, and (d) spherand.

an intramolecular property of the host, and exist both in solution and solid state. Conversely, clathrands possess extramolecular cavities, often a gap between two or more hosts, that are relevant in crystalline or solid state (Figure 2.2) [35–37].

2.2.1 Cyclodextrins as Supramolecular Hosts

Cyclodextrin-based supramolecular structures are nontoxic and seminatural products, which can be produced from relatively simple enzymatic conversion of potato or starch through environmentally friendly technologies at an affordable price. The most amazing property of cyclodextrin-based supramolecular structures is elimination of toxic effect of guest, which has made them more attractive for various applications such as drug delivery, food industry, and cosmetics [58–61].

2.3 Cyclodextrin-Containing Supramolecular Structures

Inclusion complex formation between hydrophobic cavities of cyclodextrins and hydrophobic guest molecules is affected by identity of host and guest, which leads to chemical synthesis with new structures and designs. Some of these structures are introduced in the following sections.

2.3.1 Cyclodextrins

Cyclodextrins (CDs), as they are known today, were called cellulosines when first described by Villiers in 1891, which were isolated by digesting starch from certain enzymes of a bacterial origin, Bacillus macerans being the earliest source. Later, Schardinger identified naturally occurring analogues, referred to as Schardinger sugars [62–64].

Cyclodextrins with six to eight D-glucopyranoside units, linked by a 1,4-glycosidic bond, are the main CDs, which are cyclic oligosaccharides and named α-cyclodextrin, β-cyclodextrin, and γ-cyclodextrin, respectively. The special geometry of OH groups of CDs provide torus or truncated funnel CD ring shape, in which primary hydroxyl groups are located on the narrow end of the torus and secondary hydroxyl groups on the broader side. Cyclodextrins have hydrophobic cavity with hydrophilic surface, so they can form inclusion complexes with a wide range of hydrophobic and hydrophilic guest molecules. Schematic of cyclodextrin shape is shown in Figure 1.2 [42].

2.3.2 Cyclodextrin Shape and Inclusion Complex Formation

Various parameters affect formation of inclusion complexes between CD and guest molecules such as steric and interactional complementarity, strong hydrogen bonding by various OH groups, and large contact areas [61–63]. Number of guest molecules that can be incorporated in cyclodextrin cavity will vary and is related to guest size; suitable guest molecules can also be used to thread one, two, or many CD rings in solution, resulting in certain supramolecular structures. Threading guest molecule and cyclodextrins formed mutually interlocked and intertwined molecular architectures such as rotaxanes and catenanes, the existence of which had not been realized until relatively recent times [64–66].

2.4 Supramolecular Chemistry

Self-assembled supramolecular structures open a new concept on formation of new polymolecular structures such as catenanes, rotaxanes, pseudorotaxanes, and knot compounds in nanometer scale (Figure 2.3) [7].

Formation of self-assembled supramolecular structures is based on several interactions such as hydrophobic interactions, hydrogen bonding, donor–acceptor interactions, and transition metal coordination so show some unique properties. Cyclodextrin is one of the most attractive

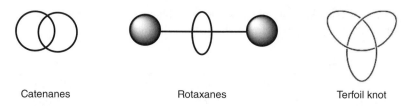

Figure 2.3 Self-assembled supramolecular structures.

monomers in self-assembled supramolecular compounds because of its simultaneous hydrophobic and hydrophilic properties, which allow formation of inclusion complexes in aqueous media and organic solvents with various functionalities [67–70].

2.4.1 Cyclodextrin Rotaxanes

Polyrotaxane is a necklace-like supramolecule in which many cyclic molecules are threaded into a single polymer chain and can be cross-linked to produce functional polymeric materials. Rotaxane is a Latin word and meaning wheel and axle which wheel is consist of a macrocycle such as cyclodextrins and the axle is usually a linear polymer that threaded through the macrocycle, so for hindered the dethreading of the wheel, bulky blocking groups attached the end of axle [71, 72].

Polyrotaxane formation is based on versatile noncovalent cross-linking (mainly hydrogen bonding and van der Waals), and these interactions make them unique from other polymers based on covalent interactions. Macrocycle may slip off the axle because there are no bulky terminal groups on the axle [73–75] (Figure 2.4).

The important point in formation of polyrotaxanes is that the bulky blocking groups must be bigger than cyclodextrin cavity; if the stoppers at the ends of the axle are smaller than the cyclodextrin cavity, the polymer

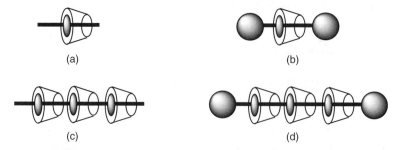

Figure 2.4 Schematic presentation of (a) pseudorotaxane, (b) rotaxane, (c) poly pseudorotaxane, and (d) polyrotaxane.

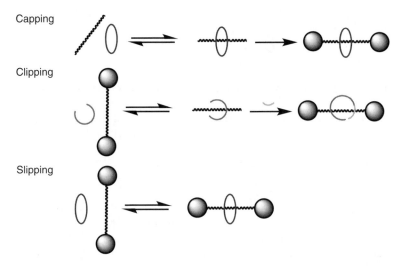

Figure 2.5 Polyrotaxanes synthesis method based on nature and location of the bonding.

is called pseudorotaxane. Because of unique structures of polyrotaxanes and their specific topology, their properties tend to be distinct from other polymeric architectures for specific applications for lateral and translational motion of the cyclic molecules relative to the linear chain [73–75].

Polyrotaxanes based on the nature and location of the bonding can be synthesized by various methods such as capping, clipping, and slipping (Figure 2.5). Capping is strongly related to thermodynamically driven template effect and threading is based on noncovalent interactions, and also obtained structure is changeable. Clipping method is similar to capping reaction, but in this reaction the macrocycle undergoes a ring-closing reaction around the dumbbell-shaped molecule, forming rotaxane. In slipping method, kinetic stability of rotaxane and size of end groups are important and affect reversibly thread of rotaxane [76–78].

Cross section of an axle plays an important role in the formation of these supramolecular structures; the cross section must be smaller than the CD cavity and the length must be greater than that of CD [51, 53]. If inclusion complex is rigid and stable, it prevents the hemolytic activity of free unmodified CD molecules and, therefore, it can be used in biological systems for delivery purposes; but in some cases capping agent must be used to cap the end groups of the complexed CD/polymer system as shown in Figure 2.5 [79–82].

In formation of inclusion complexes between CD and guest molecules and formation of polyrotaxanes based on CDs, hydrogen bonding has an important role and obtained macrostructures display various structures,

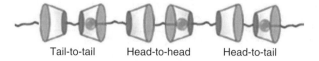

Tail-to-tail   Head-to-head   Head-to-tail

Figure 2.6 Orientation between CD and polymer chain based on hydrogen bonding.

which are named channel or cage structure and formed via head-to-head, tail-to-tail, or head-to-tail orientations (Figure 2.6) [51].

2.4.2 Studies on Responsive CD-Based Polymers

2.4.2.1 Cyclodextrin Dimers

The moderate size of the hydrophobic molecular cavities of α-, β-, and γ-cyclodextrin prevents the accommodation of large substrates by one cyclodextrin. However, efficient binding of large substrates can be achieved in 2:1 inclusion compounds. In this way, a type of dimer is obtained. The gain in enthalpy due to the hydrophobic binding of a substrate by a cyclodextrin is partly compensated by the loss of entropy because of the molecular association [57–59, 83]. By the addition of cyclodextrins to inclusion complexes of CD/guest molecules, binding constant decreased compared to that of the first cyclodextrin to the substrate, enthalpy decreased, and entropy increased due to geometrical interferences between two cyclodextrins [60–62].

In the case of large guest molecules, one cyclodextrin is not sufficient for formation of inclusion complexes, and cyclodextrin dimers can be used for the complexation and guest molecules can be entrapped between CDs and entropy will be lower [84–86]. Some of this dimerization is caused by covalent interaction via long bridges between cyclodextrins and polymers such as aromatic guests. By modifying primary or secondary hydroxyl faces or both hydroxyl faces of cyclodextrins, micelles or vesicles can be obtained via inclusion complex formation with guest molecules. Vesicle is a small structure within a cell, which can be formed via entrapment of guest molecules within the individual cavities or in the interior of the bilayer. Guest molecules may be lipophilic hydrocarbon chains or hydrophilic compounds, within the aqueous vesicle core (Figure 2.7) [86–88].

2.4.2.2 Catenanes

By mechanical interaction of two or more interlocked macrocycles, an interlocked supramolecular structure is formed and cannot be separated without breaking the covalent bonds of the mechanically linked molecule and it is named catenane. The name catenanes is derived from the Latin word catena, meaning chain. Catenanes show significantly different

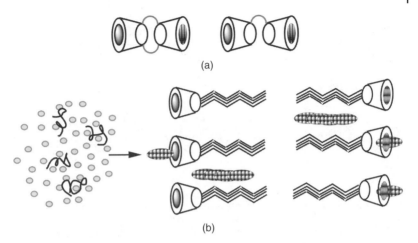

Figure 2.7 Schematic presentation of (a) CD dimers and (b) molecular inclusion complex by a cyclodextrin vesicle.

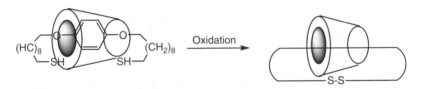

Figure 2.8 Synthesis of catenane by linking both end of a thiol threaded into α-CD in the presence of copper(II) ions and oxidation in the presence of air.

properties from individual macrocycle components. The first attempt to synthesize a catenane by ring closure of a long guest molecule included α-CD was done in 1957, but the result was frustrating (Figure 2.8) [89]. The first successful synthesis of [2]- and [3]catenanes obtained from heptakis (2,6-di-O-methyl)-β-CD was reported by Mirzoian [89].

2.4.2.3 Rotaxanes

The name rotaxane is derived from the Latin words rota, meaning wheel, and axle, meaning axis. Rotaxanes are comprised of a dumbbell-shaped component in the form of a rod and two bulky stopper groups, around which macrocyclic components are encircled. By attaching stoppers blocking agent, macrocycles dethreading is prevented from the rod. If the size of bulky groups is small or insufficient to stop equilibration of the bead, pseudorotaxanes are formed (Figure 2.9) [90–92].

Rotaxanes are supramolecular structures based on noncovalent bonds between two individual components, in which only mechanical bonds are responsible for the linking of the components. The simplest rotaxane

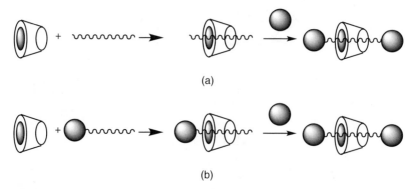

Figure 2.9 Schematic synthesis of rotaxane via two methods, (a) synthesis of a pseudorotaxane and then reaction with bulky blocking group and (b) formation of an inclusion complex containing a guest molecule with one blocking agent at the end and reaction with another blocking agent.

is based on threading of an axle molecule through the cavity of CDs with the assistance of noncovalent intermolecular attractive forces or one dumbbell-shaped molecule and one macrocycle are interlocked. Bulky blocking groups prevented dethreading of the macrocycle in rotaxane chain; if the dethreading occurs, the supramolecular structure is called a pseudorotaxane [93–96]. Cyclodextrin-based rotaxanes can be synthesized via entrapment of cyclodextrin between bulky blocking groups, which caused relatively high-yield production. Polyrotaxane or cyclodextrin is complexed on a chain molecule, already blocked by a bulky blocking group at one end. The final rotaxane can be obtained by blocking the other end of the chain molecule. A schematic representation of this route is shown in Figure 2.9 [97–99]. Based on the type of cyclodextrin and its cavity diameter, different blocking groups can be used, and the size of the blocking group should at least be as large as the diameter of the cyclodextrin at the narrow side [98, 99].

2.5 Cyclodextrin-based Rotaxanes and Pseudorotaxanes

Because of cyclodextrin's unique cone shape, properties, structure, and formation of inclusion complexes with a wide range of hydrophobic and hydrophilic guest molecules in aqueous solutions, various rotaxanes and pseudorotaxanes are obtained [70, 71]. In these rotaxanes, an asymmetric CD is threaded by a nonsymmetrical dumbbell and different structures may be formed. The idea of self-assembling rotaxanes, incorporating CDs as the ring components, was implemented for the first time by using transition metal complexes as "stoppers" at the ends of bisfunctionalized threadlike guests encircled by the CDs. Polyrotaxanes/polypseudorotaxanes differ from conventional linear polymers in that they are composed of a macrocyclic species and a linear species. They are different from polymer blends as well as block copolymers because the two components are mechanically interlocked. Due to the aforementioned structural differences, these species exhibit unique properties depending on the composition of the cyclic and linear components, the interactions between these components, and the density of the cyclic components [100–103].

For example, the solubility of crown-ether-based polyrotaxanes/polypseudorotaxanes in methanol and water was enhanced because of the hydrogen bonding between the crown ethers and the solvents. Ultimately, polyrotaxanes/polypseudorotaxanes are believed to have various potential applications due to their novel properties. In 1981, Ogino reported the use of the reaction between cis-[CoCl$_2$(en)$_2$]Cl (en denotes ethylenediamine) and α,ω-diaminoalkanes as a way to construct the dumbbell components of rotaxanes by incorporating either α-CD or β-CD (Figure 2.10) [50].

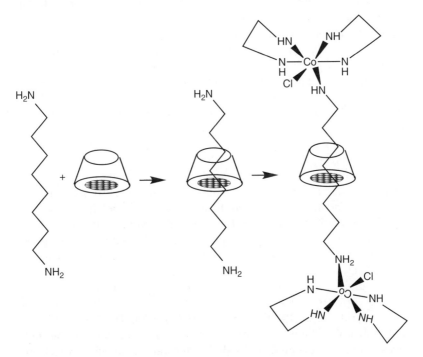

Figure 2.10 Self-assembly of a rotaxane using threading procedure.

Chen used bisimidazolyl compounds as axle guests in polyrotaxanes [104]. In another study, Park reported pseudorotaxanes based on the long alkylene chain that thread two β-CDs, while the shorter chain can thread only one β-CD ring to become the pseudorotaxanes and investigated thermodynamic threading of α-CD of obtained pseudorotaxane [105]. Harada reported a kinetic threading of cyclodextrin ring to realize the unidirectional preparation of nonsymmetric pseudorotaxane [106].

Easton reported synthesis of polyrotaxane by using succinic anhydride protected amino-α-CD treated with stilbene guests and then capped by trinitrobenzene. Incorporation of methoxyl group and succinimide restricts the rotational motion of the cyclodextrin around the stilbene axle, and this system is analogous to a ratchet's tooth and pawl to restrict rotational motion [107].

Yamamoto reported slipping of β-cyclodextrin rings to a polythiophene polymer chain to obtain a water-soluble polypseudorotaxane [108].

Polypseudorotaxane based on triblock polymers (PPO-b-PEO-b-PPO) and α-CD under ultrasonic water bath with a channel structure, while the flanking PPO blocks are uncovered and remain amorphous, was reported by Li et al. [109]. Polyrotaxane based on poly(ethylene glycol) (PEG) and α-CD and β-CD via slipping of CD rings in aqueous solution in which polyrotaxane formed by end-capping with dimethylbenzene by etherification was reported by Zhao and Harada, respectively [110, 111].

Formation of polyrotaxane between aromatic compounds and CDs were reported by Harada and Farcas and rotaxane polymer was synthesized by homopolymerization of bithiophene in water using $FeCl_3$ as an oxidative initiator and copolymerization of a diformylcarbazole derivative with a benzenediamine pseudorotaxane in DMF, respectively [37, 38].

Liu et al. also developed a simple strategy to prepare polyrotaxane-like supramolecules based on inclusion complex formation between 4,4-dipyridine and β-CD with nickel ions in aqueous solution [112]. Dimerization and photodimerization also can be used for rotaxane synthesis. Harada et al. reported the preparation of a poly(polyrotaxane) by photodimerization and PEG, β-CD, and 9-anthracene groups were subjected to irradiation by visible light, and the anthracene stoppers underwent photoinduced dimerization and the poly(polyrotaxane) formed [113].

The dimerization is reversible and the polyrotaxane monomer can be converted back in more than 90% conversion by heating the poly(polyrotaxane) DMSO solution 393 K. Harada [111] also reported the poly(polypseudorotaxane), which is quite similar to the poly (polyrotaxane), except that the β-CD rings are changed to γ-CD and the cavity of γ-CD ring is big enough to pass over the anthracene unit.

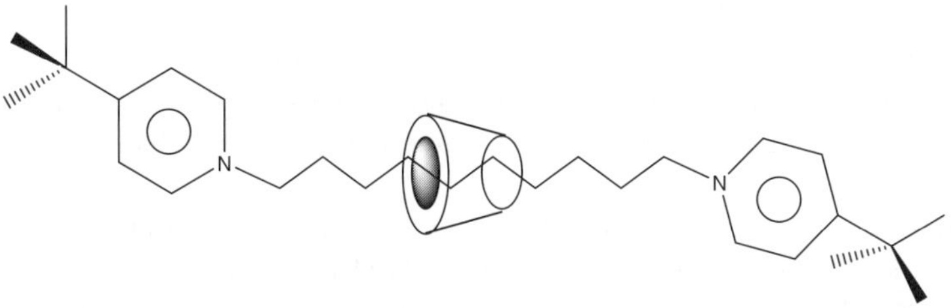

Figure 2.11 CD-based rotaxane prepared using a slippage method.

Formation of inclusion complexes between guest molecules and CDs is based on reversible hydrogen or van der Waals forces, although stoppers are attached to the chain by covalent linkage and are stable, which are groups with relatively high nucleophilicities – such as NH_2 or S—, ferrocenyl methyl unit or potassium 5-amino-2-naphthalensulfonate [50, 51, 72, 73].

Using steric bulk groups as blocking groups caused inclusion formation via slippage mechanism and prevented dethreading of polyrotaxane chain because of their relatively similar size with diameters of the CD cavities (Figure 2.11) [74, 75].

A rather unique kind of superstructure has been observed in the solid state for a monosubstituted β-CD derivative that behaves both as a host and as a guest, such that a —$CH_2NH(CH_2)_6NH_2$ side chain on the primary face of one molecule enters into the cavity of the β-CD ring of a neighboring molecule and so on in a linear manner. The monosubstituted β-CD derivatives are arranged spirally, forming a polymer-like superstructure [43, 71].

2.5.1 Pseudopolyrotaxanes

Synthesis methods for rotaxane and pseudorotaxane are similar and are convertible, and attaching a bulky substituent group to the open ends of the threaded molecule converts pseudorotaxane into rotaxane and it displays relatively similar properties. Solubility of rotaxane and pseudorotaxane is similar at the same temperature and reaction conditions. In other words, by using a pseudopolyrotaxane consisting two interlocked macrocycles that blocked by a blocking agent, a polyrotaxane was obtained with a supramolecular interlocked architecture at the macromolecular level [43].

Various factors such as synthesis method, kind of cyclic units, and how cyclic and linear units are connected affect the shape and classification of polyrotaxanes. Main- and side-chain polyrotaxanes are obtained due to location of the rotaxane unit which rotaxane unit is located on the main and the side chains, respectively [99–102].

2.5.1.1 Synthetic Route

Synthetic methods containing statistical and template or direct methods which reported by Harrison via a purely statistical threading, and there is no attractive force between the linear segment and cyclic moieties of the rotaxane [114].

In statistical threading method, the rotaxane yield is directly dependent on the size of macrocycle. Rotaxanes are made by slippage of the macrocycle over blocking groups, which was reported by Schill *et al.* In this reaction, linear segment with reactive end groups was chemically bonded to the cyclic moieties, and dethreading of cyclic species prevented by blocking groups. Then, the chemical bond between the linear segment and the macrocycle underwent a cleavage to yield rotaxane [115, 116].

In another method, macrocycle may be threaded onto the presynthesized polymer backbone via clipping route through which the polymer chain was incorporated in macrocycle. In formation of inclusion complexes, the important parameter is threading derived by enthalpy. The intermolecular attractive force between the linear segment and the macrocycle is a driving force leading to negative ΔH for threading [115, 116]. Self-assembly threading is another method that was reported by Stoddart's research group and is based on the reaction between an electron-rich and an electron-poor species, between macrocycle and linear segment [117]. The last method is catalytic self-threading in which the macrocycle has catalytic ability to catalyze the 1,3-dipolar cycloaddition reaction between functionalized monomers to produce polymer threaded by macrocycles with control on the number of macrocycles per repeating units [118].

2.5.2 Polyrotaxanes and Pseudopolyrotaxanes Based on Cyclodextrins

CDs formed inclusion complexes by axial polymeric chain with blocking groups or by polymer-analogous reactions of polymers with monomers. If rotaxane axis lies along the polymer main chain, make branch in its side chains (see Figure 2.12).

Main-chain polyrotaxanes based on α-CD and β-CD and (poly(iminoundecamethylene) and poly(iminotrimethylene–iminodecamethylene) via reaction of the amino groups with nicotinoyl chloride was reported by Wenz and Keller [119, 120]. Cavity size of β-CD is bigger than that of nicotinoyl groups, so inclusion complexes did not form and blocking groups such as 2,4-dinitro-5-aminophenyl groups were used instead [119, 120]. Also, Harada reported a main-chain polyrotaxane based on reaction of the terminal amino groups of a polyethylene oxide and CD, which is shown in Figure 2.13 [76]

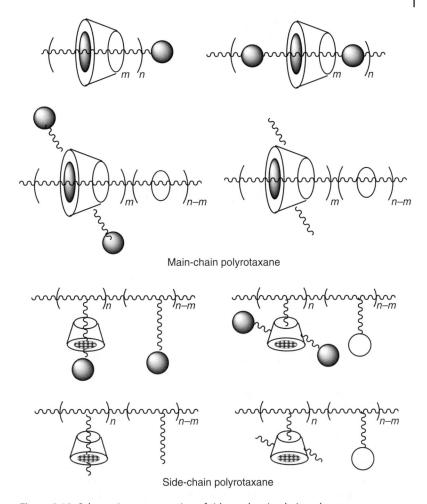

Figure 2.12 Schematic representation of side- and main-chain polyrotaxanes.

Ritter reported the first kind of side-chain polyrotaxanes using cyclodextrin, in which 2,6-di-O-methyl-β-CD formed inclusion complexes with 4′-triphenylmethyl-4-aminobutananilide and 4′-triphenylmethyl-1,1-aminoundecananilide stopper group as guest, in water and chloroform to afford a monorotaxane (Figure 2.14) [79].

Harada and coworkers developed inclusion complex formation between α-CD and various polymeric backbones including poly(ethyleneoxide), poly(propyleneoxide), and polyisobutylene in the form of column structure of stacked CD rings. In these reactions, dinitrophenyl, amino moieties were introduced at the chain ends as the stopper

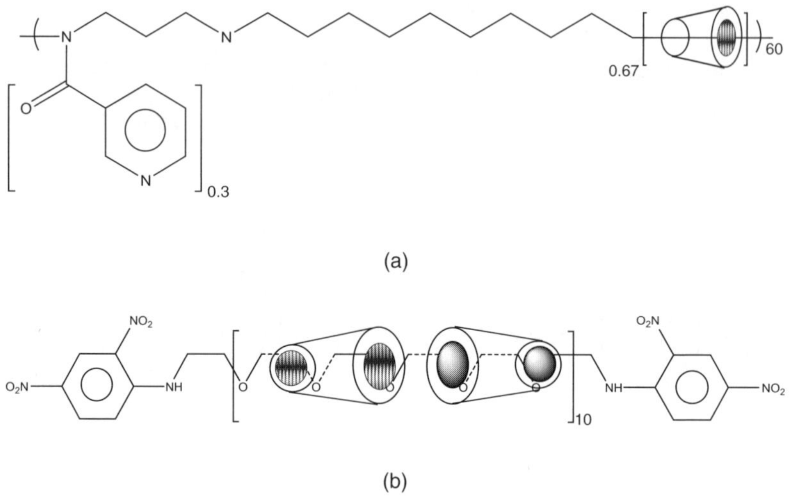

Figure 2.13 Main chain polyrotaxane by using (a) small blocking group and (b) a larger blocking group.

to prevent threaded rings from slipping off the backbone [78, 121]. These supramolecular structures were used as tubular polymers by treating with epichlorohydrin and subsequently cleaving the stopper function.

2.5.3 Cyclodextrin-based Polyrotaxanes

When the guest molecules are hydrophobic incorporated in the CDs cavity and hydrophobic interactions between cyclodextrins and polymer chains led to the formation of pseudopolyrotaxanes (Figure 2.15). Pseudopolyrotaxanes can be obtained via in situ polymerization of a monomer complexed inside a cyclodextrin [40, 122]. Radiation polymerization of vinylidene chloride and β-CD formed a crystalline inclusion compounds with high stability, which is based on nonchemical grafting or chain transfer to the β-CD molecule [60].

The formation of pseudopolyrotaxanes composed of α-CD and poly(ethylene glycol) was the first report on threading cyclodextrins on polymers. When aqueous solutions of poly(ethylene glycol) were added to a saturated aqueous solution of α-CD, the solution became turbid [123–125]. Cyclodextrin-based polyrotaxanes are synthesized by threading many cyclodextrins on a polymer chain and bulky blocking groups are attached at the end of polymer chain to prevent unthreading along a part of the chain. This supramolecular structure then can be reacted with bulky group substituents along the polymer chain and forms pseudopolyrotaxane based on cyclodextrins (Figure 2.16) [126].

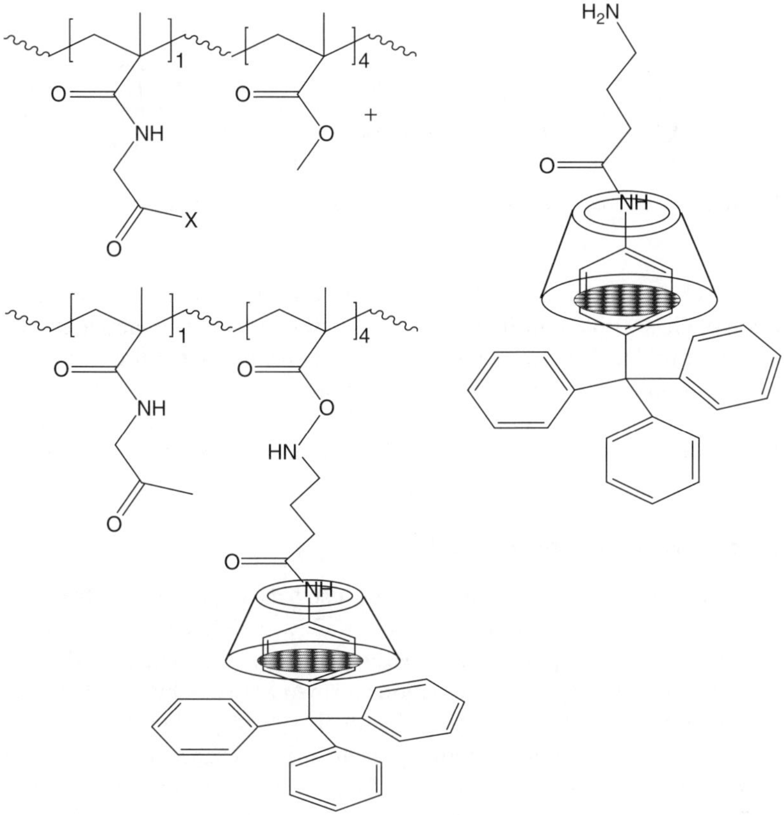

Figure 2.14 Side-chain polyrotaxane from methylated β-CD (X = O—COOEt).

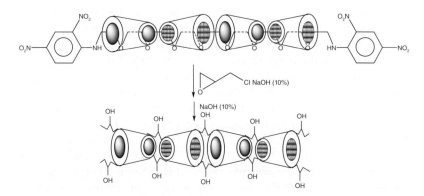

Figure 2.15 Tubular polyrotaxane based on α-CD from stacked polyrotaxane rings.

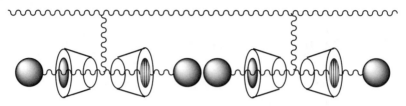

Figure 2.16 Schematic presentation of Tandem side-chain polyrotaxane.

Side-chain polyrotaxanes based on polyrotaxanes formed by complexation of CDs with comb-like polymers and the unblocked reactive end of the guest molecules can then react with an appropriate main-chain polymer to produce the desired side-chain polyrotaxanes. Also, side chains of the comb polymer and the activated ends of the side chains can be blocked with bulky blocking groups and tandem polyrotaxane is formed [19, 127].

2.5.4 Cyclodextrin Molecular Tubes

Rod-like rotaxanes containing can be used as nanotubular structure and called nanotubes and obtained by CDs which CDs arranged in the direction of the polyene axis without linkage together, so hydrogen bonding or van der Waals interactions formed between hydroxyl groups of neighboring γ-CD. When the edge hydroxyls of γ-CD were replaced by nonhydrogen bonding methoxy groups, in the absence of polyene nanotubes and when hydrogen bonding cannot form because of high pH solvent, nanotubes cannot form. Diphenylhexatrienes also formed nanotube aggregates by formation of inclusion complex with β-CD and γ-CD [128].

The simplest way to synthesize a real tube in which the cyclodextrins are linked together is by first synthesizing a polyrotaxane in which the cyclodextrins are closely packed on the polymer chain. Then, the adjacent cyclodextrins in the polyrotaxane can be cross-linked to create the nanotube. Finally, the bulky ends of the "template polymer" can be removed and an empty tube is obtained. These tubes are expected to be able to form inclusion complex with host molecules in an identical manner to that single cyclodextrins can form inclusion compounds with small molecules [129–132]. Using this general strategy, α-CD tubes have been synthesized by first threading many α-CDs on PEG. The α-CDs are closely packed from end to end on the polymer chain. Then, the polyrotaxane is synthesized by reaction of the end groups of the "template polymer" chain with bulky blocking groups to prevent the unthreading of the cyclodextrin from the polymer chain. Subsequently, the cyclodextrin molecular tube is synthesized by removing the bulky blocking groups and release of the "template polymer" [129].

2.5.4.1 Cyclodextrin-Based Nanotube Structure

Supramolecular complexes consisting of cyclic molecules, such as cyclodextrins (CDs), and polymeric chains form molecular tubes. Geometry and specific properties of CDs, linear or branched CD hosts with multiple cavities can interact with guest molecules with high affinity and specificity by taking advantage of the cooperative effect. Substituting all primary hydroxyl groups of a CD with guest molecules such as disulfide linkages forms nanotubes, which is so important for physicists, chemists, materials scientists and biology and other areas of research because of potential biodegradability of the bonds [132, 133].

Nanotubes based on CDs have various applications such as building blocks in the formulation of novel materials, molecular devices such as molecular reactors, molecular nanotubes, and molecular wires. Nanotubes based on CDs are obtained by self-assembly process of cyclodextrins threading on a polymer chain by noncovalent bonding such as hydrogen bonds or van der Waals. Guest molecules fit in CD cavities and give supramolecular entities (Figure 2.17) [4–6].

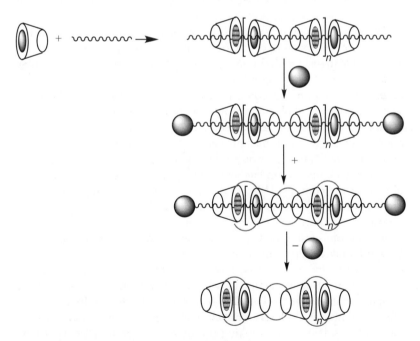

Figure 2.17 Synthesis of cyclodextrin molecular tubes.

CD molecular tubes are able to form inclusion complexes with various kinds of guests, based on hydrogen bonding, and building blocks show excellent complexation activity. Polyrotaxanes and pseudopolyrotaxane can be formed by threading CD molecules onto linear polymer chains. Various kinds of linear polymers can be used for threading of cyclodextrin rings onto polymer chains and incorporating in the CD cavities [134, 134a–c, 135, 135a,b, 136].

Polyrotaxane based on CDs can be obtained as molecular necklace, in which carbohydrate units are threaded on a polymer chain, and three conformations are possible. These alignments are defined depending on whether two equal or two different rims are facing each other and are named head-to-head (HH), tail-to-tail (TT), or head-to-tail (HT), among which head-to-head and tail-to-tail orientation are the most favorable. In head-to-head and tail-to-tail orientations, higher hydrogen bonds can be formed and in the case of head-to-tail orientation, formation of hydrogen bonds becomes more difficult because of the larger distance between the hydroxyl groups [51, 58, 134, 135]. The important parameter is that the ends of the polymer chain should be able to react with bulky blocking groups. Polymers such as poly(ethylene glycol), poly(propylene glycol), poly(isobutene), and poly(methyl vinyl ether) can be used for formation of pseudopolyrotaxane by parent CDs, which are closely packed on the polymer chains [134, 135].

The ends of polymer chain must be blocked with bulky groups, which prevent the unthreading of cyclodextrins from the polymer chain and also are able to react with the end groups of the polymer chain. Molecular nanotubes show various applications, so blocking groups must be stable under the cross-linking conditions but in some specific applications such as delivery field blocking groups are removed in order to release the cyclodextrin molecular tubes. Pseudopolyrotaxanes and polyrotaxanes are similar and show similar properties, but solubility of the former is lower than that of the latter one. Pseudopolyrotaxanes are in equilibrium with their components and depend on the solvent type, which is used for dissolving the complex [134, 135]. Cyclodextrins have many hydroxyls groups, which should not react with blocking groups in pseudopolyrotaxanes structure and end groups must be more reactive than these hydroxyl groups to be stable under various conditions. For compatibility of pseudopolyrotaxanes and polyrotaxanes in various conditions must be reacted with bifunctional cross-linking and fit between two close-packed cyclodextrins; also, interaction between cross-linking and polymer chain should be stable when the blocking groups are removed. Hydroxyl groups of two neighbor cyclodextrins are close to each other and can react with the reactive side of a cross-linking compound and both hydroxyl groups contain dangling cross-linking groups [134, 135].

Also, length of the cross-linking compound must be appropriate for connections between adjacent cyclodextrins and probably intermolecular cross-links and bonds between the blocking groups, and the polymer should be stable under the reaction conditions used for cross-linking the cyclodextrins [134, 135].

References

1 Kabanov, A.V. and Vinogradov, S.V. (2009) Nanogels as pharmaceutical carriers: finite networks of infinite capabilities. *Angew. Chem., Int. Ed.*, **48**, 5418–5429.
2 Motornov, M., Roiter, Y., Tokarev, I. and Minko, S. (2010) Stimuli-responsive nanoparticles, nanogels and capsules for integrated multifunctional intelligent systems. *Prog. Polym. Sci.*, **35**, 174–211.
3 Albrecht, K., Moeller, M. and Groll, J. (2011) Nano- and microgels through addition reactions of functional oligomers and polymers. *Adv. Polym. Sci.*, **234**, 65–93.
4 Pich, A. and Richtering, W. (2011) Microgels by precipitation polymerization: synthesis, characterization and functionalization. *Adv. Polym. Sci.*, **234**, 1–37.
5 Moya-Ortega, M.D., Alvarez-Lorenzo, C., Sigurdsson, H.H. *et al.* (2012) Cross-linked hydroxypropyl-cyclodextrin and cyclodextrin nanogels for drug delivery: physicochemical and loading/release properties. *Carbohydr. Polym.*, **87**, 2344–2351.
6 Del Valle, E.M.M. (2004) Cyclodextrins and their uses: a review. *Process Biochem.*, **39**, 1033–1046.
7 Lehn, J.M. (1978) Cryptates: inclusion complexes of macropolycyclic receptor molecules. *Pure. Appl. Chem.*, **50**, 871–892.
8 Steed, J.W. and Atwood, J.L. (2009) *Supramolecular Chemistry*, 2nd edn, John Wiley & Sons, Ltd., Chippenham, England.
9 Pedersen, C.J. (1988) The discovery of crown ethers (noble lecture). *Angew. Chem., Int. Ed.*, **27**, 1021–1027.
10 (a) Christy, F.A. and Shrivastav, P.S. (2011) Conductometric studies on cation-crown ether complexes: a review. *Crit. Rev. Anal. Chem.*, **41**, 236–269; (b) Zheng, B., Wang, F., Dong, S.Y. and Huang, F.H. (2012) Supramolecular polymers constructed by crown ether-based molecular recognition. *Chem. Soc. Rev.*, **41**, 1621–1636.
11 (a) Ashton, P.R., Philp, D., Spencer, N. and Stoddart, J.F. (1991) The self-assembly of [*n*]pseudorotaxanes. *Chem. Commun.*, **23**, 1677–1679; (b) Ashton, P.R., Philp, D., Reddington, M.V. *et al.* (1991) The self-assembly of complexes with [2]pseudorotaxane

superstructures. *J. Chem. Commun.*, **23**, 1680–1683; (c) Ashton, P.R., Baxter, I., Fyfe, M.C.T. *et al.* (1998) Rotaxane or pseudorotaxane? That is the question. *J. Am. Chem. Soc.*, **120**, 2297–2307.
12 Willner, I., Liondagan, M., Marxtibbon, S. and Katz, E.J. (1995) Bioelectrocatalyzed amperometric transduction of recorded optical signals using monolayer-modified Au-electrodes. *J. Am. Chem. Soc.*, **117**, 6581–6592.
13 Petty, M. (2008) *Molecular Electronics: From Principles to Practice*, John Wiley & Sons Ltd., Chichester.
14 (a) Lehn, J.M. (1990) Perspectives in supramolecular chemistry—from molecular recognition towards molecular information processing and self-organization. *Angew. Chem., Int. Ed.*, **29**, 1304–1319; (b) Menger, F.M.P. (2002) Supramolecular chemistry and self-assembly. *Natl. Acad. Sci. USA*, **99**, 4818–4822; (c) Lehn, J.M. (2002) Toward complex matter: supramolecular chemistry and self-organization. *Proc. Natl. Acad. Sci. USA.*, **99**, 4763–4768.
15 Ciferri, A. (2005) *Supramolecular Polymers*, 2nd edn, Taylor & Francis Group, New York.
16 Brunsveld, L., Folmer, B.J.B., Meijer, E.W. and Sijbesma, R. (2001) Supramolecular polymers. *Proc. Chem. Rev.*, **2001** (101), 4071–4098.
17 De Greef, T.F.A., Smulders, M.M.J., Wolffs, M. *et al.* (2009) Supramolecular polymerization. *Chem. Rev.*, **109**, 5687–5754.
18 Liu, Y., Wang, Z. and Zhang, X. (2012) Characterization of supramolecular polymers. *Chem. Soc. Rev.*, **2012** (41), 5922–5932.
19 Aida, T., Meijer, E.W. and Stupp, S.I. (2012) Functional supramolecular polymers. *Science*, **335**, 813–817.
20 Yan, X., Wang, F., Zheng, B. and Huang, F. (2012) Stimuli-responsive supramolecular polymeric materials. *Chem. Soc. Rev.*, **41**, 6042–6065.
21 Appel, E.A., Del Barrio, J., Loh, X.J. and Scherman, O.A. (2012) Supramolecular polymeric hydrogels. *Chem. Soc. Rev.*, **41**, 6195–6214.
22 Whittell, G.R., Hager, M.D., Schubert, U.S. and Manners, I. (2011) Functional soft materials from metallopolymers and metallo supramolecular polymers. *Nat. Mater.*, **10**, 176–188.
23 Moore, T.S. and Winmill, T.F.J. (1912) The state of amines in aqueous solution. *Chem. Soc.*, **101**, 1635–1676.
24 Jorgensen, W.L. and Pranata, J. (1990) Importance of secondary interactions in triply hydrogen bonded complexes: guanine-cytosine vs uracil-2,6-diaminopyridine. *J. Am. Chem. Soc.*, **112**, 2008–2010.
25 Pranata, J., Wierschke, S.G. and Jorgensen, W.L. (1991) OPLS potential functions for nucleotide bases. Relative association constants of hydrogen-bonded base pairs in chloroform. *J. Am. Chem. Soc.*, **113**, 2810–2819.

26 Berg, J.M., Tymoczko, J.L. and Stryer, L. (2002) *Biochemistry*, 5th edn, Freeman, New York.
27 Isaacs, N. (1995) *Physical Organic Chemistry*, 2nd edn, Longman Scientific & Technical, Essex, UK.
28 (a) Larson, J.W. and Mcmahon, T.B. (1984) Gas-phase bihalide and pseudobihalide ions. An ion cyclotron resonance determination of hydrogen bond energies in XHY-species. *Inorg. Chem.*, **23**, 2029–2033; (b) Emsley, J. (1980) Very strong hydrogen bonding. *Chem. Soc. Rev.*, **9**, 91–124.
29 Murray, T.J. and Zimmerman, S.C. (1992) New triply hydrogen bonded complexes with highly variable stabilities. *J. Am. Chem. Soc.*, **114**, 4010–4011.
30 Drew, H.R., Mccall, M.J. and Calladine, C.R. (1988) Recent studies of DNA in the crystal. *Annu. Rev. Cell Biol.*, **4**, 1–20.
31 Chandler, D. (2005) Interfaces and the driving force of hydrophobic assembly. *Nature*, **437**, 640–647.
32 (a) Voskuhl, J. and Ravoo, B. (2009) Molecular recognition of bilayer vesicles. *J. Chem. Soc. Rev.*, **38**, 495–505; (b) Stoddart, J.F. (2009) The chemistry of the mechanical bond. *Chem. Soc. Rev.*, **2009** (38), 1802–1820.
33 Blight, B.A., Hunter, C.A., Leigh, D.A. et al. (2011) An AAAA–DDDD quadruple hydrogen-bond array. *Nat. Chem.*, **3**, 244–248.
34 Corbin, P.S. and Zimmerman, S.C. (2000) Diphosphine oxide–Brϕnsted acid complexes as novel hydrogen-bonded self-assembled molecules. *J. Am. Chem. Soc.*, **122**, 3779–3780.
35 Desiraju, G.R. (1995) Supramolecular synthons in crystal engineering—a new organic synthesis. *Angew. Chem., Int. Ed. Engl.*, **34**, 2311–2327.
36 Powell, H.M. (1973) Electrophilic and nucleophilic organoselenium reagents. New routes to. alpha.beta.-unsaturated carbonyl compounds. *J. Chem. Soc.*, **95**, 6137–6139.
37 Pedersen, C.J. (1967) Cyclic polyethers and their complexes with metal salts. *J. Am. Chem. Soc.*, **89**, 7017–7036.
38 Pedersen, C.J. and Frensdorff, H.K. (1972) Makrocyclische Polyäther und ihre Komplexe. *Angew. Chem., Int. Ed. Engl*, **11**, 16–26.
39 Szetjli, J. (1998) Introduction and general overview of cyclodextrin chemistry. *Chem Rev.*, **98**, 1743–1753.
40 Szejtli, J. and Osa, T. (1996) *Comprehensive Supramolecular Chemistry*, vol. **3**(Cyclodextrins), Pergamon, Oxford, p. 693.
41 Pringsheim, H. (1932) *Chemistry of the Saccharides*, McGraw-Hill, New York, p. 280.

42 Wenz, G. (1994) Cyclodextrins as building blocks for supramolecular structures and functional units. *Angew. Chem., Int. Ed. Engl.*, **33**, 803–822.
43 Amabilino, D.B. and Stoddart, J.F. (1995) Interlocked and intertwined structures and superstructures. *Chem. Rev.*, **95**, 2725–2828.
44 Connors, K.A. (1997) The stability of cyclodextrin complexes in solution. *Chem. Rev.*, **97**, 1325–1358.
45 Inoue, Y. (1998) Complexation thermodynamics of cyclodextrins. *Chem. Rev*, **98**, 1875–1918.
46 Schneider, H.-J. (1998) NMR studies of cyclodextrins and cyclodextrin complexes. *Chem. Rev.*, **98**, 1755–1786.
47 Croft, A.P. and Bartsch, R.A. (1983) Synthesis of chemically modified cyclodextrins. *Tetrahedron*, **39**, 1417–1474.
48 D'Souza, V. (1998) Methods for selective modifications of cyclodextrins. *Chem. Rev.*, **98**, 1977–1996.
49 Werts, M.P.L. (2001) *Mechanically Linked Oligorotaxanes*, Ph.D. thesis, University of Groningen, NL.
50 Ogino, H. (1983) Relatively high-yield syntheses of rotaxanes. Syntheses and properties of compounds consisting of cyclodextrins threaded by.alpha.,.omega.-diaminoalkanes coordinated to cobalt(III) complexes. *J. Am. Chem. Soc.*, **103**, 1303–1304.
51 Harada, A., Li, J. and Kamachi, M. (1992) The molecular necklace: a rotaxane containing many threaded [alpha]-cyclodextrins. *Nature*, **356**, 325–327.
52 Loethen S. (2008). Cyclodextrin Based Pseudopolyrotoxanes and Polyrotoxanes for Biological Application. Doctor of Philosophy presented to Purdue University.
53 Gupta, U., Agashe, H.B., Asthana, A. and Jain, N.K. (2006) A review of in vitro–in vivo investigations on dendrimers: the novel nanoscopic drug carriers. nanomedicine: nanotechnology. *Biol. Med.*, **2**, 66–73.
54 Loftsson, T. and Duchêne, D. (2007) Cyclodextrins and their pharmaceutical applications. *Int. J. Pharm.*, **329**, 1–11.
55 Van, d.M., Vermonden, T., van Nostrum, C.F. and Hennink, W.E. (2009) Cyclodextrin-based polymeric materials: synthesis, properties, and pharmaceutical/biomedical applications. *Biomacromolecules*, **10**, 3157–3175.
56 Wenz, G., Han, B. and Miller, A. (2006) Cyclodextrin rotaxanes and polyrotaxanes. *Chem. Rev.*, **106**, 782–817.
57 Zhang, B. and Breslow, R. (1993) Enthalpic domination of the chelate effect in cyclodextrin dimers. *J. Am. Chem. Soc.*, **115**, 9353–9354.

58 Harada, A., Furue, M. and Nozakura, S. (1977) Interaction of cyclodextrin-containing polymers with fluorescent compounds. *Macromolecules*, **10**, 676–681.
59 Xu, W., Demas, J.N., DeGraff, D.A. and Whaley, M. (1993) Interactions of pyrene with cyclodextrins and polymeric cyclodextrins. *J. Phys. Chem.*, **97**, 6546–6554.
60 Martel, B., Leckchiri, Y., Pollet, A. and Morcellet, M. (1995) Cyclodextrin-poly(vinylamine) systems—I. Synthesis, characterization and conformational properties. *Eur. Polym. J.*, **31**, 1083.
61 Ravoo, B.J. and Darcy, R. (2000) Cyclodextrin bilayer vesicles. *Angew. Chem., Int. Ed.*, **39**, 4324–4326.
62 Ravoo, B.J., Darcy, R., Mazzaglia, A. *et al.* (2001) Supramolecular tapes formed by a catanionic cyclodextrin in water. *Chem. Commun.*, **9**, 827–828.
63 Mazzaglia, A., Donohue, R., Ravoo, B.J. and Darcy, R. (2001) Novel amphiphilic cyclodextrins: graft-synthesis of heptakis(6-alkylthio-6-deoxy)-beta-cyclodextrin 2-oligo(ethylene glycol) conjugates and their omega-halo derivatives. *Eur. J. Org. Chem.*, **9**, 1715–1721.
64 Coleman, A.W. and Kasselouri, A. (1993) Supramolecular assemblies based on amphiphilic cyclodextrins. *Supramol. Chem.*, **1**, 155–161.
65 Kasselouri, A. and Coleman, A.W. (1996) Mixed monolayers of amphiphilic cyclodextrins and phospholipids: I. Miscibility under dynamic conditions of compression. *J. Coll. Interf. Sci.*, **180**, 384–397.
66 Jullien, L., Lazrak, T., Canceill, J. *et al.* (1993) An approach to channel-type molecular structures. Part 3. Incorporation studies of the *bouquet*-shaped B_M and B_{CD} in phosphatidylcholine vesicles. *J. Chem. Soc., Perkin Trans*, **2**, 1011–1020.
67 Lin, J., Creminon, C., Perly, B. and Djedaïni-Pillard, F. (1998) New amphiphilic derivatives of cyclodextrins for the purpose of insertion in biological membranes: the "Cup and Ball" molecules. *J. Chem. Soc., Perkin Trans.*, **2**, 2639–2646.
68 Gulik, A., Delacroix, H., Wouessidjewe, D. and Skiba, M. (1998) Structural properties of several amphiphile cyclodextrins and some related nanospheres: an X-ray scattering and freeze-fracture electron microscopy study. *Langmuir*, **14**, 1050–1057.
69 Zhu, S.S. and Swager, T.M. (1997) Conducting polymetallorotaxanes: metal ion mediated enhancements in conductivity and charge localization. *J. Am. Chem. Soc.*, **119**, 12568–12577.
70 Armspach, D., Ashton, P.R., Moore, C.P. *et al.* (1993) The self-assembly of catenated cyclodextrins. *Angew. Chem., Int. Ed. Engl.*, **32**, 854–858.

71 Asakawa, M., Ashton, P.R., Ballardini, R. et al. (1997) The slipping approach to self-assembling [n]rotaxanes. *Am. Chem. Soc.*, **119**, 302–310.
72 Isnin, R. and Kaifer, A.E. (1991) Electronic structures of exciplexes and excited charge-transfer complexes. *J. Am. Chem. Soc.*, **113**, 8188–8199.
73 Isnin, R. and Kaifer, A.E. (1993) A new approach to cyclodextrin-based rotaxanes. *Pure Appl. Chem.*, **65**, 495–498.
74 Macartney, D.H. (1996) The self-assembly of a [2]pseudorotaxane of α-cyclodextrin by the slippage mechanism. *J. Chem. Soc., Perkin Trans.*, **2**, 2775–2778.
75 Dimitrius, M., Terzis, A., Coleman, A.W. and de Rango, C. (1996) The crystal structure of 6I-(6-aminohexyl) amino-6I-deoxycyclomaltoheptaose. *Carbohydr. Res.*, **282**, 125–135.
76 Ogata, N., Sanui, K. and Wada, J. (1976) Novel synthesis of inclusion polyamides. *J. Polym. Sci., Polym. Lett. Ed.*, **14**, 459–462.
77 Maciejewski, M.J. (1979) Polymer inclusion compounds by polymerization of monomers in β-cyclodextrin matrix in DMF solution. *Macromol. Sci. Chem.*, **A13** (1), 87–109.
78 Wenz, G. and Keller, B. (1992) Threading cyclodextrins on polymer chains. *Angew. Chem., Int. Ed. Engl.*, **31**, 197–199.
79 Born, M. and Ritter, H. (1991) Comb-like rotaxane polymers containing non-covalently bound cyclodextrins in the side chains. *Macromol. Rapid Commun.*, **12**, 471–476.
80 Anderson, S., Aplin, R.T., Claridge, T.D.W. et al. (1998) An approach to insulated molecular wires: synthesis of water-soluble conjugated rotaxanes. *J. Chem. Soc., Perkin Trans.*, **1**, 2383–2398.
81 Taylor, P.N., O'Connell, M.J., McNeill, L.A. et al. (2000) Insulated molecular wires: synthesis of conjugated polyrotaxanes by Suzuki coupling in water. *Angew. Chem., Int. Ed.*, **39**, 3456–3460.
82 Lagrost, C., Lacroix, J.C., Chane-Ching, K.I. et al. (1999) Host–guest complexation: a strategy to form sexithiophene exhibiting self-assembly properties. *Adv. Mater.*, **11**, 664–667.
83 Hollas, M., Chung, M.A. and Adams, J. (1998) Complexation of pyrene by poly(allylamine) with pendant β-cyclodextrin side groups. *J. Chem. Phys. B*, **102**, 2947–2953.
84 Bergamini, J.F., Lagrost, C., Chane-Ching, K.I. et al. (1999) Host–guest complexation: a general strategy for electrosynthesis of conductive polymers. *Synth. Met.*, **102**, 1538–1539.
85 Harada, A. and Kamachi, M. (1990) Complex formation between poly(ethylene glycol) and α-cyclodextrin. *Macromolecules*, **23**, 2821–2823.

86 Harada, A. and Takahashi, S. (1984) Preparation and properties of cyclodextrin–ferrocene inclusion complexes. *Chem. Lett.*, **10**, 645–646.
87 Shimomura, T., Yoshida, K.i., Ito, K. and Hayakawa, R. (2000) Insulation effect of an inclusion complex formed by polyaniline and β-cyclodextrin in solution. *Polym. Adv. Technol.*, **11**, 837–839.
88 Hermann, W., Keller, B. and Wenz, G. (1997) Kinetics and thermodynamics of the inclusion of ionene-6,10 in α-cyclodextrin in an aqueous solution. *Macromolecules*, **30**, 4966–4972.
89 Mirzoian, A. and Kaifer, A.E. (1997) Reactive pseudorotaxanes: inclusion complexation of reduced viologens by the hosts β-cyclodextrin and heptakis(2,6-di-*O*-methyl)-β-cyclodextrin. *Chem. Eur. J.*, **3**, 1052–1058.
90 Weickenmeier, M. and Wenz, G. (1997) Association thickener by host guest interaction of a β-cyclodextrin polymer and a polymer with hydrophobic side-groups. *Macromol. Rapid Commun.*, **18**, 1117–1123.
91 Harada, A., Li, J. and Kamachi, M. (1994) Sequential interconnected interpenetrating polymer networks of polyurethane and polystyrene. 1. Synthesis and chemical structure elucidation. *Macromolecules*, **27**, 4538–4594.
92 Born, M. and Ritter, H. (1996) 25th Anniversary article: a soft future: from robots and sensor skin to energy harvesters. *Adv. Mater.*, **8**, 149–162.
93 Wang, J.S., Greszta, D. and Matyjaszewski, K. (1995) Atom transfer radical polymerization (ATRP): a new approach towards well-defined (co) polymers. *Polym. Mater. Sci. Eng.*, **73**, 416–417.
94 Haddleton, D.M., Kukulj, D., Kelly, E.J. and Waterson, C. (1999) Organosoluble star polymers from a cyclodextrin core. *Polym. Mater. Sci. Eng.*, **80**, 145–146.
95 Li, J., Xiao, H., Li, J. and Zhong, Y. (2004) Drug carrier systems based on water-soluble cationic β-cyclodextrin polymers. *Int. J. Pharm.*, **278**, 329–342.
96 Li, J., Xiao, H., Young, S.K. and Lowe, T.L. (2005) Synthesis of water-soluble cationic polymers with star-like structure based on cyclodextrin core via ATRP. *J. Polym. Sci. Part A: Polym. Chem.*, **43**, 6345–6354.
97 Chamberlain, M.P., Sturgess, N.C., Lock, E.A. and Reed, C.J. (1999) Methyl iodide toxicity in rat cerebellar granule cells in vitro: the role of glutathione. *Toxicology*, **139**, 27–37.
98 Karaky, K., Reynaud, S., Billon, L. *et al.* (2005) Organosoluble star polymers from a cyclodextrin core. *J. Polym. Sci. A Polym. Chem.*, **43**, 5186–5194.

99 Kakuchi, T., Narumi, A., Matsuda, T. *et al.* (2003) Glycoconjugated polymer. 5. Synthesis and characterization of a seven-arm star polystyrene with a α-cyclodextrin core based on TEMPO-mediated living radical polymerization. *Macromolecules*, **36**, 3914–3920.

100 Miura, Y., Narumi, A., Matsuya, S. *et al.* (2005) Synthesis of well-defined AB_{20}-type star polymers with cyclodextrin-core by combination of NMP and ATRP. *J. Polym. Sci. Part A: Polym. Chem.*, **43**, 4271–4279.

101 Yamaguchi, I., Osakada, K. and Yamamoto, T. (1996) Polyrotaxane containing a blocking group in every structural unit of the polymer chain. Direct synthesis of poly(alkylenebenzimidazole) rotaxane from Ru complex-catalyzed reaction of 1,12-dodecanediol and 3,3′-diaminobenzidine in the presence of cyclodextrin. *J. Am. Chem. Soc.*, **118**, 1811–1812.

102 Cramer, F. and Hettler, H. (1967) Inclusion compounds of cyclodextrins. *Naturwissens chaften*, **54**, 625–632.

103 Bender, M.L. and Komiyama, M. (1977) in *Bioorganic Chemistry* (ed. E.E. van Tamelen), Academic Press, New York, USA.

104 Zhang, Y.H., Gao, Z.X., Zhong, C.L. *et al.* (2007) An inexpensive fluorescent labeling protocol for bioactive natural products utilizing Cu(I)-catalyzed Huisgen reaction. *Tetrahedron*, **63**, 6813–6821.

105 Park, J.W. and Song, H.J. (2004) Isomeric [2]rotaxanes and unidirectional [2]pseudorotaxane composed of α-cyclodextrin and aliphatic chain-linked carbazole-viologen compounds. *Org. Lett.*, **6**, 4869–4872.

106 Oshikiri, T., Takashima, Y., Yamaguchi, H. and Harada, A. (2005) Kinetic control of threading of cyclodextrins onto axle molecules. *J. Am. Chem. Soc.*, **127**, 12186–12187.

107 Onagi, H., Blake, C.J., Easton, C.J. and Lincoln, S.F. (2003) Installation of a ratchet tooth and pawl to restrict rotation in a cyclodextrin rotaxane. *Chem. Eur. J.*, **9**, 5978–5988.

108 Yamaguchi, I., Kashiwagi, K. and Yamamoto, T. (2004) β-Cyclodextrin pseudopolyrotaxanes with π-conjugated polymer axles. *Macromol. Rapid Commun*, **25**, 1163–1166.

109 Li, J., Ni, X. and Leong, K. (2003) Block-selected molecular recognition and formation of polypseudorotaxanes between poly(propylene oxide)–poly(ethylene oxide)–poly(propylene oxide) triblock copolymers and alpha-cyclodextrin. *Angew. Chem., Int. Ed.*, **42**, 69–72.

110 Zhao, T. and Beckham, H.W. (2003) Direct synthesis of cyclodextrin-rotaxanated poly(ethylene glycol)s and their self-diffusion behavior in dilute solution. *Macromolecules*, **36**, 9859–9865.

111 Okada, M. and Harada, A. (2004) Organolanthanide-catalyzed cyclization/boration of 1,5- and 1,6-dienes. *Org. Lett.*, **6**, 361–363.

112 Ying-Wei, Y., Chen, Y. and Liu, Y. (2006) Linear polypseudorotaxanes possessing many metal centers constructed from inclusion complexes of α-, β-, and γ-cyclodextrins with 4,4′-dipyridine. *Inorg. Chem.*, **45**, 3014–3022.

113 Okada, M. and Harada, A. (2003) Poly(polyrotaxane): photoreactions of 9-anthracene-capped polyrotaxane. *Macromolecules*, **36**, 9701–9703.

114 Harrison, I.T. and Harrison, S. (1967) Synthesis of a stable complex of a macrocycle and a threaded chain. *J. Am. Chem. Soc.*, **89**, 5723–5724.

115 Li, J.J., Zhao, F. and Li, J. (2011) Polyrotaxanes for applications in life science and biotechnology. *Appl. Microbiol. Biotechnol.*, **90**, 427–443.

116 Schill, G. and Luttringhaus, A. (1964) The preparation of catena compounds by directed synthesis. *Angew. Chem., Int. Ed. Engl.*, **3**, 546–547.

117 Philp, D. and Stoddart, J.F. (1996) Self-assembly in natural and unnatural systems. *Angew. Chem., Int. Ed. Engl.*, **35**, 1154–1196.

118 Tuncel, D. and Steinke, J.H.G. (2004) Catalytic self-threading: a new route for the synthesis of polyrotaxanes. *Macromolecules*, **37**, 288–302.

119 Ogino, H. and Ohata, K. (1984) Synthesis of tin-tellurium polyanions: structure of $(Me_4N)_4Sn_2Te_6$. *Inorg. Chem.*, **23**, 2312–2315.

120 Wendorff, J.H. and Schartel, B. (1999) Molecular composites for molecular reinforcement: a promising concept between success and failure. *Poly. Eng. Sci.*, **39**, 128–151.

121 Harada, A., Li, J. and Kamachi, M. (1993) Synthesis of a tubular polymer from threaded cyclodextrins. *Nature*, **364**, 516–518.

122 Raymo, F.M. and Stoddart, J.F. (1996) Polyrotaxanes and pseudopolyrotaxanes. *Trends Polym. Sci.*, **4**, 208–211.

123 Mazzaglia, A., Donohue, R., Ravoo, B.J. and Darcy, R. (2001) Novel amphiphilic cyclodextrins: graft-synthesis of heptakis(6-alkylthio-6-deoxy)-β-cyclodextrin 2-oligo(ethylene glycol) conjugates and their ω-halo derivatives. *Eur. J. Org. Chem.*, 1715–1721.

124 Mazzaglia, A., Ravoo, B.J., Darcy, R. et al. (2002) Aggregation in water of nonionic amphiphilic cyclodextrins with short hydrophobic substituents. *Langmuir*, **18**, 1945–1984.

125 Natoli, M., Pagliero, C., Trotta, F. and Drioli, E. (1997) A study of catalytic *p*-cyclodextrin carbonate membrane reactor performance in PNPA hydrolysis. *J. Mol. Catal. A Chem.*, **121**, 179–186.

126 Evers, R.C., Dang, T.D. and Moore, D.R. (1991) Synthesis and characterization of graft copolymers of rigid-rod poly(*p*-phenylenebenzobisimidazole). *J. Polym. Sci. Part. A-Polym. Chem.*, **1**, 121–125.

127 Tsai, T.T., Arnold, F.E. and Hwang, W.F. (1989) Synthesis and properties of high strength high modulus ABA block copolymers. *J. Polym. Sci. Chem.*, **27**, 2839–2848.

128 Li, G. and McGown, L.B. (1994) Molecular nanotube aggregates of β- and γ-cyclodextrins linked by diphenylhexatrienes. *Science*, **264**, 249–251.

129 Ishizu, K., Kuwabara, S., Chen, H. *et al.* (1986) Kinetics of crosslinking reactions. 1. Reaction rate of intramolecular crosslinkings by using a soluble microgel. *J. Polym.Sci., Part A: Polym. Chem.*, **24**, 1735.

130 Schartel, B., Stiirnpflen, V., Wendling, J. and Wendorff, J.H. (1996) Hydrophobic Fluorescent Polymer Synthesis Mediated by Cyclodextrin. *Polym. Adu. Tech.*, **7**, 160.

131 Braun, D., Hartig, C., Reubold, M. *et al.* (1993) Molecular reinforcement by λ-shaped molecules. *Makrornol. Chern.-Rapid Corn.*, **14**, 663.

132 French, D. (1957) The Schardinger dextrins. *Adv. Carbohydrate Chem.*, **12**, 189–260.

133 Tabushi, I. (1982) Cyclodextrin catalysis as a model for enzyme action. *Acc. Chem. Res.*, **15**, 66–72.

134 (a) Ooya, T., Eguchi, M. and Yui, N. (2003) Supramolecular design for multivalent interaction: maltose mobility along polyrotaxane enhanced binding with concanavalin A. *J. Am. Chem. Soc.*, **125**, 13016–13017; (b) Nelson, A. and Stoddart, J.F. (2003) Dynamic multivalent lactosides displayed on cyclodextrin beads dangling from polymer strings. *Org. Lett.*, **5**, 3783–3786; (c) Nelson, A., Belitsky, J.M., Vidal, S. *et al.* (2004) A self-assembled multivalent pseudopolyrotaxane for binding galectin-1. *J. Am. Chem. Soc.*, **126**, 11914–11922.

135 (a) Gong, C. and Gibson, H.W. (1998) Self-assembly of novel polyrotaxanes: main-chain pseudopolyrotaxanes with poly(ester crown ether) backbones. *Angew. Chem. Int., Ed. Engl.*, **37**, 310–314; (b) Gong, C., Balanda, P.B. and Gibson, H.W. (1998) Supramolecular chemistry with macromolecules: new self-assembly based main chain polypseudorotaxanes and their properties. *Macromolecules*, **31**, 5278–5289.

136 Fred, E.A. and Fred, E.A. (2005) Rigid-rod polymers and molecular composites. *Adv. Polym. Sci.*, **117**, 257–295.

3

Smart Polymers

3.1 Introduction

The development and applications of novel biomaterials play an important role in improving the treatment of diseases and the quality of health care. Recent research on polymeric biomaterials focuses on producing new materials, with improved biocompatibility, mechanical properties, and responsiveness. Polymeric biomaterials have been used in medicine, including controlled drug delivery systems (DDSs), tablet coatings, artificial organs, tissue engineering, polymer-coated stents, dental implants, and sutures [1, 2]. This is a major step toward the development of polymeric therapeutic devices from newly synthesized polymeric materials with desired properties.

Various polymerization systems can be used for polymers such as free radical polymerization (FRP), controlled radical polymerization (CRP), and living polymerization techniques. Living polymerization is a method in which growth of polymer chain is continuous without termination until all monomers are consumed. In these reactions, rate of chain initiation is much larger than that of chain propagation, so chain termination and chain transfer reactions are negligible, leading to growth of polymer chains at a constant rate, and chains with very similar lengths are obtained. Polymers obtained from this method have predetermined molar mass and control over end groups so is desirable for materials design and also synthesis of block copolymers [1, 2].

Various groups of supramolecular structures can be synthesized by living polymerization, and obtained polymers undergo reversible large physical or chemical changes in response to small external changes in the environmental conditions, such as temperature, pH, light, magnetic or electric field, ionic factors, biological molecules, and smart materials or stimuli-sensitive or stimuli-responsive materials. These supramolecular structures can be used in a wide range of applications such as drug delivery, gene delivery, bio-separation devices and membranes, tissue engineering scaffolds, cell culture supports, artificial sensors, and actuators [1–3].

3.2 Supramolecular Self-Assembly

Smart or stimuli-sensitive or stimuli-responsive polymers with well-defined architectures and appropriate physical and chemical properties can be produced by inclusion complex (IC) formation of guest molecules with cyclodextrins and self-assembly reaction [1–3]. These supramolecular polymers open a new route to design novel supramolecular nanostructures, which are reversible under external stimuli, such as temperature, pH, and complexation and have potential for numerous applications, such as biomaterials, DDSs, and chemical separations [4–6]. The formation of stimuli-responsive polymers via reaction between CDs and polymers yields well-defined architectures with tailored physical properties and allows external stimuli to change their conformation and solubility and modify the hydrophobic/hydrophilic balance of the system [7–9]. Cyclodextrins (CDs) are interesting materials in this field. CDs are macrocyclic oligosaccharides joined together by α-1,4-glycosidic linkages. A specific geometry of CDs with hydrophobic cavity and hydrophilic surface enables them to form ICs with a wide range of organic and inorganic chemical species and induces self-assembly of polymers [1–3]. Self-assemblies consist of reversible noncovalent interactions and are thus considered dynamic processes [10].

Supramolecular self-assemblies are based on reversible hydrogen bonding or van der Waals interactions, which may be induced by either changing the external environment such as pH, temperature, and electropotential, or induced mechanical motion or by transforming the physical characteristics of one or more building blocks. These reversible interactions are controlled by external stimuli so are receiving attention in the scientific community [11, 12]. Stimuli-responsive polymers form ICs with CDs and induce self-assembly of polymer chain/CD complexes, and supramolecular structures are obtained with different morphologies and physical properties [12, 13].

3.3 Synthesis of Block Copolymers

3.3.1 Free and Living Radical Polymerization

One of the simplest and most widely employed polymerization techniques for the synthesis of high-molecular-weight polymers is FRP. FRP has some advantages such as insensitivity to impurities, moderate reaction temperatures, and availability of multiple polymerization processes, for example, bulk, solution, precipitation, or emulsion polymerization [14–16].

The fast chain growth and rapid irreversible termination reaction pose some limitations on the degree of molecular weight distribution of the polymer and polymer structure. Poor control of the molecular weight and the molecular weight distribution and difficulty of preparing well-defined copolymers or polymers with a predetermined functionality or molecular weight are disadvantages of FRP [14, 15].

Conventional radical polymerization cannot be used for synthesis of pure block copolymers and polymers with controlled architecture. To overcome these disadvantages, living radical polymerizations based on reversible deactivation of polymer radicals or a degenerative transfer process were developed, which are called CRPs [16, 17].

Living radical polymerization is involves fast exchange between active and dormant species; a majority of the growing chains are in the dormant state and only a small fraction is present as propagating free radicals. The propagation rate is independent of the degree of polymerization, and the polymerization reaction should be free of termination and transfer reactions. The important parameter is concentration of the propagating radicals, which should be sufficiently low to enable chain growth on the one hand and reduce termination events on the other. In living/CRPs, lifetime of growing chains is very short and participation of dormant species and alternative reversible activation [14–17].

In these systems, initiation is very fast and initiator is consumed, which leads to instant growth of all chains and exerts control on chain architecture. Balance between activation and deactivation rates leads to a steady radical concentration based on the persistent radical effect. Living controlled radical polymerizations (LRPs) have great potential for the production of polymers of lower molecular weight and high degrees of functionality. The polymer produced contains different functional groups related to monomers and functional groups at the head and tail of living radicals [15].

Living/controlled radical polymerizations are classified into nitroxide-mediated polymerization (NMP) [18–20], atom transfer radical polymerization (ATRP) [21–23], and reverse addition–fragmentation chain transfer polymerization (RAFT) [24, 25] (Figure 3.1).

NMP have regained increasing interest because the increasing number of monomers can now be polymerized in a controlled manner through this technique with a significant decrease in the reaction times. This system can be controlled by reversible activation and deactivation of active radical polymer chains by a nitroxide radical. The concentration of the growing radical species is decreased and the chains grow with a uniform speed and side reactions are kept at a minimum and polymers

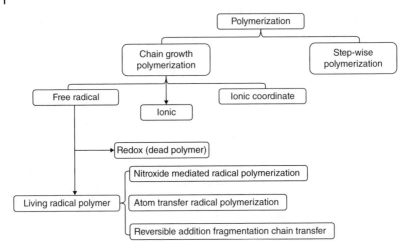

Figure 3.1 Classification of polymerization methods.

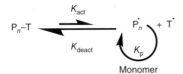

Scheme 3.1 Nitroxide-mediated polymerization mechanism.

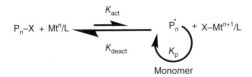

Scheme 3.2 Atom transfer radical polymerization mechanism.

with relatively narrow molecular weight distributions are produced [18–20] (Scheme 3.1).

ATRP is based on the reversible transfer of halogen atoms, between a dormant species (P_n–X) and a transition metal catalyst (M_nt/L). The alkyl halides are reduced to active radicals and transition metals are oxidized via an inner sphere electron transfer process [21–23] (Scheme 3.2).

Viscosity of the polymerization system increases and termination rate decreases significantly by the increase in the chain length, and reaction conditions are controlled to prepare polymers having a wide range of architectures including blocks, grafts, gradient copolymers, stars, combs, branched, and hyperbranched. The RAFT is based on radical addition–fragmentation chain transfer agents in radical polymerization,

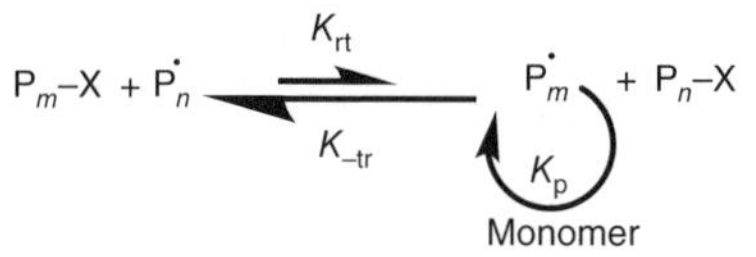

$$P_m\text{-X} + \dot{P}_n \xrightleftharpoons[K_{-tr}]{K_{rt}} \dot{P}_m + P_n\text{-X}$$

with Monomer / K_p

Scheme 3.3 Reversible addition–fragmentation chain transfer mechanism.

which controls the molecular weight and end functionalities of the obtained polymer. In RAFT process, the important parameter is a good transfer agent such as thiocarbonyl thio derivatives [24, 25] (Scheme 3.3).

The most significant advantage is compatibility of the technique with a wide range of monomers, such as styrene, acrylates, methacrylates and derivatives. This large number of monomers provides the opportunity of creating well-defined polymer libraries by combining different monomeric units. RAFT is among the most rapidly developing areas of polymer chemistry, with the number of publications approximately multiplying each year [21–25].

3.3.2 Block Copolymers

Homopolymers are formed when one kind of repeating unit thread whole backbone, and nature of repeating units determine molecular weight and weight distribution [26–28]. A polymer is a molecule composed of repeating smaller units; a block copolymer is two or more chemically distinct homopolymers linked together, and its physical properties are dependent upon molecular characteristics such as chemical structure, size, dispersity, and block ratios. Various architectures can be made, such as di, tri, and tetra block copolymers or star and mixed-star block copolymers [29–31].

Polymer A can be linked to another polymer B or polymer B can be grown from polymer A using B monomers. These are two methods for creating a diblock copolymer. There are, however, other architectures that can be made, such as tri and tetra block copolymers, or star and mixed-star block copolymers [32–34]. Copolymers are divided into random, block, and graft copolymers, in which polymeric side chains of one repeating unit are grafted to a strand of another repeating unit, among others (Figure 3.2) [35, 36].

3.4 Self-Assembly of Amphiphilic Block Copolymers

Building blocks of block copolymers can be different with similar solubility characteristics, being either polar or nonpolar or, in reference to

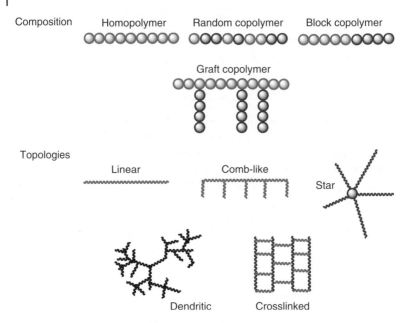

Figure 3.2 Classification of polymers and copolymers based on repeating unit.

water, hydrophilic or hydrophobic. If the building blocks have differing solubilities in water, the molecule is called an amphiphilic block copolymer and increase the capacity for self-assembly structure formation. Poor affinity of the hydrophobic block in aqueous medium induces the assembly of like blocks into an energetically favorable construct that will hide this part of the polymer, which can occur in the bulk, thin-film, or solution phases, each having its own applications [37, 38]. Amphiphilic block copolymers spontaneously self-assemble into ordered structures because the hydrophobic segments associate to form a hydrophobic inner core, while the hydrophilic segments shield the core from the solvent by forming the corona. Hydrophilic volume fraction of the polymer determines morphology of self-assembled structures such as spherical micelles, worm-like micelles, vesicles, and other nanoparticles for drug delivery applications (Figure 3.3) [34–38].

Self-assembly supramolecules are ideal for drug delivery due to their size and ability to encapsulate both hydrophobic and hydrophilic drugs and release them in a controlled manner and at the same time to prolong their circulation time and protect them from premature degradation [37, 38].

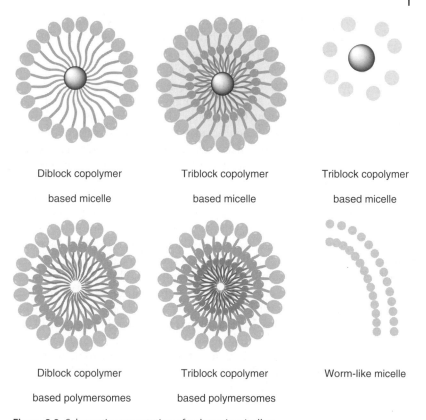

Figure 3.3 Schematic presentation of polymeric micelles.

3.4.1 Smart Polymers Synthesized Based on Living Controlled Radical Polymerization

Linear amphiphilic block copolymers may be aggregated due to their amphiphilicity and the proportions, aspect ratio, of their individual components, and the overall dimensions are much larger than among small molecule assemblies [31–33].

This aggregation alters properties of block copolymers such as optical and electrical properties, absorption and conduction behaviors, and quenching emission, which is caused by π–π-interactions between packed backbones [34–37]. Using flexible segments in block copolymers may reduce the intermolecular contact and aggregation and increase the overall solubility of block copolymers [39–42].

Copolymer chains obtained from LRP have functional groups and can be used for synthesis of polymers, which respond to external stimuli and adapt themselves to changing conditions, and are used for producing smart polymers [41, 42]. Smart polymers undergo reversible physical or chemical or ionic changes in response to small external changes in the system such as temperature, pH, light, magnetic or electric field, ionic factors, biological molecules. If the changes are small, smart polymers may overcome dramatic property changes and return to initial state [43, 44]. These polymers can be used in pharmaceutical applications such as chemical agents or DDSs, encapsulation of drugs, tissue engineering scaffolds, cell culture supports, bioseparation devices, stimuli-responsive sensors or actuator systems, biomaterials, and chemical separations [41–45].

Some smart polymers respond to two or more external stimuli, called dual-stimuli-responsive materials, which are interesting candidates because of their high potential in the biomedical field. Stimuli-sensitive polymers are synthesized via supramolecular self-assembly polymeric ICs consisting of cyclodextrins, which respond to external stimuli, and their conformation, solubility, and hydrophobic/hydrophilic balance of the system changes also determine appropriate physical properties. This behavior is attributed to formation of ICs based on reversible noncovalent interactions. These interactions can be controlled by external stimuli such as pH, temperature, and electropotential, ionic or mechanical motion. By changing the external environment or by transforming the physical characteristics of one or more building blocks, dynamic self-assemblies may occur [6, 45, 46].

3.4.2 Definition of Self-Assembly

Macromolecular self-assembly refers to the assembly of synthetic polymers, biomacromolecules, and supramolecular polymers. In self-assembled macromolecules, building blocks are at different nanometer-size scales and lead to various properties and applications such as nanoreactors, nanocarriers, biosensors, microelectronics, template surfaces for crystal growth, microstructured materials with molecular switches, and so on. Self-assembled structures are thermodynamically more stable than single, unassembled building blocks and are sensitive to disturbances exerted by the external environment [1, 47, 48].

Understanding the structure/property relationships and control of self-assembly process lead to systems due to a combination of forces such as hydrophobic interactions, van der Waals forces, and hydrophilic interactions with unique properties [33, 41].

Self-assembled structures can introduce specific functionalities to the synthetic polymers and design desired morphologies, or provide other properties on demand, such as responsiveness, to new systems

[43–45]. Self-assembled structures can be synthesized from block copolymers, which have both long-range repulsive and short-range attractive forces and called amphipathic and amphiphilic (in the case of water). Self-assembly principle is based on the poor solubility of one of the blocks in a given solvent, while the other block has good solvent–solute interactions [42–45].

When solvent is water, a large energy barrier causes disturbance in the water lattice surrounding chains. Weak interaction is formed between hydrophobic chains with the water molecules and strong solvent–solute interactions improve solubility of hydrophilic chains [33, 41, 47, 48]. Self-assembled structures are formed by these forces. The hydrophobic chains aggregate into a core to avoid contact with solvent molecules, and the hydrophilic chains interacting with solvent molecules form the corona at the surface and form self-assembled structures [33, 41].

3.4.3 Self-Assembled Structures Based on Block Copolymers

Total free energy and thermodynamic considerations of the system such as decrease in interfacial tension at the hydrophilic/hydrophobic interface and the entropy loss from polymer chains are key factors in the formation of self-assembled structures based on amphiphilic block copolymers [13, 30].

From amphiphilic block copolymers, self-assembled structures such as micelles, spheres, vesicles, or rods are formed in polar solvents at concentrations exceeding the critical micelle concentration (CMC). Hydrophobic inner core is formed by the hydrophobic segments and hydrophilic segments form the corona by shielding the core from the solvent [36–38].

Self-assembly and disassembly rates are controllable by tuning the amphiphilicity of the building blocks, leading to fabrication of new functional supramolecular assemblies and materials. Formation of covalent interactions or dynamic covalent bonds between functional groups converts amphiphiles into supra-amphiphiles, for which building blocks can be either small organic molecules or polymers [6, 46–48].

Amphiphilic characteristics are imparted to water-soluble polymers by the grafting of hydrophobic blocks, which is used for formation of various stimuli-responsive polymeric hydrogels to a variety of external factors [2, 49, 50]. Temperature-sensitive microporous hydrogels are synthesized by cross-linking acrylamide-based polymers such as poly(N-isopropylacrylamide). These hydrogels show fast and reversible phase transitions in response to changes in temperature due to changes in the affinity of the polymer chains to the aqueous medium. These smart hydrogels can control diffusion and permeation of drugs because their large volume changes with external stimuli and they have found many applications in medicine, biotechnology, environmental systems, and drug carriers [51, 52].

Novel supramolecular nanostructures have been synthesized using double hydrophilic block copolymers, which are soluble in aqueous solutions under normal conditions and will self-assemble into various types of nanostructures when subjected to external stimuli, such as pH and temperature. Vesicles disassemble and release their encapsulated species. Self-assembled structures have various morphologies, among which polymeric micelles and vesicles, dendrons and dendrimers are the most important [41–45].

3.4.3.1 Micelles

Self-assembly into micelles formation is due to the hydrophobic interaction between surfactant alkyl chains, which are strong, attractive interaction between water molecules. These interactions are energetically less favorable than interactions between water molecules, and the system tends to minimize its free energy by eliminating alkyl-chain–water molecule contacts and micelles core is formed. The balance between attractive and repulsive interactions results in micelles of finite size [53–55].

Micelles obtained by hydrophobic segments (A) and water-soluble segments (B) form the hydrophobic core and a hydrophilic corona, respectively (Figure 3.4). Length and nature of polymer segments have important role in controlling the micelles size, and morphology of micelles is mostly influenced by the hydrophilic block architecture. Simultaneous hydrophobic and hydrophilic properties make micelles an important candidate in pharmaceutical applications, which can encapsulate hydrophobic drugs and release them in the target section [41–43].

When polymer chains are fully stretched both in the corona and the core, the guest must be smaller than the core though encapsulation is done. Micelle formation can occur in a wide variety of polar organic solvents of high cohesive energy density, such as formamide and alkane diols, although the aggregate stability is generally lower than that in water (Figure 3.4) [44, 45].

3.4.3.2 Vesicles

Another class of supramolecular nanostructures is vesicles, which are formed via self-assembly of block copolymers into small spherical micelles. Vesicles are amphiphilic supramolecular nanostructures surrounded by an aqueous solution containing entangled rod-like micelles and are used in drug delivery. Building blocks of vesicles are amphiphilic molecules, and the issue is to adjust the structures of these building

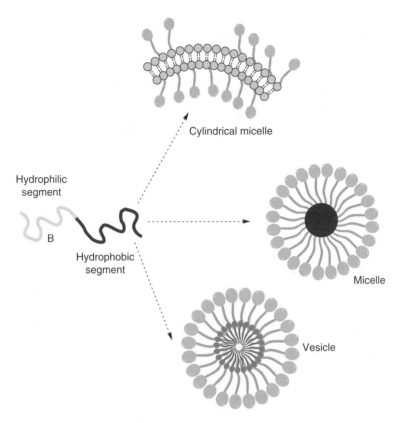

Figure 3.4 Self-assembled structure in diluted regime.

blocks so that aggregates of the desired morphology and properties are obtained, but stability of the aggregates is an important factor determining their size [56, 57].

Block copolymer vesicles are widely used as nanocarriers for drug delivery purposes, which are formed by assembly of a number of hydrophilic or hydrophobic functional molecules (molecular receptors, labels, etc.) and pave the way toward the elaboration of complex multifunctional assemblies capable of controlling regulatory circuits (Figure 3.5) [56–58].

Polymeric vesicles have been used in nanomedicines, in vivo imaging, and drug delivery because of their controllable size, shape, and in vivo circulation time. In the formation of vesicles, a wide range of building blocks can be used and can introduce features such as biocompatibility, inherent or induced permeability, and triggered release and affect their biological behavior [56–58].

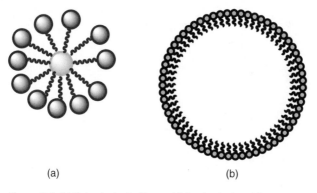

Figure 3.5 (a) Spherical micelles and (b) spherical vesicles.

Vesicles are often spherical and may not be thermodynamically stable. Various methods are used for preparing vesicles such as turbid dispersions of appropriate sparingly water-soluble surfactants or phospholipids or solubilizing an appropriate amphiphile in an organic solvent and evaporating the organic solvent, and exposing the resulting film to water. Formation of vesicles improves solubility of water-insoluble compounds in water with long-term stability. Vesicles can take different shapes and can be unilamellar or multilamellar. Methods such as sonication, extrusion have been used to transform large multilamellar vesicles into small unilamellar vesicles [59–61].

3.4.3.3 Dendrons and Dendrimers

Dendrons and dendrimers are highly branched, three-dimensional architectures, and grown generation-by-generation. They consist of three regions: the core or focal point, the repeating backbone branches, and the peripheral functional groups. They are synthesized stepwise so have very low dispersities and thus well-defined molecular weights (Figure 3.6) [62, 63].

Dendrimer and dendrons are able to encapsulate both hydrophobic and hydrophilic cargo molecules and release them in response to a stimulus and exhibit monodisperse nature, high loading capacities, large-scale production, and bioconjugation capabilities. A specific opportunity, with these amphiphile assemblies, involves the incorporation of functional groups that make them respond to an environmental change such as photo-, temperature-, light-, and pH-responsive supramoleculars [59–61]. New generation of stimuli-responsive assemblies is developed by structure–property relationships derived from these well-defined macromolecular structures and has implications in a variety of areas including drug delivery, biomedical applications, sensing, and diagnostics [62–64].

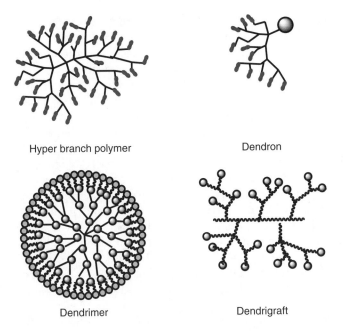

Figure 3.6 Classification of dendrimer supramolecular structures.

The most important class of stimuli-responsive systems is temperature-responsive systems, which show a considerable change in the material's physical properties by a slight change in temperature. Hydrogen bonding plays an important role in temperature-responsive supramolecules. By increasing the temperature of the solution, hydrogen-bonding network is interrupted and their solubility in the aqueous phase is reduced until the solution becomes turbid and this temperature is often referred to as the lower critical solution temperature (LCST). Therefore, hydrogen bond with water is a source of this responsiveness to changes in temperature [62–65].

3.5 Stimuli-Sensitive Supramolecular Structures

Supramolecular assemblies are in nature formed as a result of spontaneous organization of inherently disordered components into stable organized structures through noncovalent interactions (e.g., hydrogen bonding, metal coordination, hydrophobic forces, van der Waals forces, π–π interactions, and/or electrostatic) as well as electromagnetic interactions [66, 67].

The important parameter for stimuli-sensitive structures is a dramatic physicochemical change caused by stimuli, which can be altered in different ways for polymer chains such as changes in hydrophilic to hydrophobic balance, conformation, solubility, degradation, or bond cleavage, which will, in turn, cause considerable behavioral changes in self-assembled structures. Many designs with the location of responsive moieties or functional groups are possible, including side chains on one of the blocks, chain end groups, or junction points in between blocks, and lead to response [67–69].

Supramolecular assemblies demonstrate that a variety of different shapes and sizes can be obtained using molecular self-assemblies such as micelles; liposomes, fibers, and thin films have profound implications in the field of sensing, drug nanocarriers, drug delivery, and diagnostics. Amphiphilic molecules can form a micelle by interactions between the assembly and surrounding solvent, which constituent functional groups provided by the hydrophilic–lipophilic balance [68–70].

Stimuli are of physical, chemical, or biological nature [5, 71, 72]. Physical stimuli such as light, temperature, ultrasound, magnetic, mechanical, electric usually modify chain dynamics and the energy level of the polymer/solvent system. Chemical stimuli such as solvent, ionic strength, electrochemical, pH modulate molecular interactions, in between polymer and solvent molecules, or between polymer chains. Enzymes and receptors are biological stimuli, which are related to functions of molecules: enzymes corresponding to chemical reactions, receptor-recognition of molecules [5, 71, 72].

The most important classes of stimuli-sensitive polymers are pH-sensitive and temperature-sensitive, which have been used widely in biomedical field and drug delivery [59, 60]. Sudden rearrangement with small ranges of pH or temperature is the most important parameter for pH-sensitive and temperature-sensitive polymers, which is caused by reversible interaction between polymer–polymer and polymer–solvent and show a transition between extended and compacted coil states of polymer chains. They are useful for DDSs or tissue engineering applications [63–67].

Hydrophobic–hydrophilic balance and electrostatic repulsions may extend coiled chains and switch on–off the receptor of temperature-sensitive and pH-sensitive polymers, respectively. Electrostatic repulsions of the generated charges (anions or cations) from ionizable weak acidic or basic moieties attached to a hydrophobic backbone may extend coiled chains, and this transition causes to switch on–off the receptor [73–76].

Hydrophobic–hydrophilic balance in temperature-sensitive structures is changed by small temperature changes around the critical T, and new

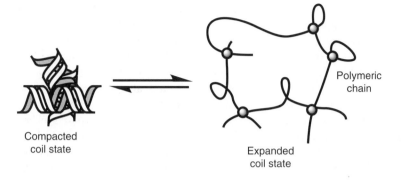

Figure 3.7 Equilibrium state between compacted and expanded state of polymer chains.

arrangement of hydrophobic and hydrophilic interactions is formed between the polymeric chains and the aqueous media. The solution will change from monophasic to biphasic due to polymer precipitation when the transition occurs [75–78]. This transition is related to the reversible sol–gel transition in micellization or micelle aggregation between a collapsed and an expanded state (Figure 3.7). By surface modification of stimuli-sensitive systems, different behaviors may be obtained from sensitive structures such as converting hydrophobic surface to hydrophilic surface [73–75].

3.5.1 Stimuli-Responsive Polymers Based on Cyclodextrins

In stimuli-responsive polymers or gels, conformation, solubility, and hydrophobic/hydrophilic balance of the system modified by external stimuli lead to well-defined supramolecular self-assemblies with appropriate physical properties. Dynamic self-assemblies based on reversible noncovalent interactions may be induced by transforming the physical characteristics of one or more building blocks or changing the external environment such as changes in pH, temperature, and electropotential, or induced mechanical motion [79–83].

One of the reversible noncovalent interactions for synthesis of stimuli-responsive polymers is IC formation between polymers and cyclodextrins, which leads to self-assembly of polymer/CD complexes with different morphologies and physical properties [84–88]. Various kinds of polymers such as linear polymers, random and block copolymers, and selective polymers could form ICs with CDs via hydrogen bonding between the CD and the guest molecule and between the hydroxyl groups on the rims of neighboring CDs [89, 90].

Formation of ICs greatly modifies the physical and chemical properties and solubility of the guest molecules. CDs are used to encapsulate hydrophobic compounds to enhance the scrubbing efficiency of low-polarity volatile organic compounds and also encapsulate toxic substances by converting them into nontoxic ICs and have been used for a wide range of applications in organic and polymer synthesis [78–80], catalysis [78], pharmaceuticals [91–94], biotechnology [95], supramolecular chemistry [96–98], and environmental mediation [99]. The self-assembled threading of CDs onto linear polymer chains depends on electrostatic, hydrophobic, and hydrogen-bonding interactions. Supramolecular structures ranging from aggregates to reversible networks could be formed simply by mixing amphiphilic polymers with CDs in water, which have been used for drug delivery [100–102].

Stimuli-responsive nanogels change their behavior when exposed to external signals and introduce new class of environment-sensitive or smart structures [5, 68–70]. Stimuli-responsive structures are used for controlled drug release in vivo when a specific stimulus (such as variations in solubility, macromolecular structure, surface properties, swelling, disassembly, or a chemical reaction) is triggered at the target site (Figure 3.8) [4, 76].

3.5.2 pH-Responsive Systems

pH-sensitive polymers are polyelectrolytes obtained via protonation or deprotonation of a polybase or polyacid, pH-induced conformational changes, or selective cleavage of pH-sensitive bonds. They have weak acidic or basic groups in their structure that either accept or release protons in response to pH-induced conformational changes or selective

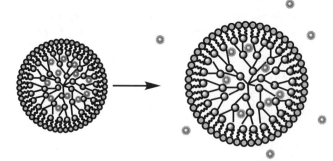

Figure 3.8 Multistimuli nanogel tend which released drug under pH or temperature differences.

cleavage of pH-sensitive bonds. The pendant acidic or basic groups on polyelectrolytes undergo ionization similarly to acidic or basic groups of monoacids or monobases. Electrostatic repulsion exerted by other adjacent ionized groups results in increase in the hydrodynamic volume of the polymer and makes the complete ionization on polyelectrolytes more difficult [103]. Parameters that modify electrostatic repulsion, such as pH, ionic strength, and type of counter ions lead to transition between tightly coiled (collapsed state) and expanded state [5, 68–70].

At low pH, the acidic groups will be protonated and unionized, so polyacidic polymers will unswell , and by increasing the pH, a negatively charged polymer will swell. In polybasic polymers, the opposite behavior is found – increasing pH decreases ionization of the basic groups. Incorporating hydrophobic moieties into the polymer backbone and controlling their nature, amount, and distribution is important for reversible phase transition. Typical examples of pH-sensitive polymers with anionic groups are poly(carboxylic acids) such as poly(acrylic acid) (PAA) or poly(methacrylic acid) (PMA), poly(*N*,*N*-dialkyl aminoethyl methacrylates), poly(lysine), poly(ethylenimine), and chitosan (Figure 3.9) [104–106].

Poly(acrylic acid) (PAA) is generally used as a block for synthesis of pH-sensitive vesicles and undergoes a reversible pH-triggered inversion of the membrane with other blocks such as polystyrene or poly(4-vinyl pyridine) (P4VP). For the synthesis of a pH-sensitive copolymer, one chain has positive charge, water-soluble, and located on the outside corona, such as P4VP, and the other block is protonated, water insoluble, and located inside such as PAA [104–106].

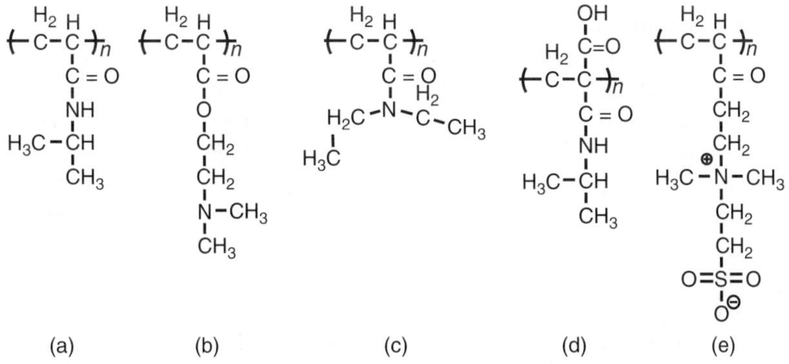

Figure 3.9 (a) Poly(*N*-isopropylacrylamide) (PNIPAAm), (b) poly(*N*,*N*-diethylaminoethyl methacrylate) (PDEAEMA), (c) poly(*N*,*N*'-diethylacrylamide) (PDEAAm), (d) poly(2-carboxyisopropylacrylamide) (PCIPAAm), and (e) poly[2-(methacryloyloxy)ethyl]-dimethyl(3-sulfopropyl) ammonium hydroxide (PMEDSAH).

By changing pH, ionizable groups become neutral and electrostatic repulsion forces disappear within the polymer network, so hydrophobic interactions dominate and conformation in the uncharged state is obtained and phase transition occurs. Hydrophobicity or hydrophilicity of pH-sensitive polymers can be controlled by the copolymerization of hydrophilic ionizable monomers with more hydrophobic monomers with or without pH-sensitive moieties, such as 2-hydroxyethyl methacrylate, methyl methacrylate, and maleic anhydride [85–89].

At low pH, the acidic groups will be protonated and unionized, so polymers will unswell at low pH. With pH increase, a negatively charged polymer will swell. When hydrogel is in swollen state, each polymer chain is isolated by solvent molecules and is exposed as a single molecular unit to tension and shear forces produced in gel deformation process. Most polyelectrolyte gels exhibit a decrease in modulus with increasing swelling degree. Solute permeability of pH-sensitive polymers and hydrogels is affected by the nature of ionizable groups, polymer composition, hydrophobicity of the polymer backbone, and cross-linking density, which is important for bioactive compounds release applications such as drug delivery to solid tumors. pH of extracellular environment of solid tumors is acidic (pH = 6.5), normal tissues have a pH around 7.4, and endosomes and lysosomes have a pH = 5.0–5.5 [85–89, 104–106].

Based on body pH, stable nanogels at physiological pH but sensitive to pH changes toward lower values for a more efficient delivery of small drugs, nucleic acids, or proteins in target sites have been investigated. When pH-responsive hydrogels have weak acid or base groups, such as carboxylic acids, phosphoric acids, and amines, can present a change in the ionization state by varying the pH, which controlled and accelerated drug release through dissociation and association with protons in the aqueous environment [104–106].

If pH-responsive hydrogels have acid-degradable linkages, these polymers may support induced cleavage in acidic conditions and porosity, and possibly polymer dissolution, is increased and response to changes in pH is volume deformation [104–106]. The applications of pH-responsive nanogels are very important in drug delivery, particularly in the delivery of anticancer therapeutics, gene delivery systems, and glucose sensors [107–109].

Eisenberg *et al.* reported the first pH-sensitive vesicles made of polystyrene-*b*-poly(acrylic acid) in 1995 and then poly(acrylic acid)-*b*-poly(styrene)-*b*-poly(4-vinyl pyridine) (PAA-*b*-PS-*b*-P4VP) vesicles were formed, which underwent a reversible pH-triggered inversion of the membrane [109, 110]. Protonation/deprotonation cycle of the two amphiphilic blocks: at pH 1, P4VP chains are positively charged and water soluble and located on the outside corona, while PAA chains are protonated, water insoluble, and located inside. By increasing pH to 14, the PAA chains are negatively charged, the P4VP chains are deprotonated and insoluble, and the vesicles have an inverted membrane: PAA outside, P4VP inside. Armes and coworkers reported PDPA-*b*-PMPC as a highly biocompatible and pH-sensitive block copolymer and PDPA block is insoluble and deprotonated at pH above 7 (pK_a around 5.8–6.6) [108–110].

This copolymer is used to encapsulate water-soluble Doxorubicin, which was released upon lowering of the solution pH. Several groups reported formation of polypeptide-based pH-responsive polymersomes, usually triggered by a conformational change of one of the blocks such as poly(L-glutamic acid)-*b*-poly(L-lysine) copolypeptides (PGA–PLys) [48]. At pH < 4, the polybase is in a coil conformation, while the polyacid is protonated and the hydrogen bonds induce a conformational change from a coil structure to an α-helical structure. The PGA block is then water insoluble and forms the hydrophobic part of the membrane. At pH > 10, the structure is inverted, with α-helices of PLys forming the core of the membrane and PGA charged chains on the outside [111–113].

3.5.3 Temperature-Responsive Systems

Temperature-responsive polymers showing a phase transition in solution due to temperature change passing through a critical temperature, known as critical solution temperature (CST), are the most studied class of environmentally sensitive polymers as they have potential applications in the biomedical field. This type of systems exhibit a CST (typically in water) at which the phase of polymer and solution change according to their composition. Polymer solutions and hydrogels appear as monophasic below a specific temperature called LCST and are swollen below it, and are deswollen below a given temperature exhibiting an upper critical solution temperature (UCST) [105, 114, 115].

In temperature- responsive polymers, small temperature variations across the CST lead to either contraction or expansion of the polymer chain structure, and hydrophobic and hydrophilic interactions between polymer chains and aqueous solution should be optimized and show a volume change at a certain temperature, at which a sharp alteration in the solvation state occurs. Presence of additives in solution, such as salts, cosolvents, and surfactants affect thermoreversibility of polymers, and they can alter the solvent quality and modify the interactions between polymer and solvent [112–114].

These types of polymers have various applications in biomedical and DDS and are applicable both in vitro and in vivo because of their change in temperature is relatively easy to control. Due to dissolution process of thermoreversible hydrogels or polymers, ordered state of water molecules is near the polymer, and entropic effects balanced between intra- and intermolecular forces and due to solvation, hydrogen bonding and hydrophobic interaction. At low temperature, hydrogen bonding between hydrophilic segments of the polymer chains and water molecules becomes stronger, thereby enhancing swelling or dissolution in water [105, 114–116].

When temperature increases, hydrogen bonding becomes weaker and hydrophobic interactions among hydrophobic segments become stronger and lead to contraction of hydrogels. Many polymers can be used for synthesis of thermoresponsive polymer hydrogels such as poly(N-isopropylacrylamide) (PNIPAAm), poly(N,N'-diethylacrylamide), poly(N-acryloyl-N'-propylpiperazine), poly(N-vinylcaprolactam) (PNVCL), and poly(methylvinylether) (PMVE), combination of acrylamide (AAm) and acrylic acid (AAc), poly([N-(3-aminopropyl)-methacrylamide hydrochloride] (PAMPA), and Pluronics (Figure 3.10) [104–107].

Among these polymers, PNIPAAm shows an LCST of 32 °C close to the body temperature, so extensively used for DDSs, which are soluble below this temperature due to hydrogen bonding and hydrophilic interaction. So a phase separation occurs above the LCST (cloud point) due to the predomination of hydrophobic interactions. Below 32 °C, PNIPAAm dissolves in water and above 32 °C hydrophobicity of compound is changed. Copolymers based on PNIPAAm such as PAMPA-b-PNIPAM, PEO–PNIPAM are used as vesicles, which can encapsulate cancer drugs such as doxorubicin, PNIPAM forming the membrane core. The LCST

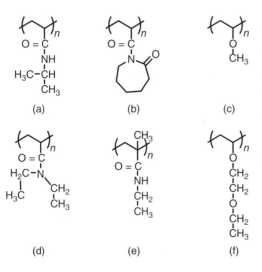

Figure 3.10 (a) Poly(N-isopropyl acrylamide)PNIPAM, (b) poly(N-vinylcaprolactam) PNVCL, (c) poly(methyl vinyl ether) PMVE, (d) poly(N,N-dimethylacryl amide) PDEAM, (e) poly(N-methyl-methacrylamide) PNEMAM, and (f) poly(2-ethoxyethyl vinyl ether) PEOVE.

of the PNIPAAm can be also changed by incorporating ionic groups into the gel network or by changing solvent composition. Some of thermoreversible polymers are made of hydrophobic and hydrophilic groups such as pluronics composed of polyethylene oxide–polypropylene oxide–polyethylene oxide copolymers (PEO–PPO–PEO) [103–105].

At low temperatures in aqueous solutions, a hydration layer surrounds pluronics molecules. At high temperatures, hydrogen bonds between the solvent and hydrophilic chains break down. Hydrophobic interactions among the polypropylene oxide lead to the formation of micelle structures above the critical micelle temperature (CMT), which is stable at low temperatures and transforms into the cubic structure with the increase in temperature. Dehydration of the polymer leading to increased chain friction and entanglement, producing a hydrophobic association, and forms a gel [103–107].

Thermoresponsive polymers are used for synthesis of temperature-responsive nanogels for biomedical applications, which incorporate hydrophilic and hydrophobic molecules, and show a reversible swelling–deswelling behavior, passing through a critical temperature known as CST. Such hydrogels can be loaded with an anticancer drug, and, at the target location, by moderately increasing the temperature above the LCST, the nanogel can change volume and the drug release can be accelerated. Reversible swelling–deswelling behavior of polymers represents smart behavior originating from the architecture of the polymer and the balance between the hydrophilic/hydrophobic fragments (Figure 3.11) [103–107].

Synthesis of temperature-responsive materials without amide functions for in vivo delivery of various bioactive compounds remains a challenge, such as poly(ethylene glycol)-based polymers, which undergo a phase transition upon heating; the LCST ranges from 99 to 176 °C, depending on the molecular weight but limits its use as a temperature-responsive material in medical applications. Poly(ethylene glycol) (PEG)

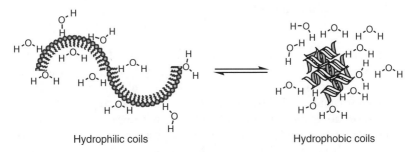

Figure 3.11 Phase transfer of thermoresponsive polymers.

is a neutral, water-soluble, biocompatible, nontoxic, nonimmunogenic, FDA-approved, and, probably the most widely used polymer in biomedical applications [99–102, 104–107].

3.5.4 Redox-Responsive Systems

The most important stimuli-responsive systems are temperature- and pH-responsive hydrogels. Other stimuli systems used for various applications are applied stress, magnetic and electromagnetic field, ionic strength, bioactive species, redox-responsive systems, and so on. Redox-responsive hydrogels are obtained by incorporation of redox-responsive centers in the polymer main chain, in the side groups or as cross-linking moieties and undergo reversible oxidation–reduction reactions. These hydrogels are capable of responding reversibly in a controllable and predictable manner and may be applied chemically or electrochemically [117–119]. Sometimes, a reductive environment could be a powerful stimulus for drug delivery in the case of nanogels containing reducible bonds, such as disulfide bonds [118–120].

The intracellular and the extracellular environment have a huge difference in terms of reduction potential, and it is being explored for triggering the intracellular delivery of drugs. Redox active gels include ferrocene, conjugated polymer, tetrathiafulvalene, transition metal ions, and disulfide-based gels with various dimensions, including macroscopic gels, microgels, and nanogels. These copolymers spontaneously self-assemble into vesicular structures, which can be rapidly disrupted in the presence of cysteine in intracellular concentrations. The latter is responsible for the separation of the blocks within 10 min, subsequently triggering the destruction of the vesicles [118–120].

3.5.5 Other Stimuli-Responsive Hydrogels

Stimuli can also be classified into external stimuli such as light, ultrasound, and magnetic field and internal stimuli such as temperature, pH, redox potential, and glucose level.

3.5.5.1 Light-Sensitive Materials

In the drug delivery and other delivery fields, one of the most attractive factors is light because it provides precise temporal (when the light source is switched on) and spatial control (where the light is directed to), and also light-responsive systems do not require additional substances to trigger release [120, 121]. Liposomes, micelles, hybrid materials, microcapsules, and few types of vesicles are light-responsive systems. These compounds do not require additional substances to release and do not modify the immediate environment around the hydrogel, as is the case for systems sensitive to internal stimuli, such as pH. Azobenzene derivatives, spirobenzopyran derivatives, nitrocinnamate, and O-nitrobenzyl derivatives are used for synthesis of light-sensitive molecules. Azobenzene derivatives are used for synthesis of light-sensitive molecules, which respond to both near-UV and visible light, and lead to a reversible isomerization between the cis and trans conformation upon light irradiation [120–124].

3.5.5.2 Photoresponsive Polymers

Photoresponsive polymers have one or more properties that can be significantly changed in a controlled manner on receiving an external stimulus. Photoresponsive polymers containing azobenzenes and other chromophores in the side or main chains have been extensively studied as advanced materials, which are based on photoresponsive behaviors, including photo-isomerization, photo-crosslinking (or photodimerization), photoalignment, and photoinduced cooperative motions. Spiropyrans, spirooxazines, fulgides, and diarylethenes undergo a reversible *trans*-to-*cis* isomerization, which leads to a fast change in the geometric shape and polarity of the molecule [122–125].

Photoresponsive block copolymers are prepared by incorporating photoresponsive constituent segments into block copolymers, and if they show a lysogenic phase, the regular periodicity of liquid-crystalline (LC) ordering will influence the microphase-separated nanostructures, making it possible to self-assemble into periodic nanostructures on a macroscopic scale (Figure 3.12) [122–125].

3.5.5.3 Photoresponsive Liposomes

There are a few types of photosensitive liposomes and lipids. Light may also serve as a trigger via the photothermal conversion of energy of absorbed light. Chandra designed photocleavable lipids composed of charged amino acids (Asp, Glu, and Lys) as polar head groups linked to alkyl chains of stearyl amine as tail groups by a photocleavable O-nitrobenzyl linker [126]. Troutman *et al.* designed thermoresponsive liposomes coated with gold, forming a plasmon resonant coating. Irradiation with IR-light (1094 nm) triggered the release of model encapsulated compounds. UV-induced isomerization of the azo chromophore results in changes in membrane permeability and exploits the alteration of the membrane to release encapsulated acridine orange and doxorubicin (Figure 3.13) [125].

3.5.5.4 Photoresponsive Micelles

Photoinduced isomerization of azobenzenes can also be used to induce changes in macromolecular self-assemblies dispersed in a liquid medium,

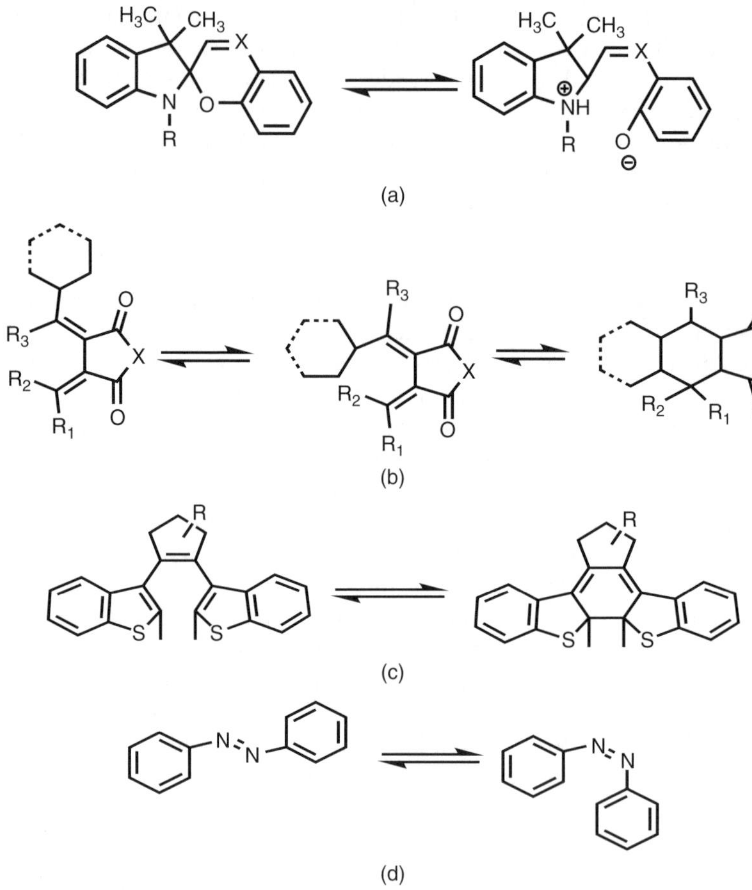

Figure 3.12 (a) Spiropyrans, (b) flugides, (c) diarylethenes, and (d) azobenzene.

such as micelle dissociation or vesicle deformation. Photoresponsive polymeric micelles have received increasing attention in both academic and industrial fields due to their efficient photosensitive nature and unique nanostructure and are promising for applications in different areas such as DDSs to release hydrophobic drugs and nanocarriers. Due to photoreaction mechanism, photoresponsive polymeric micelles are classified into photoisomerization, photoinduced rearrangement, photoinduced cross-linkable, photocleavage, and photoinduced energy conversion polymeric micelles [105, 115–117, 122–125].

Self-assembly of an amphiphilic block copolymer with a functional photochromic chromosphere leads to formation of photoresponsive micelles. In these systems, photochromic molecules absorb optical

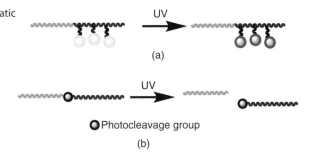

Figure 3.13 Schematic representation of (a) photochromic containing block copolymers and (b) photodegradable block copolymers.

signal, and photoreactions such as isomerization, rearrangement, cleavage, dimerization, and energy conversion chromophores in the photoreceptor convert photoirradiation to a chemical signal. Various self-assembled block copolymer micelles were reported as photoresponsive polymeric micelles [122–124]. Poly(ethylene oxide)-*b*-poly([7-(diethylamino)coumarin-4-yl]methyl methacrylate) (PEO-*b*-PDEACMM) was used for synthesis of self-assembled block copolymer micelles by Babin [127].

Coumarin pendant groups are cleaved upon NIR absorption, converting hydrophobic PDEACMM polymer chains to poly(methacrylic acid) (PMA) hydrophilic chains and lead to destruction of micellar structures and release of Nile red in solution. Matyjaszewski and coworkers developed PEO-*b*-SP (poly ethylene oxide–spiropyran) reversible micelles with light responsiveness. Hydrophobic spiropyran undergoes a reversible isomerization in response to light, yielding merocyanine hydrophilic structures. Micelles formed from PEO-*b*-SP can be disrupted by UV irradiation, when PEO-*b*-SP is converted into PEO-*b*-ME (methyl), a fully water-soluble diblock. They are used to encapsulation, release, and partial reencapsulation of a hydrophobic dye [127].

Zhao *et al.* have developed a block copolymer poly(ethylene oxide)-*b*-poly(2-nitrobenzyl methacrylate) (PEO–PNBM), where the ONB (*O*-nitrobenzyl) has been placed as a side chain. Chromophores are cleaved upon UV irradiation and the polarity balance is changed, causing the micellar structures to fall apart. Poly(ethylene oxide)-*b*-poly-(ethoxytri(ethylene glycol) acrylate-*co*-*O*-nitrobenzyl acrylate) (PEO-*b*-P(TEGEA-*co*-NBA)) triblock copolymer were synthesized, which are responsive to both UV-trigger and temperature-trigger due to the thermoresponsive PTEGEA block [127, 128].

3.5.5.5 Photoresponsive Vesicles

Photoresponsive polymers have several advantages over other stimuli, but only a few relevant systems have been described such as photoresponsive lipids and micelles. Azobenzene is used for synthesis of photodegradable vesicles. PAzo-b-P(tBA-AA) copolymers based on PAzo (which is a hydrophobic methacrylate-based azobenzene containing side-chain liquid crystalline polymer) and poly-($tert$-butyl acrylate-co-acrylic acid) polymer (which is a weakly hydrophilic polymer) were synthesized by Zhao and coworkers [129]. PAzo block from hydrophobic to hydrophilic causes a change in the hydrophilic/hydrophobic balance of the copolymer inducing vesicle dissociation. Upon UV irradiation, hydrophilic/hydrophobic balance of the copolymer changed and PAzo became hydrophilic, inducing vesicle dissociation and aggregates formed were coined as vesicles. In another study, Han *et al.* reported poly(N-isopropylacrylamide)-*block*-poly{6-[4-(4-pyridylazo)phenoxy]hexylmethacrylate} (PNIPAM-b-PAzPy) block copolymer, which is sensitive to UV and temperature [130].

Swelling of the vesicles occurs by direct light to the solution or hydration, which is because of isomerization flip of the azobenzene chromophores in the membrane. Swelling-shrinking is applicable to the transformation energy from light to mechanical energy by deformation of the membrane [130].

Lin reported a novel photoresponsive polymersome, based on hydrophilic poly(ethylene oxide) (PEO) and hydrophobic azopyridine containing poly(methacrylate) (PAP) self-assembly, which is sensitive to UV exposure [131].

These changes are obtained by deformation of the membrane structure due to the isomerization of azopyridine moieties disturbing the tight packing of the polymer chains in the membrane. Photoresponsive vesicles can be used for encapsulation and controlled release of drugs such as poly(ethylene glycol)-b-polybutadiene (PEG-b-PBD), and a liquid crystal-based copolymer, PEG-b-PMAazo (PAzo), which was reported by Mabrouk [132]. The PAzo polymer adopts a rod-like structure or coil conformation when the azo moieties are in the trans form or cis form, respectively.

3.5.5.6 Electroresponsive Polymers

Electroresponsive polymers swell, shrink, or bend in response to electric field, which are usually made of polyanions, polycations, or amphoteric polyelectrolytes and are thus pH-responsive as well. Chemical cross-linking of water-soluble polymers or free-radical polymerization of monomers yields electroresponsive polymers. Natural polymers such as hyaluronic acid, chondroitin sulfate, agarose, xanthan gum, and calcium alginate and synthetic polymers such as methacrylate and acrylate derivatives such as partially hydrolyzed PAAm and poly(dimethylaminopropyl acrylamide) are the most used polymers for synthesis of electroresponsive polymers [121, 133–135].

The electrical response of polyelectrolyte hydrogels is influenced by many parameters, such as the shape and the orientation of the gel, its composition (charge density, nature and hydrophilicity of cross-links, monomers, and pendant groups), the nature of the aqueous conducting medium, the eventual presence of electrolytes in the medium, and the experimental setup. An electrical field in the form of an external stimulus offers numerous advantages and allows precise control over the magnitude of the current, the duration of electrical pulses, and the interval between pulses, which is important for DDSs, artificial muscle, or biomimetic actuators. Controlled and predictable release rates are obtained by altering the magnitude of the electric field between the electrodes [121, 133, 134].

The drug ions are bound in one redox state and released from the other [135]. Swelling, shrinking, or bending of electroresponsive polymers depend on the electrostatic attraction that exists between the anode surface and the negatively charged groups, creating a uniaxial stress along the gel axis (Figure 3.14) [136, 137].

Depending on electroresponsive polymers shape and position relative to the electrodes, deswelling or bending is obtained by an electric field when the hydrogels lie perpendicular to the electrodes or when the main axis of the gel lies parallel to the electrodes, respectively [134–136]. Deswelling has mainly been used for the production of mechanical devices, such as artificial tendons/fingers/hands, soft actuators, and molecular machine, while the hydrogel deswelling has been used in DDSs. Bending in electroresponsive polymers is obtained by swelling on one side and deswelling on the other side. Concentration of the electrolytes influences deformation [98–101].

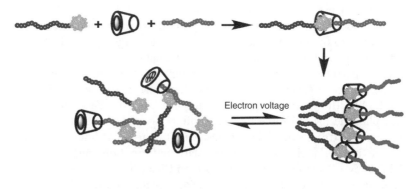

Figure 3.14 Schematic representation of electroresponsive polymers.

When the concentration of electrolytes is low, an electric field causes the hydrogel to collapse, which is due to migration of Na^+ to the cathode, resulting in changes in the carboxyl groups of the polymer chains from COONa to COOH. When the concentration of electrolytes in solution is high, more Na^+ enter the hydrogel than in migration from the hydrogel to the cathode. Swelling is more prominent at the hydrogel side facing the anode. This results in bending of the hydrogels. Deswelling of an electroresponsive polymer hydrogel may depend on increase in gel opacity, magnitude of the electric field, and amount of charge transported through the gel [98–101].

Electroresponsive hydrogels have been mainly applied in controlled DDSs. Kwon *et al.* prepared hydrogels of poly(2-acrylamido-2-methylpropane sulfonic acid-*co-n*-butylmethacrylate) that are able to release edrophonium chloride and hydrocortisone in a pulsatile manner using electric current [115].

Another application of electrosensitive hydrogels is in biomedical field for the production of artificial muscles and tissues. These types of hydrogels are able to convert chemical energy into mechanical energy [138–141]. All living organisms move by the isothermal conversion of chemical energy into mechanical work, for example, muscular contraction, and flagellar and ciliary movement. Electrically driven motility has been demonstrated using weakly cross-linked poly(2-acrylamido-2-methylpropanesulfonic acid), PAMPSA, hydrogels. In the presence of positively charged surfactant molecules, the surface of the polyanionic hydrogel facing the cathode is covered with surfactant molecules, reducing the overall negative charge. This results in local deswelling of the hydrogel, leading to its bending. Application of an oscillating electrode polarity could lead the hydrogel to quickly repeat its oscillatory motion, resulting in a worm-like motion [138–141].

3.5.5.7 Magnetic-Responsive Polymers

These kinds of polymers are obtained by incorporating colloidal magnetic solution into the polymer network, and they undergo a volume change after the application of an external magnetic field.

Field–particle interaction and particle–particle interaction are responses of a magnetic-responsive hydrogel to an external field [142, 143]. The particles undergo a dielectrophoretic (DEP) or a magnetophoretic (MAP) force by a nonuniform field and are attracted to regions of stronger field intensities. Due to the cross-linking bridges in the network, changes in molecular conformation, due to either DEP or MAP forces, can accumulate and lead to macroscopic shape changes and/or motion. In uniform fields, because of the lack of a field gradient, there are no attractive or repulsive field–particle interactions;

thus, particle–particle interactions become dominant and field induces electric or magnetic dipoles [142, 143].

In this situation, if the particles are so closely spaced that the local field can influence their neighbors, mutual particle interactions occur, which can be very strong, leading to a significant change in the structure of particle ensembles. The particles attract each other when aligned end to end and repel each other in the side-by-side situation [139–141, 144].

Magnetic-responsive hydrogels have been found useful in biomedical applications, such as cell separation, gene and drug delivery, and magnetic intracellular hyperthermia treatment of cancer. Rapid and controllable shape changes of these gels would be expected to mimic muscular contraction [139–141, 144].

3.6 Polymers with Dual-Stimuli Responsiveness

It is possible that obtained polymeric structures will be sensitive to multiple stimuli such as light and temperature, pH and temperature, and light and electric field by the simple combination of ionizable and hydrophobic (inverse thermosensitive) functional groups. These kinds of polymers can be obtained from the incorporation of different functional groups responding to different stimuli such as free radical copolymerization of N-isopropylacrylamide (responsive to temperature). A polymerizable spiropyran derivative (responsive to light) with a N,N'-methylenebisacrylamide cross-linker produces a thermo- and light-responsive polymer [133, 142]. Dual-stimuli-responsive polymers act as smart polymers, which are capable of changing their chemical and physical, or conformational, properties in response to small external signals and encapsulate or release molecules and alter their shape and are singularly applicable to gene delivery [143, 145].

Core–shell microgels based on PNIPAAm, MBAAm, chitosan, or poly(ethyleneimine) were synthesized via graft copolymerization in the absence of surfactants and were reported by Leung *et al.*, which consists of temperature-sensitive cores (based on PNIPAAm) with pH-sensitive shells (based on cationic water-soluble polymers) [125].

Gan *et al.* prepared and studied a new water-soluble pH- and temperature-sensitive polymer based on poly(acryloyl-N-propyl piperazine) (crNP) (Figure 3.10) that exhibited a LCST of 37 °C in water. Our group has also been working in this field, developing new polymers based on N-methylpyrrolidone methacrylate (EPyM) that present pH- and temperature-sensitivity [146, 147].

The homopolymer (PEPyM) (Figure 3.3c) exhibited a phase separation transition temperature in water at 15 °C, being also sensitive to pH changes (the polymeric structure was collapsed at basic pH and highly swollen at acidic pH). PEPyM presented a pulsatile swelling–deswelling behavior when the stimuli were removed or reversed (on–off). LCST of the polymer can be modulated [143, 145–147].

3.6.1 Cyclodextrins for Synthesis of Responsive Supramolecules

Polymeric hydrogels have unique bulk and surface properties and have been an interesting topic in both fundamental and applied research as smart materials that are capable of responding to external stimuli such as temperature and pH. Supramolecular hydrogels based on IC formation of cyclodextrins with small organic molecules or polymers have attracted much attention due to their potential applications as novel DDSs and tissue engineering [148, 149].

In recent years, supramolecular self-assembly structure formation from cyclodextrins threaded onto the polymer chains and formation of polyrotaxane/pseudopolyrotaxane have been investigated. Supramolecular hydrogels are formed by inclusion complexation between CDs and the hydrophobic interactions or by cross-linking of CD rings lying on the polymer chains with switchable and controllable properties [85].

CDs have hydrophobic cavity and hydrophilic outer surface. Guest molecules of varying shapes and sizes can be included into the cavities and ICs are formed in different stoichiometric ratios. Stimuli-responsive supramolecular hydrogels can also be constructed by interactions between CDs and side chains of polymers, and an external stimulus does not induce directional motion of the macrocycle, but rather changes the noncovalent intercomponent interactions. The system can be returned to its original state by using a second chemical modification. In suitably designed rotaxanes, the switching process can be controlled by reversible chemical reactions (protonation–deprotonation, reduction–oxidation, isomerization) from chemical, electrochemical, or photochemical triggers [84–88].

3.6.2 pH-Responsive Inclusion Complexes

Stimuli-responsive polymers modulate interactions between the CD cavity and its guest and play a key role in controlling the assembly and disassembly processes. pH-responsive polymers are polyelectrolytes containing acid or basic groups in their structure and are able to accept or release protons in response to pH changes in the surrounding environment, and electrical charge of the polymer molecule is changed. Decreasing pH, neutralizing the electric charge, reducing hydrophilicity, or increasing hydrophobicity of the polymeric macromolecules lead to transition from a soluble state to an insoluble state [69, 150, 151].

Cyclodextrins are used for synthesis of pH-responsive polymers because of their specific structure. Cyclodextrin is often complexed with a polyelectrolyte and forms polyrotaxanes or polypseudorotaxanes. Becuwe *et al.* reported pyridine-4-yl indolizin/β-CD IC as a pH-dependent supramolecule; at neutral pH values, the molecule displayed a stable self-included conformation of the aromatic groups and the hydrophobic cavity, and in acidic pH conditions, the pyridyl group became protonated, causing a conformational change in the structure. Poly(iminooligomethylene), polyethyleneimine (PEI) also formed pH-responsive ICs with cyclodextrins [152].

The complexation ratios, at high pH values, of PEI units to α-CD were 2:1 and 4:1 for γ-CD. Single linear PEI chains threaded onto α-CD, while two linear PEI chains threaded onto γ-CD in the parallel and antiparallel directions. At lower pH values (<8), the secondary amine on the PEI chains became protonated, and the ionization of these groups led to dethreading. PEI- and PEI-based copolymers such as PEG-*b*-PEI, PEG-*b*-PEI-*b*-PEG, PEI-*b*-PEG-*b*-PEI formed ICs with cyclodextrins at both high and low pH values [151].

At high pH values, α-CD threaded onto both PEG and PEI blocks and viscoelasticity of the system increased, but at lower pH values, CD only threaded onto the neutral PEG block and viscoelasticity of the system decreased. Furthermore, Karaky *et al.* investigated pH-switchable supramolecular sliding gels composed of α-CD and PEI-*b*-PEG-*b*-PEI. At basic pH, the CD rings were homogeneously distributed along the polymer chains, and at acidic pH, the EI units of the microgel deprotonated, leading to an increase in swelling. Due to unfavorable electrostatic interactions, the α-CD molecules moved from the PEI block to the central PEG block [152–154] (Figure 3.15).

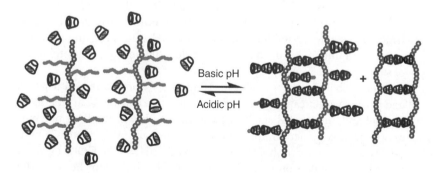

Figure 3.15 Schematic design for pH-responsive CD–polypseudorotaxane and polyrotaxane hydrogels.

Another report was about pH-responsive microgels of epichlorohydrin (EPI), poly(vinyl alcohol) (PVA) with cyclodextrin. Sensitivity of these complexes was checked by the association constant of methyl orange (MO) guest molecule, which is thirteen times greater at pH 7.4 than at 1.4. At low pH values, the protonation of MO molecules produced ammonium and azonium tautomers that greatly weakened the hydrophobic interactions between MO and β-CD. Due to low complexation at low pH values, MO readily diffused out of the microgel, resulting in a higher release rate [154].

Effects of pH and composition on associative properties of amphiphilic acrylamide/acrylic acid terpolymers and relation between conformation and the degree of ionization of the polymeric chains of polymers were reported by McCormick et al., and the results showed that at low pH values, compact polymer coils were formed, and at high pH values, the repulsion of ionized groups produced extended structures [155]. Moine et al. reported pH-responsive ICs as thickening agent, where the viscosity of the solution changed with pH, such as poly(β-malic acid-co-β-ethyladamantyl malate) (PMLA–Ad)/β-CD [156].

Adamantine groups are hydrophobic and form ICs with apolar cavities of β-CD. At low pH values of 2.1–2.5, carboxylic acid groups were protonated and led the polymer to exhibit a compact conformation with hydrophobic microdomains, and the mobility and accessibility of the adamantyl groups decreased and the probability of complexation was lowered. Decrease in the mobility and accessibility of the adamantyl groups lowered the probability of complexation , so at pH values greater than 5, the COOH groups dissociated, allowing the copolyester to swell, which increased the connectivity of the system and induced better accessibility for the adamantine groups, and the formation of the complex increased the viscosity of the system [156].

In another research, Huh et al. reported a pseudorotaxane based on poly(ε-lysine) (PL) and α-CD in aqueous solutions at high pH values, which is a stable crystalline precipitates at pH values ranging from 8.5 to 11.5 and undergoes decomposition at low pH because of protonation of side groups [157, 158].

Pyrene-labeled poly(acrylic acid) (PAAMePy) formed ICs with γ-CD, and pyrene was encapsulated. Pyrene units were capable of arranging in parallel and antiparallel stacking patterns. At pH = 3, the PAA polymer existed in a compact conformation and free pyrenes formed ICs with γ-CD due to the hydrophobic interactions. At pH = 8 the COOH groups of PAA became deprotonated, so the repulsive interactions between the PAA polymers decreased the proximity of pyrene units, and dimer formation was not observed [159].

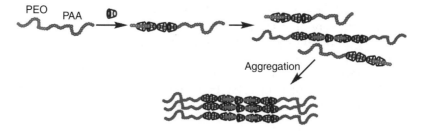

Figure 3.16 Vesicle self-assembly PEO-*b*-PAA induced by the complexation with CD.

Poly(ethylene oxide)-*b*-poly(acrylic acid) (PEO-*b*-PAA) block copolymer was studied by Liu *et al.* [149]. With increasing pH, the PAA groups were ionized, and by adding α-CD, PEO blocks were encapsulated and self-assembled into spherical nanostructures. At low pH, ICs of polymer/CD aggregated to produce a turbid solution due to protonation of PAA groups. A schematic of the PEO-*b*-PAA/CD self-assembly is shown in Figure 3.16 [160].

3.6.3 Temperature-Responsive Inclusion Complexes

Due to a wide range of applications of thermoreversible supramolecular structures of CDs with desired architectures, synthesis of this class of compounds is highly regarded. Formation of ICs of polymers and CDs is due to van der Waals interaction and hydrogen bonding between guests and the cavities of cyclodextrins. Cavity size, molecular geometrical shape of CDs, and rotation of guest molecules cause a steric barrier, so inclusion–dissociation behavior is dependent on temperature [108, 154, 161].

Polymers such as poly(alkyloxazoline), poly(propylene glycol), poly(*N*,*N*-dimethylacrylamide), poly(*N*-isopropylacrylamide) copolymers (PNIPAM), hydroxyl propyl methyl cellulose (HPMC), poly(ethylene oxide) (PEO), and poly(propylene oxide) (PPO) (poloxamers, Tetronics, or Pluronics), poly(acrylic acid) (PAA) and polyacrylamide (PAAm), poly(ethylene oxide)/poly(D,L-lactic acid-*co*-glycolic acid) are used for synthesis of purpose thermoresponsive systems based on CD–polymer complexes [162].

Various studies were conducted in synthesis of CD-based thermoresponsive polymers. Conformational structure of a thermoresponsive α-CD/PEO-*b*-PNIPAM complex was studied by Tu *et al.* and they showed the self-assembled structure of α-CD/PEO-*b*-PNIPAM IC into long-range ordered lamellar structures containing alternating layers of α-CD/PEO ICs with coiled segments of free PEO/PNIPAM and rod segments of α CD/PEO units [163].

Ooya et al. used Pluronic consisting of PEG-b-PPG-b-PEG for synthesis of thermoresponsive polyrotaxane by β-CDs. Many β-CDs were threaded onto PEG-b-PPG-b-PEG capped with fluorescein diisothiocyanate (FITC). These supramolecules were self-assembled and β-CDs showed sliding motion between PEG and PPG segments by changing the temperature. Also in acidic pH, the terminal hydrazone bond divided and is used as a delivery carrier in biological or catalyzed acidic conditions [164].

Nozaki et al. reported a hydrophobic environment produced by complex formation between 8-anilino-1-naphthalene sulfonic acid ammonium salt (ANS) and CD. The LCST decreased and the polymer chain shrank and became insoluble. The steric hindrance due to shrinkage and crowding of the polymer chains destabilized the CD/ANS complex, resulting in a reduction in complex formation. When ANS was decomplexed from CD, the LCST of PNIPAM increased due to the hydrophilic environment. The hydrophilic environment caused the polymer to swell, and the polymer resolubilized. When the polymer swelled, there was an increase in CD/ANS complexation due to the decrease in steric hindrance around the CD/ANS complex [165–167].

The LCST of the system with the ANS guest was lower than that of the polymer in water. As more number of ANS guests were encapsulated, the LCST decreased. The hydrophobic phenyl groups of the ANS were found outside the CD cavity after encapsulation, which induced a sudden phase transition with increasing temperature. In systems using NS (2-naphthalenesulfonic acid) as the guest molecule, the opposite LCST effects were observed. The complexation between CD and NS enlarged the hydrophilic moiety of the polymers. Wang et al. reported a thermoreversible block copolymer based on IC formation of mono-6-deoxy-6-ethylene diamino-β-CD (ECD) or mono-6-deoxy-6-hexane diamino-β-CD (HCD) with poly(N-isopropylacrylamide-co-glycidyl methacrylate) (P(NIPAM-co-GMA)) [168].

In another study, a thermosensitive copolymer was synthesized via ATRP of PNIPAM polymer with an insoluble pyrene end group and β-CD by Duan et al. At low temperatures, the solution was clear due to the formation of micelles and with increasing temperature, the solution became turbid as the PNIPAM chains collapsed and agglomerated due to the increased hydrophobicity. Formation of ICs disturbed PNIPAM aggregates and increased the LCST of the system, and the hydrophobic pyrene group affected the polymer's liquid phase transition [169].

Pluronic systems comprising PEO-b-PPO-b-PEO and heptakis (2,6-di-O-methyl)-β-CD (hβ-CD) were investigated over a temperature range of 5–70 °C by Dreiss et al. At low temperatures, Pluronics dispersed as unimers, and at ambient temperatures, the increased hydrophobicity caused the PPO segments to aggregate, forming spherical micelles. The aggregation number was found to increase with increasing temperature. β-CD formed pseudopolyrotaxanes with the polymers in their unimeric or micellar form by preferentially threading onto the PPO segment of the polymer backbone. This threading improved the water solubility of the block and disrupted aggregate formation. However, at temperatures above 50 °C, solubility decreased, aggregation became irreversible, and large-scale aggregation of the system occurred [170].

Poly(N-isopropylacrylamide-co-2-hydroxyethylacrylate) (PNIPAM–PHEAc–β-CD) brilliant colloidal crystals exhibited a variety of brilliant colors was reported by Mathews et al. in 2009, which was thermoreversible and reversibly transformed from invisible to visible materials with changes in temperature. As this complex was biocompatible with good water stability and solubility, it was evaluated as a possible drug delivery carrier such as antitumor cancer drug [171].

Zhang reported thermoresponsive hydrogels based on incorporation of water-soluble epichlorohydrin–β-CD (EP–β-CD) into a PNIPAM hydrogel, and release of the ibuprofen (IBt) guest was prolonged due to the IC between the CD and the drug [172]. Wang et al. also reported on poly(acrylic acid)-graft-β-CD (PAAc-g-β-CD) and polyacrylamide (PAAM) polymer network, which showed a UCST of approximately 35 °C, and the IBU drug guest was found to diffuse at a faster rate at 37 °C than at 25 °C [173].

3.6.4 Photoresponsive Inclusion Complexes

In photoresponsive systems, energy and electron transfer are governed by noncovalent interactions such as hydrogen bonds and aromatic π-stacking and control supramolecular self-assemblies through external stimuli. Photoresponsive polymers have found useful applications in photofunctional systems, photonics, and photoswitchable materials due to low cost, fast response, high sensitivity, and reversibility and repeatability [174].

ICs of cyclodextrins with polymers such as polyacrylamide (PAAm), poly(acrylic acid) (PAA) and derivatives, poly(methacrylic acid) (PMAA), poly(2-diethylaminoethyl methacrylate) (PDEAEMA), poly(ethylene imine), poly(L-lysine), and poly(N,N-dimethylaminoethylmethacrylate) (PDMAEMA) display photoresponsive properties. Tamura et al. reported a polyrotaxane consisting of various proportions of α-CD and naphthalene, and α-CD threaded onto a PEG chain bearing anthracene moieties at each end. Naphthalene and anthracene moieties act as energy donors and energy acceptors, respectively [174].

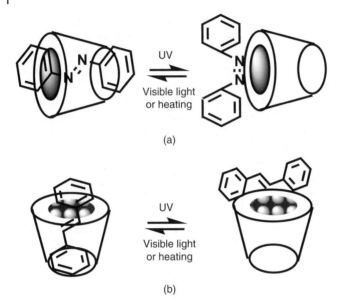

Figure 3.17 Photoresponsive inclusion complexes containing (a) azobenzene and (b) stilbene.

Hu *et al.* studied the thermoresponsive property and formation of ICs of some hydrophobically modified (HM) polymers with α-CD. An azobenzene functionalized hydroxypropyl methylcellulose (AZO-HPMC) was used as a photoresponsive trigger and showed an exothermic enthalpy-driven reaction with a stoichiometric ratio close to 1.0 and trans–cis isomerization of the azobenzene groups affected the stoichiometric number [175].

Reversible changes to physical and chemical properties are caused by photoisomerization in monomers such as azobenzene (azo) and azo derivatives, stilbene and 4,4′-azodibenzoic acid (ADA) [175]. Azobenzene in trans state is the most stable due to better configuration and more electron delocalization. Dark or visible light induces trans state, but irradiation of ultraviolet (UV) light isomerizes to the cis state (Figure 3.17) [176].

3.6.5 pH-Sensitive Polyrotaxane

Remarkable changes of pH in the human body make pH-sensitive polymers the ideal pharmaceutical systems to the specific delivery of therapeutic agents to a specific body area, tissue, or cell compartment. In other words, pH-sensitive polymers are polyelectrolytes having acid or basic groups that can accept or release protons in response to pH changes in the surrounding environment [4, 66, 67, 73, 76].

Natural polymers such as albumin, gelatin, and chitosan are used for synthesis of pH-sensitive polymers through formation of ICs with cyclodextrins. These compounds have amino polysaccharides and are water soluble at pH 6.2 and form gel above this value, so can be used in oral or mucosal administration. Hudson and Gil used chitosan as DNA carrier because of the positive charge of amino groups [177].

3.7 Stimuli-Sensitive Polyrotaxane for Drug Delivery

Drugs usually have poor solubility in water, but combination of CDs and hydrogels can overcome this limitation and improve loading of drugs into hydrogels and controlled delivery features of the hydrogels by changing the drug–polymer interactions [178]. ICs of CD provide a larger hydrodynamic radius of the drug–CD, which causes an additional diffusion barrier to drug release, extending the release period of drugs, improving the loading of drugs, and controlling the release rate of drugs from the hydrogel matrix, and reduces the rate of drug diffusion and controls the delivery [94, 179, 180].

CDs can act as binding points when incorporated into hydrogels, and these cross-linked networks effectively limit the entrance of physiological fluids, and, as a result, the covalently attached CDs cannot move apart from each other, so drug–CD affinity becomes the driving force to retain the drug and control the delivery. When the hydrogels come in contact with the physiological fluids, they swell but the volume of water taken up is limited by the polymer network, and, consequently, the polymeric chains do not dissolve [179, 180].

This creates a microenvironment rich in cavities available to interact with the guest drug molecules. When the microenvironment is rich in cyclodextrin cavities and interaction of CD–guest drug molecules is available, affinity of the drug molecules for the CD cavities drives the drug delivery. Decomplexation of a drug molecule from one CD cavity makes the drug available to form complexes with the neighboring empty CD cavities, and the likelihood of recomplexation is strongly dependent on the drug–CD affinity until the drug reaches the surface (Figure 3.18). The higher the drug–CD affinity, the slower is the drug release [94, 179, 180].

CDs can form ICs with hydrogels through different synthetic strategies in which CDs form a part of the network structure that can be obtained

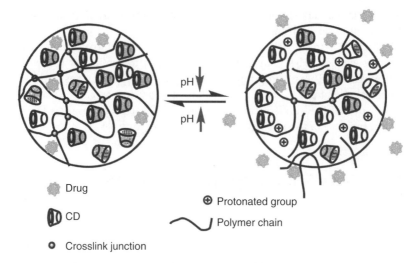

Figure 3.18 Drug released from pH-sensitive hydrogel containing CD and polymer chain.

by direct cross-linking of the CDs or by copolymerization of the CDs with other monomers [179–181]. Condensation reactions of the hydroxyl groups of the CDs with various cross-linking agents such as aldehydes, ketones, epoxides, and epichlorohydrin (EPI) form ICs between CDs and hydrogels [181]. EPI has two reactive functional groups that react with the hydroxyl groups of CDs or with other EPI molecules under alkaline conditions, and mixture of cross-linked CDs joined by repeating glyceryl units of polymerized EPI is obtained and used for removal of components from food, bioremediation, separation science, and as DDSs [179–181].

Hydrogels based on PEG/β-CD IC in the presence of isocyanate groups exhibited high hydrophilicity and biocompatibility and higher loading efficacy and sustained release of estradiol, quinine, and lysozyme, and PEG-diamine networks had been created in which the tie junctions were polyrotaxanes with isocyanate-activated β-CD groups [182].

By using diisocyanates as cross-linker agent for synthesis of CD-based hydrogels, networks with smaller mesh size and a lower swelling degree in water are formed while the use of EPI provides longer bridges in CDs. Leroux and Ruel-Gariépy used chitosan/glycerophosphate, which is spongy and can promote the controlled delivery of macromolecules and drugs with low solubility in water such as paclitaxel (a hydrophobic anticancer drug). So it is a good alternative to pharmaceutical implants [183].

Polymers that form ICs with cyclodextrins can be used as anticancer drugs carrier because of high hydrostatic pressure on the tumor, which lead to difficult diffusion of the drug, low oxygen concentration, high lactic acid concentration, and a consequent pH decrease to values between 6.0 and 7.4. When drugs conjugate with pH-sensitive polymers containing cyclodextrins and are encapsulated, solubility of drugs increases, specific delivery of anticancer drugs is promoted, and are more effective in reducing the tumor cells than the drug in its original state. Shenoy *et al.* synthesized poly(β-amino ester)-based, biodegradable, pH-sensitive polymers, which formed paclitaxel nanoparticles. They are more effective in reducing the tumor cells than the drug on its original state [184]. In another study, paclitaxel nanoparticles were prepared with the polymers poly(*N,N*-dimethylaminoethyl methacrylate) (PDMAEMA) and 2-hydroxyethyl (methacrylate) (HEMA) [185].

3.7.1 Photoresponsive Inclusion Complex Application

Photoresponsive ICs of cyclodextrins and various compounds act as molecular shuttles with extra degrees of conformational freedom, which the ring can rotate around the axle and shuttling is the movement of the ring along the axle is possible. Rotaxanes are good candidates for molecular machines because of this conformational freedom [185].

Various compounds form photosensitive rotaxane with CDs such as stilbene, azobenzene derivatives, azo-modified poly(acrylic acid), fumaramide (*trans*), and maleamide (cis) isomers of the olefin units, ruthenium(II) tris(bipyridine). Photoresponsive mixture of azo-modified poly(acrylic acid) and CD was reported by Yamaguchi *et al.* (Figure 1.19a) [185].

Reversible ICs of azo unit and α-CD was formed in sol–gel. Irradiation with UV light converts the gel into a sol, while visible light or heating (60 °C) induces back-isomerization of the *cis*-azo group to the *trans*-azo group and changes the viscosity of the polymer mixture. ICs act as cross-linking points between the polymers to yield supramolecular hydrogels without any covalent cross-links [185, 186]. Nakashima used ethylene spacer length between the stopper and aromatic unit in the rotaxane structures and showed that the shuttling process can be driven by light, heating, and solvent polarity (Figure 3.19) [186].

A rotaxane containing α-cyclodextrin, azobenzene, biphenyl chain, and two different fluorescent naphthalimide units was reported, where reversible motion of the CD between azobenzene and biphenyl station takes place after irradiation at 360 and 430 nm, respectively (Figure 3.20) [187].

The CD macrocycle locates over the azobenzene unit in the trans form, and it moves to the biphenyl site when the azo motif undergoes isomerism to the cis form by reversible changes in fluorescence intensity of the two

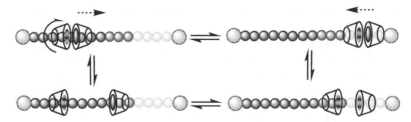

Figure 3.19 Possible shuttling process in stimuli-responsive rotaxane.

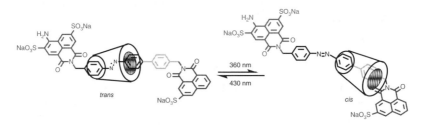

Figure 3.20 Light-driven molecular shuttle based on CD.

stoppers. Wang et al. reported photoresponsive rotaxane based on stilbene and CD and isophthalic acid stopper. Formation of hydrogen-bond interactions with CD prevents the shuttling motion [187].

Anderson et al. demonstrated directional shuttling of an asymmetric cyclodextrin along a symmetric organic axle featuring a stilbene moiety (Figure 3.21). A rapid gliding motion of the CD ring along the thread was observed by 2D NMR spectroscopy for *trans*-**80**. Irradiation at 340 nm led to trans–cis isomerization to afford *cis*-**80**. In the cis-conformer, the stilbene unit is located in close proximity to the 6-rim of the CD ring, while the wider 3-rim is able to better accommodate the bulky "stopper" group. The molecular switching cycle between the two co-conformers was completed upon irradiation of *cis*-**80** at 265 nm [188].

Leigh et al. reported the shuttles that use the interconversion of fumaramide (*trans*) and maleamide (*cis*) isomers of the olefin units by photochemical and thermal stimuli (Figure 3.22) [189].

Wang et al. developed a lockable light-driven molecular shuttle of α-CD on a NPSI track (biphenyl unit (P), stilbene unit (S), sulfonic naphthalimide disodium salt (N) and isophthalic acid (I)), while the stilbene unit was in the trans state, the CD preferred to rest over the S unit, which is shown in Figure 3.23 [187].

Stanier et al. investigated α-CD/NPSI IC formed by UV irradiation. α-CD did not move from the stilbene unit to the biphenyl unit, because carboxylic acid groups from isophthalic acid (as stopper) formed

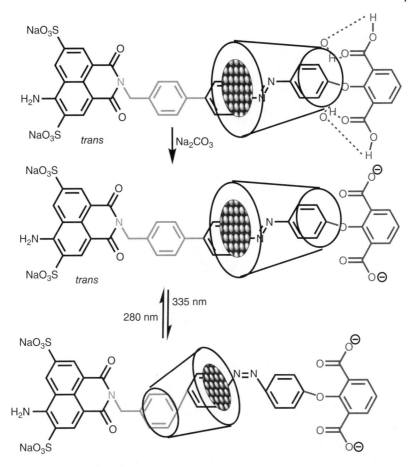

Figure 3.21 Photoresponsive rotaxane based on strong hydrogen bonding between OH groups of CD and two carbonyl groups in phthalic acid unit.

hydrogen bonding with hydroxyl groups of the α-CD and locked the α-CD to the stilbene unit [179].

By adding a base to the system, COOH groups of isophthalic acid were converted into COO- groups and allowed CD to be pushed from stilbene to the biphenyl when stilbene was transformed from trans to cis state. When visible light was irradiated, α-CD could be shifted back to the stilbene unit and relocked by addition of an acid [179].

Tribet *et al.* reported IC of azobenzene-modified polyacrylate (AMP) and β-CD or epichlorohydrin-containing β-CD (poly-CD), which act as photosensitive viscosity switches depending on the density of the photochrome and the interactions of the polymer chains [176].

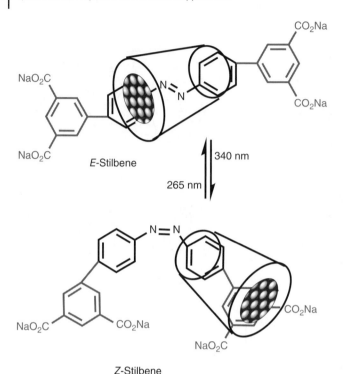

Figure 3.22 Stilbene-based photoresponsive polyrotaxane.

IC of α-CD and poly(acrylic acid) (PAA) and dodecyl side chains was reported by Tomatsu et al. via controlled gel-to-sol and sol-to-gel transitions. PAA/dodecyl solution is gel type because of hydrophobic interactions, and by addition of α-CD to the solution, hydrophobic interactions are reduced and dissociation of the network structure converted to sol state. By adding 4,4′-azodibenzoic acid (ADA) to the system, sample was reverted back to a gel state because of formation of complex between CD and ADA, left the dodecyl and reestablish the hydrophobic network structure [190, 191]. Irradiation of UV light changed ADA into the cis state. By irradiating the sample with visible and UV light, the mixture's gel–sol behavior could be altered repeatedly [190, 191].

In azobenzene hydroxypropyl methyl cellulose (HMC) network formed via hydrophobic interactions of azobenzene groups and with radiation of UV light, the trans azobenzene groups isomerized to the cis form, and the azobenzene groups no longer exhibited the same hydrophobic interactions, raising the sol–gel transition temperature to 36.5 °C [192].

When azobenzene groups formed IC with α-CD, trans form encapsulated by CD, the side groups became more hydrophilic,

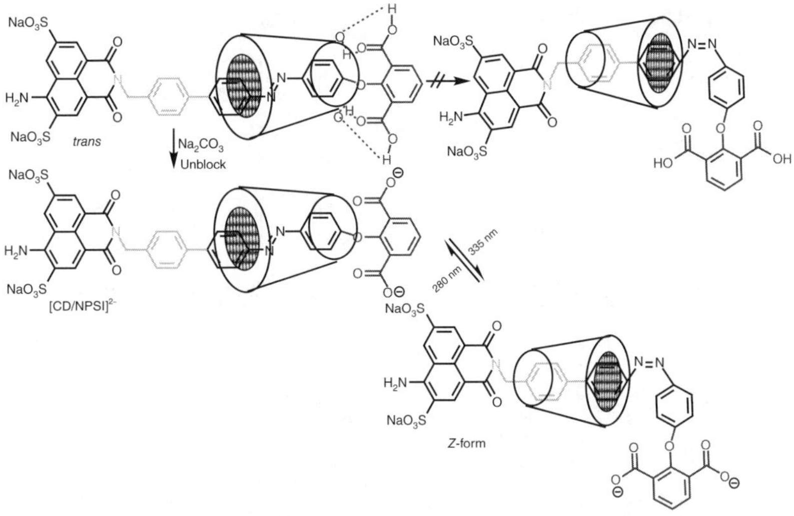

Figure 3.23 Photoresponsive α-CD/NPSI inclusion complex [187]. (*Source*: Qu 2005 [187]. Reproduced with permission of American Chemical Society.)

thereby raising the transition temperature to 57 °C but in cis conformation, the azobenzene groups could not be encapsulated by CD; thus, they no longer exhibited strong hydrophobic interactions. Transition temperatures of the two cis conformation samples should not differ by 12.5 °C due to the addition of uncomplexed CD. The azobenzene conformations are shown in Figure 3.24 [193].

Tong *et al.* investigated a variety of phenyl compounds and their ability to fit into the CD cavity, which is affected by guest molecular size, molecular shape, and structural flexibility. The guest molecule must be rigid enough to induce a good fit into the CD cavity [194]. ICs of cinnamoyl derivatives such as hydrocinnamoyl-β-CD, 6-hydrocinnamoyl-α-CD, 6-cinnamoyl-α-CD, and 6-cinnamoyl-β-CD were investigated by Harada *et al.*. Phenyl ring formed ICs with the CD groups [84–89].

3.7.2 Redox-Responsive Inclusion Complexes Applications

Redox-responsive polymers and hydrogels are attractive due to their interesting bioapplications such as actuators or artificial muscles. Oxidation states change polarity, chain stiffness, extension, and interaction with solvent. These polymers and hydrogels usually contain transition metals such as Fe in main chain or side chain groups and change the local cross-link density. Reversible cross-links between the bonding and nonbonding lead to formation of permanent covalently cross-linked polymeric hydrogel networks. Oxidation of ferrocene units by pH or

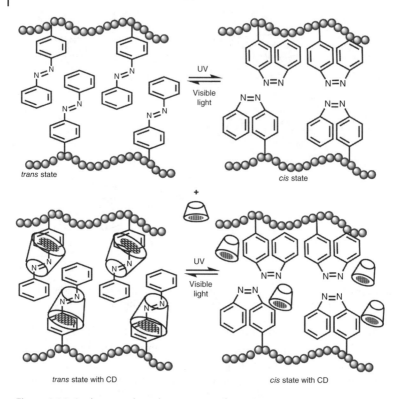

Figure 3.24 Azobenzene-based rotaxane conformations.

temperature causes hydrogen bonding to interrupt and gel formation by redox stimuli. Host–guest formation via ICs between cyclodextrins and various hydrophobic polymers such as aromatic monomers, cholesterol, ferrocene derivatives, and adamantine derivatives results in formation of responsive supramolecules. Supramolecular structures with covalent cross-links act as redox-responsive actuator and load upon oxidation and reduction [190, 191].

Based on cavity size of the CD, various metallocene compounds such as cobaltocenium, bipyridinium, aminobiphenyl, and ferrocene derivatives can form ICs with CD and form redox- responsive hydrogels [195–198]. IC formation of CDs with reduced state ferrocene is more favorable and is the most popular CD guests in electrochemically driven molecular complexes [157, 163]. Redox-responsive polymers containing ferrocene carboxylic acid, poly(acrylic acid), and dodecyl was reported by Tomatsu et al. [190, 191].

Because of hydrophobic interactions of the dodecyl chains, poly(acrylic acid)/dodecyl has high viscosity. Addition of β-CD encapsulates dodecyl

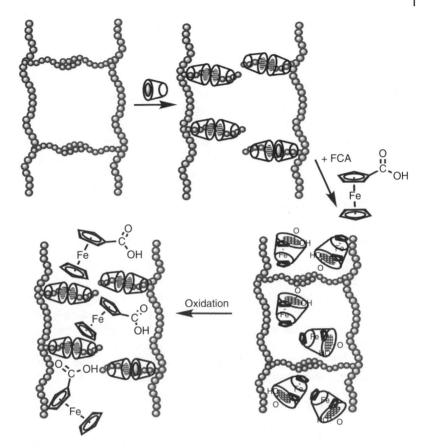

Figure 3.25 Redox-responsive gel-to-sol transition based on CD and ferrocenecarboxylic acid.

chains and viscosity decreases. After adding ferrocene carboxylic acid, cyclodextrin releases dodecyl chains and forms ICs with ferrocene carboxylic acid, thus gel behavior is regained. Addition of sodium hypochlorite to redox reagent system decreases affinity of ferrocene carboxylic acid for β-CD and the CD recomplexes with the dodecyl side chains, so redox-responsive gel-to-sol transition behavior is obtained (Figure 3.25) and is useful as gel actuators that could be controlled by applied electric potential [190–192].

Supramolecular structures of poly(propyleneimine) (PPI) dendrimer with ferrocene end units and β-CD were based on oxidation of the ferrocene units. Formation of IC of PPI dendrimer and CDs was dependent on removal of cobaltocenium end groups. In the reduced state, the dendrimers bound to the gold and SiO_2 surfaces [195, 199].

3.8 Multi-Stimuli-Responsive Inclusion Complexes

Two or more stimuli-responsive groups can be combined within a polymer via formation of ICs of CDs with polymers or radical polymerization, which is dependent on molecular architecture and composition, the preparation of systems, and responses to multiple stimuli changes in the environments [200]. Multi-stimuli-responsive polymers with multiple functionalities allow for more complex interactions and morphologies. In double-responsive polymers, temperature- and light-responsive systems are most often studied, which result in phototuning of the thermal response [201, 202].

3.8.1 pH- and Temperature-Responsive Inclusion Complexes Applications

Temperature-responsive polymers can change solubility, conformation, and hydrophobic/hydrophilic balance and promote the formation, transformation, or deformation of aggregates in response to variations in temperature; so they are the most important stimuli-responsive materials for several decades. pH and temperature control are important for biocompatible stimuli changes, so dual-response hydrogels are used as smart polymers with potential applications in biomedical field [203].

Dual-sensitive supramolecular hydrogels based on poly(ε-lysine)-CDs are constructed by unique noncovalent interactions between CDs and polymeric guests show reversible responses to changes in temperature and pH and reversible gelation induced by IC formation [81, 204]. Yamamoto *et al.* reported a dual-sensitive polymer based on β-CD-poly(ε-lysine) with 3-trimethylsilylpropionic acid (TPA) guest molecules and was shown to exhibit reversible gel–sol behavior with heating and cooling. pH sensitivity of this polymer is because of ionizable α-amino groups of poly(ε-lysine) and the COOH groups of 3-trimethylsilylpropionic acid and control molecular aggregation. By lowering the pH, trimethylsilylpropionic acid protonation led to complex dissociation and gel-to-sol transition [164].

Hamcerencu *et al.* [140] copolymerized xanthan maleate and β-CD acrylate with NIPAM hydrogels to form interpenetrating polymer networks. Carboxylic acid groups of xanthan existed at low pH and nonionic form, so hydrogen bonding formed between the polymer chains and the hydrogel deswelled [205].

The swelling of the hydrogel was found to be pH dependent and it increased with increasing pH, which is related to ionization of carboxylic acid groups and decreases the hydrogen bonding of the xanthan chains, so the electrostatic repulsion between the ionized groups caused the hydrogels to swell and released encapsulated drug or compound. ICs of PNIPAM chain, methyl orange, and β-CD were reported by Liu [206]. At low pH(=1.4), azo groups of methyl orange protonated and produced ammonium and azonium tautomers. At pH = 4, protonation of PNIPAM–β-CD led to ionic interaction between amino groups of the polymer and sulfate groups of methyl orange. Also this hydrogel is thermoreversible because of PNIPAM [64].

Hydrogels containing IC of cyclodextrin with synthetic polymers, such as poly(N,N-dimethylacrylamide-co-N-phenyl-acrylamide) and poly(N-vinylamide-co-vinylacetate), are thermoresponsive [202, 207, 208]. The concept of controlling factors such as temperature, pH, photo, or irradiation state has been applied to produce double-stimuli-responsive materials [200]. Hydrogels based on poly(N,N-dimethylaminomethylmethacrylate) and poly(acrylic acid) are pH- and thermoresponsive polymers – with increase in pH, the cloud point decreases[209].

3.8.2 pH- and Redox-Responsive Inclusion Complexes Applications

The most important systems for DDSs for cancer are systems with pH and redox potential changes and are employed to trigger sharp changes in the properties of a material and can be used for extracellular or intracellular drug delivery. pH of subcellular segment and tumor is different, where delivery vehicles will reside, change dramatically from early endosome to late endosome, and eventually to lysosomes, where the pH can reach 5.0 or even lower. On the other side, in cancer cells, levels of glutathione (GSH) are three times more than that in healthy cells, which leads to use redox-sensitive nanoparticles for drug delivery. Carriers that can respond to both pH and redox stimuli are useful for cancer drug delivery with high efficacy and selectivity [210, 211]. Zhong *et al.* reported a novel core–shell pH- and reduction dual-sensitive micelles for the anticancer drug delivery, which quickly release the loaded drugs responding to acidic and reductive stimuli [212].

3.8.3 Temperature- and Photoresponsive Inclusion Complexes Applications

Photoresponsive polymers are interesting because of easy and rapid control of light compared with traditional stimuli such as temperature, pH, electric field, and ionic strength [172–175]. Azobenzene derivatives can undergo reversible photo-isomerization upon UV and visible light irradiation and change physical and chemical properties of hydrogel, which are attractive for potential applications as drug carriers, enzymatic bioprocessing, photo-triggered targeted DDSs, photo-controlled separation/recovery systems in bioMEMs formats, and photo-driven smart surface [91, 206, 213].

Azobenzene units can be introduced into temperature-responsive polymers such as poly(N-isopropylacrylamide) (PNIPAM) and poly(N,N-dimethylacrylamide) (PDEAM), and produce dual-stimuli-responsive polymers. Although these compounds can form ICs with cyclodextrins and a reversible transition has been obtained by alternatively using UV and visible light, it is an ideal system for constructing photoresponsive polymers, whose solubility properties can be controlled by using the light-controlled molecular recognition of α-CD and azobenzene. Kungwatchakun and Irie synthesized a copolymer of PNIPAM and N-(4-phenyl(azo)-phenyl-acrylamide) and they showed that, over a narrow range of composition and temperature, photo-stimuli may be used to control the solubility of the polymer in water [214].

Luo et al. combined thermosensitive PNIPAM with photosensitive azobenzene end groups, as illustrated gold nanoparticles were capped with α-CD (Figure 3.26). α-CD can encapsulate the trans form of PNIPAM-azo polymers and PNIPAM encapsulated the gold nanoparticle. Below the LCST of PNIPAM, the solution was clear, and above the LCST, the PNIPAM chains collapsed, resulting in a turbid solution. With UV irradiation, azobenzene converted into the cis state and detached from the gold particles. These temperatures and phototransitions were found to be reversible and repeatable. These thermo- and photosensitive nanoparticles are important for applications in photoactive materials and switching devices that can be controlled with light (Figure 3.26) [109, 215, 216].

In another study, Luo et al. reported dual thermal- and photoresponsive copolymers based on IC of α-CD and thermosensitive azobenzene side chains, and poly(N,N-dimethylacrylamide-co-N-4-phenylazophenyl acrylamide), which led to light-controlled solubility behavior of the ICs and the mechanism of polymer solubility change upon photoirradiation. Hydrophobicity of the polymers increased with increase in azobenzene contents and solubility decreased, so LCST of the polymer decreased. On the other hand, α-CD encapsulated the azobenzene side groups and system became hydrophilic and LCST of the polymers largely increased. Upon UV irradiation, the azobenzene released from the α-CD cavity, resulting in a reduction in the LCST due to an increase in hydrophobicity, and is useful in light-controlled drug delivery (Figure 3.27) [108, 109].

ICs of poly(N,N-dimethylacrylamide)-co-poly(N-4-phenylazophenyl acrylamide) (PDMA-co-PAPA)/α-CD showed a photoinduced phase separation phenomenon due to a balance between hydrogen bond formation with water and hydrophobic intermolecular forces. Addition of azobenzene to the system rendered the system hydrophobic and caused

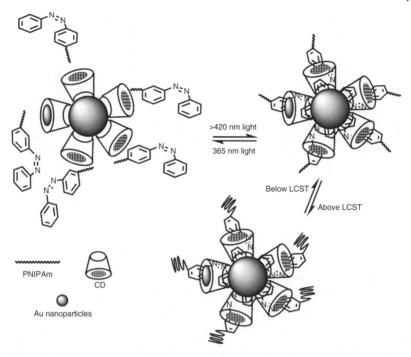

Figure 3.26 Temperature- and UV-responsive rotaxane based on AZO-PNIPAM and Au nanoparticles.

precipitation at a lower temperature, so LCST of the polymer decreased. UV irradiation converted trans form of azobenzene to the cis state, so solubility and LCST increased. IC formation of *trans*-azobenzene with CD increased polymer solubility and a large increase in LCST was demonstrated. After UV irradiation, the azobenzene was converted to the cis state and the solubility increased. In the trans state, the system was transparent due to complexation of the polymer with CD [215].

3.8.4 Temperature- and Redox-Responsive Inclusion Complexes Applications

Supramolecular materials with reversible responsiveness to environmental changes are formed by IC formation of polymers with CDs, which change solubility of guest molecule, and hydrophobic and hydrophilic interactions. Host–guest complexation between cyclodextrin and ferrocene shows reversible association–dissociation controlled by the redox state of ferrocene [108, 109]. On the other side, by addition of PNIPAM to this complex, interchain aggregation of PNIPAM to CD/ferrocene complex led to temperature- and redox-responsive hydrogels with a

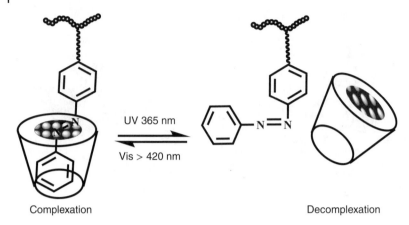

Figure 3.27 UV- and light irradiation of PDMAA-co-PAPA/CD rotaxanes.

sol–gel transition by oxidizing agent or a competitive guest to ferrocene. By addition of CD to the system, CD formed complexes with the reduced ferrocene side groups and disrupted the hydrophobic interactions between side groups, decreased viscosity, and caused an increase in LCST. When ferrocene was oxidized, there was no interaction between the ferrocene groups and CD, so no complex formed and specific viscosity or LCST did not change. Temperature- and redox-sensitive systems are noted for applications in sensors and controlled release systems [108, 109].

References

1 Chen, G. and Jiang, M. (2011) Cyclodextrin-based inclusion complexation bridging supramolecular chemistry and macromolecular self-assembly. *Chem. Soc. Rev.*, **40**, 2254–2266.
2 Li, J. and Loh, X.J. (2008) Cyclodextrin-based supramolecular architectures: syntheses, structures, and applications for drug and gene delivery. *Adv. Drug Deliv. Rev.*, **60**, 1000–1017.
3 Xu, F.J., Li, J., Yuan, S.J. et al. (2008) Thermo-responsive porous membranes of controllable porous morphology from triblock copolymers of polycaprolactone and poly(*N*-isopropylacrylamide) prepared by atom transfer radical polymerization. *Biomacromolecules*, **9**, 331–339.
4 Chen, Y., Pang, X.H. and Dong, C.M. (2010) Dual stimuli-responsive supramolecular polypeptide-based hydrogel and reverse micellar

hydrogel mediated by host–guest chemistry. *Adv. Funct. Mater.*, **20**, 579–586.
5 Meyer, D.E., Shin, B.C., Kong, G.A. et al. (2001) Drug targeting using thermally responsive polymers and local hyperthermia. *J. Control. Release*, **74**, 213–224.
6 Jonathan, D.R.T., Steed, W. and Wallace, K.J. (2007) *Core Concepts in Supramolecular Chemistry and Nanotechnology*, John Wiley & Sons, Ltd, Chichester.
7 Hoeben, F.J.M., Jonkheijm, P., Meijer, E.W. and Schenning, A. (2005) About supramolecular assemblies of pi-conjugated systems. *Chem. Rev.*, **105**, 1491–1546.
8 Gervaise, C., Bonnet, V., Wattraint, O. et al. (2012) Synthesis of lipophosphoramidyl-cyclodextrins and their supramolecular properties. *Biochimie*, **94**, 66–74.
9 Roling, O., Wendeln, C., Kauscher, U. et al. (2013) Layer-by-layer deposition of vesicles mediated by supramolecular interactions. *Langmuir*, **29**, 10174–10182.
10 Guo, R. and Wilson, L.D. (2013) Cyclodextrin-based microcapsule materials – their preparation and physiochemical properties. *Curr. Org. Chem.*, **17**, 14–21.
11 Memisoglu, E., Bochot, A., Özalp, M. et al. (2003) Direct formation of nanospheres from amphiphilic–cyclodextrin inclusion complexes. *Pharm. Res.*, **20**, 117–125.
12 Kauscher, U., Samanta, A. and Ravoo, B.J. (2014) Photoresponsive vesicle permeability based on intramolecular host–guest inclusion. *Org. Biomol. Chem.*, **12**, 600–606.
13 Memişoğlu, E., Bochot, A., Şen, M. et al. (2003) Non-surfactant nanospheres of progesterone inclusion complexes with amphiphilic β-cyclodextrins. *Int. J. Pharm.*, **251**, 143–153.
14 Matyjaszewski, K. and Davis, T. (2002) *Handbook of Radical Polymerization*, John Wiley & Sons, Hoboken, NJ.
15 Fischer, H. (2001) The persistent radical effect: a principle for selective radical reactions and living radical polymerizations. *Chem. Rev.*, **101**, 3581–3610.
16 Szwarc, M. (1956) Living polymers. *Nature*, **178**, 1168–1169.
17 Matyjaszewski, K. and Lin, C.H. (1991) Exchange reactions in the living cationic polymerization of alkenes. *Makromol. Chem. Macromol. Symp.*, **47**, 221–237.
18 Georges, M.K. (1993) Narrow molecular weight resins by a free-radical polymerization process. *Macromolecules*, **26**, 2987–2988.
19 Hawker, C.J. (2001) New polymer synthesis by nitroxide mediated living radical polymerizations. *Chem. Rev.*, **101**, 3661–3688.

20 Wayland, B.B. (1994) Living radical polymerization of acrylates by organocobalt porphyrin complexes. *J. Am. Chem. Soc.*, **116**, 7943–7944.

21 Kato, M. (1995) Polymerization of methyl methacrylate with the carbon tetrachloride/dichlorotris-(triphenylphosphine)ruthenium(II)/ methylaluminum bis(2,6-di-*tert*-butylphenoxide) initiating system: possibility of living radical polymerization. *Macromolecules*, **28**, 1721–1723.

22 Wang, J.S. and Matyjaszewski, K. (1995) Controlled/'living' radical polymerization. atom transfer radical polymerization in the presence of transition-metal complexes. *J. Am. Chem. Soc.*, **117**, 5614–5615.

23 Matyjaszewski, K. (1995) Controlled radical polymerizations: the use of alkyl iodides in degenerative transfer. *Macromolecules*, **28**, 2093–2095.

24 Chiefari, J. (1998) Living free-radical polymerization by reversible addition–fragmentation chain transfer: the RAFT process. *Macromolecules*, **31**, 5559–5562.

25 Destarac, M. (2000) Dithiocarbamates as universal reversible addition–fragmentation chain transfer agents. *Macromol. Rapid Commun.*, **21**, 1035–1039.

26 Wang, X.S. (1999) Facile synthesis of well-defined water-soluble polymers via atom transfer radical polymerization in aqueous media at ambient temperature. *Chem. Commun.*, **18**, 1817–1818.

27 Matyjaszewski, K. and Tsarevsky, N. (2002) In PCT Int. Appl., Carnegie Mellon University, Pittsburgh, WO 0228913, p. 64.

28 Sumerlin, B.S. (2001) Water-soluble polymers. 84. Controlled polymerization in aqueous media of anionic acrylamido monomers via RAFT. *Macromolecules*, **34**, 6561–6564.

29 Greszta, D. (1994) 'Living' radical polymerization. 1. Possibilities and limitations. *Macromolecules*, **27**, 638–644.

30 Goto, A. and Fukuda, T. (2004) Kinetics of living radical polymerization. *Prog. Polym. Sci.*, **29**, 329–385.

31 Tordo, P. (2002) Design and use of β-phosphorus nitroxides and alkoxyamines in controlled/'living' free radical polymerizations. *Macromol. Symp.*, **182**, 225–247.

32 Davis, K.A. and Matyjaszewski, K. (2002) Statistical, gradient and segmented copolymers by controlled/living radical polymerizations. *Adv. Polym. Sci.*, **159**, 1–169.

33 Matyjaszewski, K. (2003) The synthesis of functional star copolymers as an illustration of the importance of controlling polymer structures in the design new materials. *Polym. Int.*, **52**, 1559–1565.

34 Coessens, V., Pintauer, T. and Matyjaszewski, K. (2001) Functional polymers by atom transfer radical polymerization. *Prog. Polym. Sci.*, **26**, 337–377.

35 Tsarevsky, N.V., Tsarevsky, N.V., Pintauer, T. and Matyjaszewski, K. (2004) Deactivation efficiency of and degree of control over polymerization in ATRP in protic solvents. *Macromolecules*, **37**, 9768–9778.

36 Sumerlin, B.S., Lowe, A.B., Thomas, D.B. et al. (2004) Aqueous solution properties of pH-responsive AB diblock acrylamido–styrenic copolymers synthesized via aqueous reversible addition–fragmentation chain transfer. *J. Polym. Sci., Part A: Polym. Chem.*, **42**, 1724–1734.

37 Muehlebach, A. (2004) Annealing time dependence of structural and magnetotransport properties of $Ni_{81}Fe_{19}$ (2 nm)/Ag (4 nm) multilayers. *Polym. Mat. Sci. Eng.*, **90**, 180–184.

38 Liu, J., Sheina, E., Kowalewski, T. and McCullough, R.D. (2002) Tuning the electrical conductivity and self-assembly of regioregular polythiophene by block copolymerization: nanowire morphologies in new di- and triblock copolymers. *Angew. Chem. Int. Ed.*, **41**, 329–332.

39 Benoit, D., Benoit, D., Hawker, C.J. et al. (2000) One-step formation of functionalized block copolymers. *Macromolecules*, **33**, 1505–1507.

40 Shipp, A.D., Wang, J.L. and Matyjaszewski, K. (1998) Synthesis of acrylate and methacrylate block copolymers using atom transfer radical polymerization. *Macromolecules*, **31**, 8005–8008.

41 Matyjaszewski, K., Shipp, D.A., McMurtry, G.P. et al. (2000) Simple and effective one-pot synthesis of (meth)acrylic block copolymers through atom transfer radical polymerization. *J. Polym. Sci., Part A: Polym. Chem.*, **38**, 2023–2031.

42 Tong, J.D., Tong, J.D., Moineau, G. et al. (2000) Synthesis, morphology, and mechanical properties of poly(methyl methacrylate)-*b*-poly(*n*-butyl acrylate)-*b*-poly(methyl methacrylate) triblocks: ligated anionic polymerization vs atom transfer radical polymerization. *Macromolecules*, **33**, 470–479.

43 Terada, T., Inaba, T., Kitano, H. et al. (1994) Raman spectroscopic study on water in aqueous solutions of temperature-responsive polymers: poly(*N*-isopropylacrylamide) and poly[*N*-(3-ethoxypropyl)acrylamide]. *Macromol. Chem. Phys.*, **195**, 3261–3270.

44 Kitano, H., Hirabayashi, T., Gemmei-Ide, M. and Kyogoku, M. (2004) Effect of macrocycles on the temperature-responsiveness of poly[(methoxy diethylene glycol methacrylate)-*graft*-PEG]. *Macromol. Chem. Phys.*, **205**, 1651–1659.

45 Malmstadt, N., Hoffman, A.S. and Stayton, P.S. (2004) Smart mobile affinity matrix for microfluidic immunoassays. *Lab Chip*, **4**, 412–415.

46 Bontempo, D., Li, R.C., Ly, T. *et al.* (2005) One-step synthesis of low polydispersity, biotinylated poly(*N*-isopropylacrylamide) by ATRP. *Chem. Commun.*, **37**, 4702–4704.

47 Wang, H., Wang, S., Su, H. *et al.* (2009) Supramolecular approach for preparation of size-controlled nanoparticles. *Angew. Chem. Int. Ed.*, **48**, 4344–4348.

48 Wang, S., Chen, K.J., Wu, T.H. *et al.* (2010) Photothermal effects of supramolecularly assembled gold nanoparticles for the targeted treatment of cancer cells. *Angew. Chem. Int. Ed.*, **49**, 3777–3781.

49 Zhang, J. and Ma, P.X. (2013) Cyclodextrin-based supramolecular systems for drug delivery: recent progress and future perspective. *Adv. Drug Deliv. Rev.*, **65**, 1215–1233.

50 Sallas, F. and Darcy, R. (2008) Amphiphilic cyclodextrins – advances in synthesis and supramolecular chemistry. *Eur. J. Org. Chem.*, **2008**, 957–969.

51 Sortino, S., Mazzaglia, A., Monsù Scolaro, L. *et al.* (2006) Nanoparticles of cationic amphiphilic cyclodextrins entangling anionic porphyrins as carrier-sensitizer system in photodynamic cancer therapy. *Biomaterials*, **27**, 4256–4265.

52 Skiba, M., Morvan, C., Duchêne, D. *et al.* (1995) Evaluation of gastrointestinal behaviour in the rat of amphiphilic β-cyclodextrin nanocapsules, loaded with indomethacin. *Int. J. Pharm.*, **126**, 275–279.

53 Dong, J., Xun, Z., Zeng, Y. *et al.* (2013) A versatile and robust vesicle based on a photocleavable surfactant for two-photon-tuned release. *Chem. Eur. J.*, **19**, 7931–7936.

54 Zhao, H., Sterner, E.S., Coughlin, E.B. and Theato, P. (2012) *o*-Nitrobenzyl alcohol derivatives: opportunities in polymer and materials science. *Macromolecules*, **45**, 1723–1736.

55 Carlmark, A., Malmstroem, E. and Malkoch, M. (2013) Dendritic architectures based on bis-MPA: functional polymeric scaffolds for application-driven research. *Chem. Soc. Rev.*, **42**, 5858–5879.

56 Percec, V., Wilson, D.A., Leowanawat, P. *et al.* (2010) Self-assembly of Janus dendrimers into uniform dendrimersomes and other complex architectures. *Science*, **328**, 1009–1014.

57 Zhang, S., Sun, H.-J., Hughes, A.D. *et al.* (2014) Self-assembly of amphiphilic Janus dendrimers into uniform onion-like dendrimersomes with predictable size and number of bilayers. *Proc. Natl. Acad. Sci. USA*, **111**, 9058–9063.

58 Nazemi, A. and Gillies, E.R. (2014) Dendrimersomes with photodegradable membranes for triggered release of hydrophilic and hydrophobic cargo. *Chem. Commun.*, **50** (76), 11122–11125.

59 Astruc, D., Wang, D., Deraedt, C. *et al.* (2015) Catalysis inside dendrimers. *Synthesis*, **47**, 2017–2031.

60 Liu, Y., Tee, J.K. and Chiu, G.N.C. (2015) Dendrimers in oral drug delivery application: current explorations, toxicity issues and strategies for improvement. *Curr. Pharm. Des.*, **21**, 2629–2642.

61 Gillies, E.R. and Frechet, J.M.J. (2005) Dendrimers and dendritic polymers in drug delivery. *Drug Discov. Today*, **10** (1), 35–43.

62 Gu, W., Zhao, H., Wei, Q. *et al.* (2013) Line patterns from cylinder-forming photocleavable block copolymers. *Adv. Mater.*, **25**, 4690–4695.

63 Kloxin, A.M., Kasko, A.M., Salinas, C.N. and Anseth, K.S. (2009) Photodegradable hydrogels for dynamic tuning of physical and chemical properties. *Science*, **324**, 59–63.

64 Jiang, J., Tong, X. and Zhao, Y. (2005) A new design for light-breakable polymer micelles. *J. Am. Chem. Soc.*, **127**, 8290–8291.

65 Bertrand, O., Gohy, J.F. and Fustin, C.A. (2011) Synthesis of diblock copolymers bearing *p*-methoxyphenacyl side groups. *Polym. Chem.*, **2**, 2284–2292.

66 Schmaljohann, B. (2006) Thermo- and pH-responsive polymers in drug delivery. *Adv. Drug Deliv. Rev.*, **58**, 1655–1670.

67 Kost, J. and Langer, R. (2001) Responsive polymeric delivery systems. *Adv. Drug Deliv. Rev.*, **46**, 125–148.

68 Galaev, I.Y. and Mattiasson, B. (1999) Smart polymers and what they could do in biotechnology and medicine. *Trends Biotechnol.*, **17**, 335–340.

69 Kumar, A., Srivastava, A., Galaev, I.Y. and Mattiasson, B. (2007) Smart polymers: physical forms and bioengineering applications. *Prog. Polym. Sci.*, **32**, 1205–1237.

70 Chen, S. and Singh, J. (2005) Controlled delivery of testosterone from smart polymer solution based systems: in vitro evaluation. *Int. J. Pharm.*, **295**, 183–190.

71 Issels, R. (1999) Hyperthermia combined with chemotherapy-biological rationale, clinical application, and treatment results. *Onkologie*, **22**, 374–381.

72 Engin, K. (1996) Biological rationale and clinical experience with hyperthermia. *Control Clin. Trials*, **17**, 316–342.

73 Soppimath, K.S., Aminabhavi, T.M., Dave, A.M. *et al.* (2002) Stimulus-responsive 'smart' hydrogels as novel drug delivery systems. *Drug Dev. Ind. Pharm.*, **28**, 957–974.

74 Lozinsky, V.I., Galaev, I.Y., Plieva, F.M. et al. (2003) Polymeric cryogels as promising materials of biotechnological interest. *Trends Biotechnol.*, **21**, 445–451.

75 Roy, I. and Gupta, M.N. (2003) Smart polymeric materials: emerging biochemical applications. *Chem. Biol.*, **10**, 1161–1171.

76 Meng, H., Mohamadian, H., Stubblefield, M. et al. (2013) Various shape memory effects of stimuli-responsive shape memory polymers. *Smart Mater. Struct.*, **22**, 1–9.

77 Kaneider, N.C., Dunzendorfer, S. and Wiedermann, C.J. (2004) Heparan sulfate proteoglycans are involved in opiate receptor-mediated cell migration. *Biochemistry*, **43**, 237–244.

78 Wenz, G. (1994) Cyclodextrins as building blocks for supramolecular structures and functional units. *Angew. Chem., Int. Ed. Engl.*, **33**, 803–822.

79 Herrmann, W., Keller, B. and Wenz, G. (1997) Kinetics and thermodynamics of the inclusion of ionene-6,10 in α-cyclodextrin in an aqueous solution. *Macromolecules*, **30**, 4966–4972.

80 Wenz, G. and Keller, B. (1992) Threading cyclodextrin rings on polymer chains. *Angew. Chem., Int. Ed. Engl.*, **31**, 197–199.

81 Choi, H.S., Ooya, T., Sasaki, S. and Yui, N. (2003) Control of rapid phase transition induced by supramolecular complexation of β-cyclodextrin-conjugated poly(ε-lysine) with a specific guest. *Macromolecules*, **36**, 5342–5347.

82 Li, J., Li, X., Toh, K.C. et al. (2001) Inclusion complexation and formation of polypseudorotaxanes between poly[(ethylene oxide)-*ran*-(propylene oxide)] and cyclodextrins. *Macromolecules*, **34**, 8829–8831.

83 Li, J., Ni, X.P., Zhou, Z.H. and Leong, K.W. (2003) Preparation and characterization of polypseudorotaxanes based on block-selected inclusion complexation between poly(propylene oxide)-poly(ethylene oxide)-poly(propylene oxide) triblock copolymers and α-cyclodextrin. *J. Am. Chem. Soc.*, **125**, 1788–1795.

84 Harada, A., Li, J. and Kamachi, M. (1995) Preparation and characterization of inclusion complexes of poly(propylene glycol) with cyclodextrins. *Macromolecules*, **28**, 8406–8411.

85 Harada, A., Li, J. and Kamachi, M. (1992) The molecular necklace: a rotaxane containing many threaded α-cyclodextrins. *Nature*, **356**, 325–327.

86 Harada, A., Li, J. and Kamachi, M. (1993) Preparation and properties of inclusion complexes of polyethylene glycol with alpha-cyclodextrin. *Macromolecules*, **26**, 5698–5703.

87 Harada, A., Li, J. and Kamachi, M. (1994) Preparation and characterization of a polyrotaxane consisting of monodisperse poly(ethylene glycol) and alpha-cyclodextrins. *J. Am. Chem. Soc.*, **116**, 3192–3196.
88 Harada, A., Nishiyama, T., Kawaguchi, Y. et al. (1997) Preparation and characterization of inclusion complexes of aliphatic polyesters with cyclodextrins. *Macromolecules*, **30**, 7115–7118.
89 Ceccato, M., LoNostro, P. and Baglioni, P. (1997) α-Cyclodextrin/polyethylene glycol polyrotaxane: a study of the threading process. *Langmuir*, **13**, 2436–2439.
90 Loethen, S., Kim, J.M. and Thompson, D.H. (2007) Biomedical applications of cyclodextrin based polyrotaxanes. *Polym. Rev.*, **47**, 383–418.
91 Uekama, K., Hirayama, F. and Irie, T. (1998) Cyclodextrin drug carrier systems. *Chem. Rev.*, **98**, 2045–2076.
92 Fundueanu, G., Constantin, M., Mihai, D. et al. (2003) Pullulan-cyclodextrin microspheres: a chromatographic approach for the evaluation of the drug–cyclodextrin interactions and the determination of the drug release profiles. *J. Chromatogr. B-Anal. Technol. Biomed. Life Sci.*, **791**, 407–419.
93 Guo, J.H. and Cooklock, K.M. (1995) Bioadhesive polymer buccal patches for buprenorphine controlled delivery: solubility consideration. *Drug Dev. Ind. Pharm.*, **21**, 2013–2019.
94 Bibby, D.C., Davies, N.M. and Tucker, I.G. (2000) Mechanisms by which cyclodextrins modify drug release from polymeric drug delivery systems. *Int. J. Pharm.*, **197**, 1–11.
95 Szejtli, J. (1990) The cyclodextrins and their applications in biotechnology. *Carbohydr. Polym.*, **12**, 375–392.
96 Fyfe, M.C.T. and Stoddart, J.F. (1997) Synthetic supramolecular chemistry. *Acc. Chem. Res.*, **30**, 393–401.
97 Conn, M.M. and Rebek, J. (1997) Self-assembling capsules. *Chem. Rev.*, **97**, 1647–1668.
98 Szejtli, J. (1998) Introduction and general overview of cyclodextrin chemistry. *Chem. Rev.*, **98**, 1743–1753.
99 Orprecio, R. and Evans, C.H. (2003) Polymer-immobilized cyclodextrin trapping of model organic pollutants in flowing water streams. *J. Appl. Polym. Sci.*, **90**, 2103–2110.
100 Szczubialka, K., Jankowska, M. and Nowakowska, M. (2003) Smart polymeric nanospheres as new materials for possible biomedical applications. *J. Mater. Sci. Mater. Med.*, **14**, 699–703.
101 Kathmann, E.E.L., White, L.A. and Cormick, C.L. (1997) Electrolyte and pH-responsive zwitterionic copolymers of 4-[(2-acrylamido-2-methylpropyl)- dimethylammonio]butanoate

with 3-[(2-acrylamido-2-methyl-propyl) dimethyl ammonio] propane sulfonate. *Macromolecules*, **30**, 5297–5304.

102 Morrison, M.E., Dorfman, R.C., Clendening, W.D. et al. (1994) Quenching kinetics of anthracene covalently bound to a polyelectrolyte: effects of ionic strength. *J. Phys. Chem.*, **98**, 5534–5540.

103 Kopecek, J. (2003) Smart and genetically engineered biomaterials and drug delivery systems. *Eur. J. Pharm. Sci.*, **20**, 1–16.

104 Schild, H.G. (1992) Poly(N-isopropylacrylamide): experiment theory and application. *Prog. Polym. Sci.*, **17**, 163–249.

105 Okubo, M., Ahmad, H. and Suzuki, T. (1998) Synthesis of temperature sensitive micron-sized monodispersed composite polymer particles and its application as a carrier for biomolecules. *Colloid. Polym. Sci.*, **276**, 470–475.

106 Aoyagi, T., Ebara, M., Sakai, K. et al. (2000) Novel bifunctional polymer with reactivity and temperature sensitivity. *J. Biomater. Sci., Polym. Ed.*, **1**, 101–110.

107 Mintzer, M.A. and Simanek, E.E. (2009) Nonviral vectors for gene delivery. *Chem. Rev.*, **109**, 259–302.

108 Tang, G.P., Guo, H.Y., Alexis, F. et al. (2006) Low molecular weight polyethylenimines linked by β-cyclodextrin for gene transfer into the nervous system. *J. Gene Med.*, **8**, 736–744.

109 Luo, C.H., Zuo, F., Zheng, Z.H. et al. (2008) Temperature/light dual–responsive inclusion complexes of α-cyclodextrins and azobenzene-containing polymers. *J. Macromol. Sci. Part A-Pure Appl. Chem.*, **45**, 364–371.

110 Kabanov, A.V., Bronich, T.K., Kabanov, V.A. et al. (1996) Temperature/light dual–responsive inclusion complexes of α-cyclodextrins and azobenzene-containing polymers. *Macromolecules*, **29**, 6797–6802.

111 Chung, J.E., Yokoyama, M., Yamato, M. et al. (1999) Thermo-responsive drug delivery from polymeric micelles constructed using block copolymers of poly(N-isopropylacrylamide) and poly(butyl methacrylate). *J. Control. Release*, **62**, 115–127.

112 Chung, E., Yokoyama, M. and Okano, T. (2000) Inner core segment design for drug delivery control of thermoresponsive polymeric micelles. *J. Control. Release*, **65**, 93–103.

113 Kohori, F., Sakai, K., Aoyagi, T. et al. (1999) Control of adriamycin cytotoxic activity using thermally responsive polymeric micelles composed of poly(N-isopropylacrylamide-co-N,N-dimethylacrylamide)-β-poly(D,L-lactide). *Colloids Surf. B.*, **16**, 195–205.

114 Suwa, K., Morishita, K., Kishida, A. and Akashi, M. (1997) Synthesis and functionalities of poly(N-vinylalkylamide). V. Control of a lower

critical solution temperature of poly(N-vinylalkylamide). *J. Polym. Sci. A Polym. Chem.*, **35**, 3087–3094.

115 Chaterji, S., Kwon, K.I. and Park, K. (2007) Smart polymeric gels: redefining the limits of biomedical devices. *Prog. Polym. Sci.*, **32**, 1083–1122.

116 Aronoff, D.M. and Neilson, E.G. (2001) Antipyretics: mechanisms of action and clinical use in fever suppression. *Am. J. Med.*, **111**, 304–315.

117 Guo, C.X. and Li, C.M. (2010) Direct electron transfer of glucose oxidase and biosensing of glucose on hollow sphere-nanostructured conducting polymer/metal oxide composite. *Phys. Chem. Chem. Phys.*, **12**, 12153–12159.

118 122. Lee, M., Kim, J.E., Fang, F.F. et al. (2011) Rectangular-shaped polyaniline tubes covered with nanorods and their electrorheology. *Macromol. Chem. Phys.*, **212**, 2300–2307.

119 Li, G., Pang, S., Liu, J. et al. (2006) Synthesis of polyaniline submicrometer-sized tubes with controllable morphology. *Nanopart. Res.*, **8**, 1039–1044.

120 Miller, L.L., Smith, G.A., Chang, A. and Zhou, Q. (1987) Electro-chemically controlled release. *J. Control. Release*, **6**, 293–306.

121 Alvarez-Lorenzo, C., Bromberg, L. and Concheiro, A. (2009) Light-sensitive intelligent drug delivery systems. *Photochem. Photobiol.*, **85**, 848–860.

122 Yuan, X., Fischer, K. and Schartl, W. (2005) Photocleavable microcapsules built from photoreactive nanospheres. *Langmuir*, **21**, 9374–9380.

123 Ercole, F., Davis, T.P. and Evans, R.A. (2010) Photo-responsive systems and biomaterials: photochromic polymers, light-triggered self-assembly, surface modification, fluorescence modulation and beyond. *Polym. Chem.*, **1**, 37–54.

124 Koňák, Č. (1997) Photoregulated association of water-soluble copolymers with spirobenzopyran-containing side chains. *Macromolecules*, **30**, 5553–5556.

125 Troutman, T.S., Leung, S.J. and Romanowski, M. (2009) Light-induced content release from plasmon-resonant liposomes. *Adv. Mater.*, **21**, 2334–2338.

126 Chandra, B. (2006) Formulation of photocleavable liposomes and the mechanism of their content release. *Org. Biomol. Chem.*, **4**, 1730–1740.

127 Babin, J. (2009) A new two-photon-sensitive block copolymer nanocarrier. *Angew. Chem., Int. Ed.*, **48**, 3329–3332.

128 Jiang, X. (2008) Multiple micellization and dissociation transitions of thermo- and light-sensitive poly(ethylene

oxide)-*b*-poly(ethoxytri(ethylene glycol) acrylate-*co*-*o*-nitrobenzyl acrylate) in water. *Macromolecules*, **41**, 2632–2643.

129 Tong, X., Wang, G., Soldera, A. and Zhao, Y. (2005) How can azobenzene block copolymer vesicles be dissociated and reformed by light? *J. Phys. Chem. B*, **109**, 20281–20287.

130 Han, K. (2008) Reversible photocontrolled swelling-shrinking behavior of micron vesicles self-assembled from azopyridine-containing diblock copolymer. *Macromol. Rapid Commun.*, **29**, 1866–1870.

131 Lin, L. (2009) UV-responsive behavior of azopyridine-containing diblock copolymeric vesicles: photoinduced fusion: disintegration and rearrangement. *Macromol. Rapid Commun.*, **30**, 1089–1093.

132 Mabrouk, E. (2009) Bursting of sensitive polymersomes induced by curling. *Proc. Natl. Acad. Sci.*, **106**, 7294–7298.

133 Li, Y., Lokitz, B.S. and McCormick, C.L. (2006) Thermally responsive vesicles and their structural 'locking' through polyelectrolyte complex formation. *Angew. Chem., Int. Ed.*, **45**, 5792–5795.

134 Du, J. and O'Reilly, R.K. (2009) Advances and challenges in smart and functional polymer vesicles. *Soft Matter*, **5**, 3544–3561.

135 Li, M.H. and Keller, P. (2009) Stimuli-responsive polymer vesicles. *Soft Matter*, **5**, 927–937.

136 Napoli, A. (2004) Glucose-oxidase based self-destructing polymeric vesicles. *Langmuir*, **20**, 3487–3491.

137 Cerritelli, S., Velluto, D. and Hubbell, J.A. (2007) PEG-SS-PPS: reduction-sensitive disulfide block copolymer vesicles for intracellular drug delivery. *Biomacromolecules*, **8**, 1966–1972.

138 Qin, S. (2006) Temperature-controlled assembly and release from polymer vesicles of poly(ethylene oxide)-*block*-poly(*N*-isopropylacrylamide). *Adv. Mater.*, **18**, 2905–2909.

139 Neuberger, T., Schopf, B., Hofmann, H. *et al.* (2005) Superparamagnetic nanoparticles for biomedical applications: possibilities and limitations of a new drug delivery system. *J. Magn. Mater.*, **293**, 483–496.

140 Mizogami, S., Mizutani, M., Fukuda, M. and Kawabata, K. (1991) Abnormal ferromagnetic behavior for pyrolytic carbon under low temperature growth by CVD method. *Synth. Met.*, **43**, 3271–3274.

141 Kopelevich, Y., Esquinazi, P., Torres, J.H.S. and Moehlecke, S. (2000) Ferromagnetic- and superconducting-like behavior of graphite. *J. Low Temp. Phys.*, **119**, 691–702.

142 Liu, X. and Jiang, M. (2006) Optical switching of self-assembly: micellization and micelle–hollow-sphere transition of hydrogen-bonded polymers. *Angew. Chem., Int. Ed.*, **45**, 3846–3850.

143 Zhao, W. (2011) ABC triblock copolymer vesicles with mesh-like morphology. *ACS Nano*, **5**, 486–492.

144 Pardoe, H., Chua-anusorn, W., Pierre, T.G. and Dobson, J. (2001) Structural and magnetic properties of nanoscale magnetic particles synthesized by coprecipitation of iron oxide in the presence of dextran or polyvinyl alcohol. *J. Magn. Mater.*, **225**, 41–46.

145 Barrio, J. (2010) Self-assembly of linear-dendritic diblock copolymers: from nanofibers to polymersomes. *J. Am. Chem. Soc.*, **132**, 3762–3769.

146 Gan, L.H., Gan, Y.Y. and Roshan, D.G. (2000) Poly(N-acryloyl-N-propylpiperazine): a new stimuli responsive polymer. *Macromolecules*, **33**, 7893–7897.

147 Gonzalez, N., Elvira, C. and San, R.J. (2003) Hydrophilic and hydrophobic copolymer systems based on acrylic derivatives of pyrrolidone and pyrrolidine. *J. Polym. Sci. Part A: Polym. Chem.*, **41**, 395–407.

148 Sangeetha, N.M. and Matitra, U. (2005) Supramolecular gels: functions and uses. *Chem. Soc. Rev.*, **34**, 821–836.

149 Peng, Z., Sun, Y., Liu, X. and Tong, Z. (2010) Nanoparticles of Block Ionomer Complexes from Double Hydrophilic Poly(acrylic acid)-*b*-poly(ethylene oxide)-*b*-poly(acrylic acid) Triblock Copolymer and Oppositely Charged Surfactant. *Nanoscale Res Lett.*, **6**, 89–95.

150 Kumar, A. (2011) *Smart Polymeric Biomaterials: Where Chemistry & Biology Can Merge.* Available at: <http://www.iitk.ac.in/directions/dirnet 7/PP~ASHOK~FFF.pdf>.

151 Choi, H.S., Ooya, T., Lee, S.C. *et al.* (2004) pH Dependence of polypseudorotaxane formation between cationic linear polyethylenimine and cyclodextrins. *Macromolecules*, **37**, 6705–6710.

152 Becuwe, M., Cazier, F., Bria, M. *et al.* (2007) Tuneable fluorescent marker appended to β-cyclodextrin: a pH-driven molecular switch. *Tetrahedron Lett.*, **48**, 6186–6188.

153 Karaky, K., Brochon, C., Schlatter, G. and Hadziioannou, G. (2008) pH-Switchable supramolecular 'sliding' gels based on polyrotaxanes of polyethyleneimine-*block*-poly(ethylene oxide)-*block*-polyethyleneimine block copolymer and α-cyclodextrin: synthesis and swelling behaviour. *Soft Matter*, **4**, 1165–1168.

154 Liu, Y., Fan, X.D., Kang, T. and Sun, L. (2004) A cyclodextrin microgel for controlled release driven by inclusion effects. *Macromol. Rapid Commun.*, **25**, 1912–1916.

155 Branham, K.D., Snowden, H.S. and McCormick, C.L. (1996) Water-Soluble Copolymers. 64. Effects of pH and Composition on Associative Properties of Amphiphilic Acrylamide/Acrylic Acid Terpolymers. *Macromolecules*, **29**, 254–262.

156 Moine, L., Amiel, C., Brown, W. and Guerin, P. (2001) Associations between a hydrophobically modified, degradable, poly(malic acid) and a β-cyclodextrin polymer in solution. *Polym. Int.*, **50**, 663–676.

157 Huh, K.M., Ooya, T., Sasaki, S. and Yui, N. (2001) Polymer inclusion complex consisting of poly(ε-lysine) and α-cyclodextrin. *Macromolecules*, **34**, 2402–2404.

158 Huh, K.M., Tomita, H., Ooya, T. *et al.* (2002) pH Dependence of inclusion complexation between cationic poly(ε-lysine) and α-cyclodextrin. *Macromolecules*, **35**, 3775–3777.

159 Melo, J.S.S., Costa, T., Oliveira, N. and Schillen, K. (2007) Fluorescence studies on the interaction between pyrene-labelled poly(acrylic acid) and cyclodextrins. *Polym. Int.*, **56**, 882–899.

160 Liu, J.H., Sondjaja, H.R. and Tam, K.C. (2007) α-Cyclodextrin-induced self-assembly of a double-hydrophilic block copolymer in aqueous solution. *Langmuir*, **23**, 5106–5109.

161 Liu, Y.Y., Fan, X.D. and Zhao, Q. (2003) A novel IPN hydrogel based on poly(N-isopropylacrylamide) and β-cyclodextrin polymer. *J. Macromol. Sci. Pure Appl. Chem.*, **A40**, 1095–1105.

162 Mias, S., Sudor, J. and Camon, H. (2008) PNIPAM: a thermo-activated nano-material for use in optical devices. *Microsyst. Technol.*, **14** (6), 747–751.

163 Tu, C.W., Kuo, S.W. and Chang, F.C. (2009) Supramolecular self-assembly through inclusion complex formation between poly(ethylene oxide-b-N-isopropylacrylamide) block copolymer and α-cyclodextrin. *Polymer*, **50**, 2958–2966.

164 Choi, H.S., Yamamoto, K., Ooya, T. and Yui, N. (2005) Synthesis of poly(ε-lysine)-Grafted Dextrans and Their pH- and Thermosensitive Hydrogelation with Cyclodextrins. *Chemphyschem*, **6**, 1081–1086.

165 Nozaki, T., Maeda, Y., Ito, K. and Kitano, H. (1995) Cyclodextrins modified with polymer chains which are responsive to external stimuli. *Macromolecules*, **28**, 522–524.

166 Hirasawa, T., Maeda, Y. and Kitano, H. (1998) Inclusional complexation by cyclodextrin–polymer conjugates in organic solvents. *Macromolecules*, **31**, 4480–4485.

167 Ohashi, H., Hiraoka, Y. and Yamaguchi, T. (2006) An autonomous phase transition–complexation/decomplexation polymer system with a molecular recognition property. *Macromolecules*, **39**, 2614–2620.

168 Yang, M., Chu, L.Y., Xie, R. and Wang, C. (2008) Polymers with pendent β-cyclodextrin groups. *Macromol. Chem. Phys.*, **209**, 204–211.

169 Duan, Q., Miura, Y., Narumi, A. *et al.* (2005) Synthesis and thermoresponsive property of end-functionalized poly(N-isopropylacrylamide) with pyrenyl group. *J. Polym. Sci. Part A-Polym. Chem.*, **44**, 1117–1124.

170 Joseph, J., Dreiss, C.A., Cosgrove, T. and Pedersen, J.S. (2007) Rupturing polymeric micelles with cyclodextrins. *Langmuir*, **23**, 460–466.

171 Mathews, A.S., Cho, W.J., Kim, I. and Ha, C.S. (2009) Thermally responsive poly[N-isopropylacrylamide-co-2-hydroxyethylacrylate] colloidal crystals included in β-cyclodextrin for controlled drug delivery. *J. Appl. Polym. Sci.*, **113**, 1680–1689.

172 Zhang, J.T., Huang, S.W., Gao, F.Z. and Zhuo, R.X. (2005) Novel temperature-sensitive, β-cyclodextrin-incorporated poly(N-isopropylacrylamide) hydrogels for slow release of drug. *Colloid. Polym. Sci.*, **283**, 461–464.

173 Wang, Q.F., Li, S.M., Wang, Z.Y. et al. (2009) Preparation and characterization of a positive thermoresponsive hydrogel for drug loading and release. *J. Appl. Polym. Sci.*, **111**, 1417–1425.

174 Tamura, M., Gao, D. and Ueno, A. (2001) A polyrotaxane series containing α-cyclodextrin and naphthalene-modified α-cyclodextrin as a light-harvesting antenna system. *Chemistry*, **7**, 1390–1397.

175 Zheng, P.J., Wang, C., Hu, X. et al. (2005) Supramolecular complexes of azocellulose and α-cyclodextrin: Isothermal titration calorimetric and spectroscopic studies. *Macromolecules*, **38**, 2859–2864.

176 Pouliquen, G., Amiel, C. and Tribet, C. (2007) Photoresponsive viscosity and host–guest association in aqueous mixtures of poly-cyclodextrin with azobenzene-modified poly(acrylic)acid. *J. Phys. Chem. B*, **111**, 5587–5595.

177 Gil, E.S. and Hudson, S.M. (2004) Stimuli-responsive polymers and their bioconjugates. *Prog. Polym. Sci.*, **29**, 1173–1222.

178 Otero-Espinar, F.J., Igea, S.A., Mendez, J.B. and Jato, J.L.V. (1991) Reduction in the ulcerogenicity of naproxen by complexation with β-cyclodextrin. *Int. J. Pharm.*, **70**, 35–41.

179 Chen, P.C., Kohane, D.S., Park, Y.J. et al. (2004) Injectable microparticle–gel system for prolonged and localized lidocaine release. II. *In vivo* anesthetic effects. *J. Biomed. Mater. Res. Part A*, **70**, 459–466.

180 Barreiro-Iglesias, R., Alvarez-Lorenzo, C. and Concheiro, A. (2001) Incorporation of small quantities of surfactants as a way to improve the rheological and diffusional behavior of carbopol gels. *J. Control. Release*, **77**, 59–75.

181 Crini, G. and Morcellet, M. (2002) Synthesis and applications of adsorbents containing cyclodextrins. *J. Sep. Sci.*, **25**, 789–813.

182 Ooya, T., Ichi, T., Furubayashi, T. et al. (2007) Cationic hydrogels of PEG crosslinked by a hydrolyzable polyrotaxane for cartilage regeneration. *React. Funct. Polym.*, **67**, 1408–1417.

183 Ruel-Gariépy, E. and Leroux, J. (2004) In situ-forming hydrogels-review of temperature-sensitive systems. *Eur. J. Pharm. Biopharm.*, **58**, 409–426.

184 Shenoy, D., Little, S., Langer, R. and Amiji, M. (2005) Poly(ethylene oxide)-modified poly(β-amino ester) nanoparticles as a pH-sensitive system for tumor-targeted delivery of hydrophobic drugs: in vitro evaluations. *Mol. Pharm.*, **2**, 357–366.

185 You, J., Almeda, D., Ye, G.J.C. and Auguste, D.T. (2010) Bioresponsive matrices in drug delivery. *J. Biol. Eng.*, **4**, 1–12.

186 Murakami, H., Kawabuchi, A., Matsumoto, R. *et al.* (2005) A multi-mode-driven molecular shuttle: photochemically and thermally reactive azobenzene rotaxanes. *J. Am. Chem. Soc.*, **127**, 15891–15899.

187 Qu, D.H., Wang, Q.C., Ren, J. and Tian, H. (2005) Superparamagnetic nanoparticle-supported catalysis of Suzuki cross-coupling reactions. *Org. Lett.*, **6**, 2085–2088.

188 Stanier, C.A., Alderman, S.J., Claridge, T.D.W. and Anderson, H.L. (2002) Unidirectional photoinduced shuttling in a rotaxane with a symmetric stilbene dumbbell. *Angew. Chem., Int. Ed.*, **41**, 1769–1772.

189 Altieri, A., Bottari, G., Dehez, F. *et al.* (2003) remarkable positional discrimination in bistable light- and heat-switchable hydrogen-bonded molecular shuttles. *Angew. Chem., Int. Ed.*, **42**, 2296–2300.

190 Tomatsu, I., Hashidzume, A. and Harada, A. (2006) Redox-responsive hydrogel system using the molecular recognition of β-cyclodextrin. *Macromol. Rapid Commun.*, **27**, 238–241.

191 Tomatsu, I., Hashidzume, A. and Harada, A. (2005) Photoresponsive hydrogel system using molecular recognition of α-cyclodextrin. *Macromolecules*, **38**, 5223–5227.

192 Zheng, P.J., Hu, X., Zhao, X.Y. *et al.* (2004) Photoregulated sol–gel transition of novel azobenzene-functionalized hydroxypropyl methylcellulose and its α-cyclodextrin complexes. *Macromol. Rapid Commun.*, **25**, 678–682.

193 Arnaud, A. and Bouteiller, L. (2004) isothermal titration calorimetry of supramolecular polymers. *Langmuir*, **20**, 6858–6863.

194 Liu, Y., You, C.C., Zhang, M. *et al.* (2000) Molecular Interpenetration within the columnar structure of crystalline anilino-β-cyclodextrin. *Org. Lett.*, **2**, 2761–2763.

195 Gonzalez, B., Casado, C.M., Alonso, B. *et al.* (1998) Synthesis, electrochemistry and cyclodextrin binding of novel cobaltocenium-functionalized dendrimers. *Chem. Commun.*, **23**, 2569–2570.

196 Harada, A. and Takahashi, S. (1984) Alumina solid Lewis superacid: activated benzene and isomerization of alkanes on aluminas chlorinated at high temperature. *J. Chem. Soc.-Chem. Commun.*, **10**, 645–646.

197 Harada, A., Hu, Y., Yamamoto, S. and Takahashi, S. (1998) Preparation and properties of inclusion compounds of ferrocene and its derivatives with cyclodextrins. *J. Chem. Soc., Dalton Trans.*, **3**, 729–732.

198 Gonzalez, B., Cuadrado, I., Alonso, B. et al. (2002) Mixed Cobaltocenium−ferrocene heterobimetallic complexes and their binding interactions with β-cyclodextrin: a three-state, host−guest system under redox control. *Organometallics*, **21**, 3544–3551.

199 Castro, R., Cuadrado, I., Alonso, B. et al. (1997) multisite inclusion complexation of redox active dendrimer guests. *J. Am. Chem. Soc.*, **119**, 5760–5761.

200 Schattling, P., Jochum, F.D. and Theato, P. (2014) Multi-stimuli responsive polymers – the all-in-one talents. *Polym. Chem.*, **5**, 25–36.

201 Kurisawa, M. and Yui, N. (1998) Gelatin/dextran intelligent hydrogels for drug delivery: dual-stimuli-responsive degradation in relation to miscibility in interpenetrating polymer networks. *Macromol. Chem. Phys.*, **199**, 1547–1554.

202 Yamamoto, N., Kurisawa, M. and Yui, N. (1996) Double-stimuli-responsive degradable hydrogels: interpenetrating polymer networks consisting of gelatin and dextran with different phase separation. *Macromol. Rapid Commun.*, **17**, 313–318.

203 Mano, J.F. (2008) Stimuli-responsive polymeric systems for biomedical applications. *Adv. Eng. Mater.*, **10**, 515–527.

204 Dimitrov, I., Trzebicka, B., Muller, A.H.E. et al. (2007) Thermosensitive water-soluble copolymers with doubly responsive reversibly interacting entities. *Prog. Polym. Sci.*, **32**, 1275–1343.

205 Hamcerencu, M., Desbrieres, J., Popa, M. and Riess, G. (2009) Stimuli-sensitive xanthan derivatives/N-isopropylacrylamide hydrogels: influence of cross-linking agent on interpenetrating polymer network properties. *Biomacromolecules*, **10**, 1911–1922.

206 Khoukh, S., Oda, R., Labrot, T. et al. (2007) Light-responsive hydrophobic association of azobenzene-modified poly(acrylic acid) with neutral surfactants. *Langmuir*, **23**, 94–104.

207 Miao, M., Cirulis, J.T., Lee, S. and Keeley, F.W. (2005) Structural determinants of cross-linking and hydrophobic domains for self-assembly of elastin-like polypeptides. *Biochemistry*, **44**, 14367–14375.

208 Yin, X.C. and Stover, H.D.H. (2003) Hydrogel microspheres by thermally induced coacervation of poly(N,N-dimethylacrylamide-co-glycidyl methacrylate) aqueous solutions. *Macromolecules*, **36**, 9817–9822.

209 (a) Schilli, C.M., Zhang, M., Rizzardo, E. et al. (2004) A new double-responsive block copolymer synthesized via RAFT polymerization: poly(N-isopropylacrylamide)-*block*-poly(acrylic acid). *Macromolecules*, **37**, 7861; (b) Mu, B. and Liu, P. (2012) Temperature and pH dual responsive crosslinked polymeric nanocapsules via surface-initiated atom transfer radical polymerization. *React. Funct. Polym.*, **72**, 983.

210 Zhou, Z.X., Shen, Y.Q., Tang, J.B. et al. (2009) Charge-reversal drug conjugate for targeted cancer cell nuclear drug delivery. *Adv. Funct. Mater.*, **19**, 3580–3589.

211 Liu, T., Li, X.J., Qian, Y.F. et al. (2012) Multifunctional pH-disintegrable micellar nanoparticles of asymmetrically functionalized β-cyclodextrin-based star copolymer covalently conjugated with doxorubicin and DOTA-Gd moieties. *Biomaterials*, **33**, 2521–2531.

212 Chen, J., Qiu, X., Ouyang, J. et al. (2011) pH and reduction dual-sensitive copolymeric micelles for intracellular doxorubicin delivery. *Biomacromolecules*, **12**, 3601–3611.

213 Otsuki, N., Fujioka, N., Kawatsuki, N. and Ono, H. (2006) Photoinduced orientation and holographic recording in polyester films comprising azobenzene side-groups using 633 nm red light. *Mol. Cryst. Liq. Cryst.*, **458**, 139–148.

214 Kungwatchakun, D. and Irie, M. (1988) Photoresponsive polymers. photocontrol of the phase separation temperature of aqueous solutions of poly-[N-isopropylacrylamide-*co*-N-(4-phenylazophenyl)acrylamide]. *Makromol. Chem.-Rapid Commun.*, **9**, 243–246.

215 Luo, C.H., Zuo, F., Ding, X.B. et al. (2008) Light-triggered reversible solubility of α-cyclodextrin and azobenzene moiety complexes in PDMAA-*co*-PAPA via molecular recognition. *J. Appl. Polym. Sci.*, **107**, 2118–2125.

216 Joung, Y.K., Ooya, T., Yamaguchi, M. and Yui, N. (2007) Modulating rheological properties of supramolecular networks by pH-responsive double-axle intrusion into γ-cyclodextrin. *Adv. Mater.*, **19**, 396–400.

4

Basics of Corrosion

Coated steel and aluminum have been widely used in construction, automobile, and domestic appliance industries because of their high quality, endurance, good formability, and relatively low cost. But a drawback of metals is corrosion, a natural process, causing defects and deterioration of their mechanical properties in the very long term [1, 2].

Corrosion is caused by a shift to lowest energy states. Metals often combine with oxygen and water and form hydrated metal oxides (rust), go to lower energy states, and lead to corrosion. On the other hand, corrosion is a direct chemical or electrochemical reaction between metallic materials under the effect of the environment and occurs in several widely differing forms. Moisture and high-temperature gases are conducive to corrosion, which is classified as wet and dry corrosion, respectively [3, 4]. In other words, corrosion is caused by breaking down of essential properties in a material due to chemical reactions with its surroundings, which means loss of electrons of metals reacting with water and oxygen. In polymer and ceramic materials, weakening of polymers by the ultraviolet light from the sun and degradation leads to corrosion. Corrosion can be concentrated locally to form a pit or crack, or it can extend across a wide area to produce general decline [2–5].

With a water layer on the metal surface, a reaction starts at the interface, which is of an electrochemical nature involving an exchange of charge and matter at the metal/water interface. The surface reaction propagates progressively and corrosion takes place. In the anodic reaction, metal goes into solution and hydrolysis occurs, , and in the cathodic reaction electrons produced in the anodic reaction are consumed in the reduction of protons in hydrogen or in the reduction of the oxygen dissolved in water, among other possibilities [5–7].

Thermodynamic stability of metals in a particular environment is important and it prevents corrosion of metals such as gold, silver, and platinum, which is called immune behavior. Corrosion protection does not depend on the stability of protective films [7–10].

In another way, when a metal dissolves in solution and forms soluble corrosion products, dissolution of the metal continues in the solution and the corrosion products do not prevent subsequent corrosion, so corrosion is active mode and is determined by more weight loss of the metal. If after immersion of the metal in solution, corrosion reaction takes place and corrosion production is insoluble, corrosion is protective or inactive and slows the reaction rate to very low levels, which is called passive corrosion, for example, aluminum, chromium, titanium, nickel, and stainless steels [1–3].

Corrosion phenomenon is an interdisciplinary science and involves materials science, metallurgy, surface science, electrochemistry, metallurgical industries, automotive industries, aeronautics, chemical industries, energy production, microelectronics, polymer, and mechanics. Protection of alloys such as stainless steels, nickel-based stainless alloys, and aluminum alloys is a promising way and economical challenge to have a better resistance to corrosion and increase lifetime of products containing metallic components, which saves 25% of product cost [2, 3, 7].

Different ways have been used to protect metals against corrosion, which is critical to avoid premature destruction of buildings, which is related to economic aspects of production cost, maintenance and lifetime, ecological and environment service conditions [2, 3].

One method is coating in which materials are covered by an ultrathin layer at the surface to protect against corrosion, which are auto-protected against corrosion, but the cost is higher and their use is restricted to products with higher added value than steel structures. Various factors such as chemical composition, atomic structure, electronic properties, and usage environment affect coating performance [9–11].

4.1 Introduction to Corrosion and Its Types

4.1.1 Corrosion

Corrosion is an electrochemical reaction between the surface of a metal or an alloy and environment, which causes oxidation of metal and electron loss (Reaction 1)

$$M \rightarrow M^{n+} + ne^-$$

M, metal; M^{n+}, the metal oxidized to metal ions; ne^-, number of electrons that metal loses during the anodic process.

If reduction of metal takes place, the reaction is as follows:

$2H^+ + 2e^- \rightarrow H_2$	Hydrogen reduction reaction
$4H^+ + O_2 + 4e^- \rightarrow 2H_2O$	Oxygen reduction in acid solution
$2H_2O + O_2 + 4e^- \rightarrow 4OH$	Oxygen reduction in basic or neutral solution
$M^{3+} + e^- \rightarrow M^{2+}$	Metal ion reduction

Anode is the surface where oxidation takes place and it carries negative charges designated by the sign (−), whereas cathode is where reduction takes place, designated by the sign (+). In addition to these two electrodes, electrolyte solution and electrical connection between the anode and cathode for the flow of electron current are crucial for corrosion, and if any one of the above-mentioned components is not present or disabled,, electrochemical corrosion process will be stopped [12–15].

4.1.2 Forms of Corrosion

Corrosion is classified into eight main forms:

1) Uniform or general attack
2) Galvanic or two-metal corrosion
3) Crevice corrosion
4) Pitting
5) Intergranular corrosion
6) Selective leaching or parting
7) Erosion corrosion
8) Stress corrosion.

4.1.2.1 Uniform Corrosion

Uniform corrosion, or general corrosion, as sometimes called, is defined as a type of corrosion in which most direct chemical attacks (e.g., an acid) affect the surface, in other words, uniform corrosion proceeds at approximately the same rate over the exposed metal surface (Figure 4.1). Iron and steel undergo uniform corrosion when exposed to atmospheres, soils, and natural waters, which is easily predictable, observable, and hence easy to protect by coating or painting. In natural environment, oxygen is the primary cause of uniform corrosion of steel and other metals and alloys [16–18].

Various methods can be used for prevention of corrosion such as increasing thickness of material by paints or metallic coatings, using corrosion inhibitors, modifying the environment, or cathodic (or anodic) protection [16, 17].

Figure 4.1 Uniform corrosion attack on structural steel.

4.1.2.2 Galvanic Corrosion

Galvanic corrosion, also called bimetallic corrosion or dissimilar metal corrosion, takes place when two dissimilar metals come in direct contact because of an electrical contact (including physical contact) with a metal or nonmetallic conductor (the cathode) in a corrosive electrolyte and leads to deterioration of the anodic metal [16–18]. Drastic attack occurs when two dissimilar metals are joined and the degree of accelerated attack is reduced for bimetallic joints. Galvanic corrosion is determined by resistance to corrosion and oxidation in moist air. In a galvanic cell, the metal with high potential value acts as the anode and rusts faster than it would alone, in contrast, the other metal acts as the cathode and corrodes slower than it out of galvanic cell [19, 20].

In the same electrolyte, under the same environmental conditions (temperature, pH, flow rate, etc.), different metals and alloys have different electrochemical potentials, which is the driving force for the destructive attack on the active metal (anode). The conductivity of electrolyte will also affect the degree of attack. The cathode-to-anode area ratio is directly proportional to the acceleration factor. Prevention of galvanic corrosion is possible by selecting metals/alloys as close together as possible in the galvanic series, avoiding unfavorable area effect of a small anode and large cathode, insulating dissimilar metals wherever practical, applying coatings with caution, and avoiding threaded joints for materials far apart in the galvanic series [18–22] (Figure 4.2).

Figure 4.2 Galvanic corrosion occurred on the aluminum plate along the joint with the mild steel within 2 years due to the huge acceleration factor in galvanic corrosion.

4.1.2.3 Pitting Corrosion

Pitting corrosion, which is one of the most damaging forms of corrosion, is defined as a localized corrosion of a metal surface limited to a point or small area that leads to the creation of small holes or cavities in the metal or [16, 17]. Pitting factor is the ratio of the depth of the deepest pit resulting from corrosion and the average penetration calculated from weight loss. Pitting corrosion occurs on the surface of metals such as aluminum, stainless steel, which have a metal oxide layer. When aluminum or other passive metals and alloys are imposed in the oxygen-containing environment such as air or water, oxide layers rapidly cover the surface and prevent aluminum or aluminum alloys from further corrosion [23, 24]. The resulting pits can become wide and shallow or narrow and deep, which can rapidly perforate the wall thickness of the metal. By damaging the oxide film (by a scratch), repassivation rate is higher than the rate of corrosion, so new oxide layer will immediately form on the metal surface [24]. The pitting corrosion is initiated from a weak site in the oxide layer by a halide ion attack [25–27]. Aluminum ions are formed via oxidation of the aluminum and combine with chloride ions to form aluminum chloride, and free electrons move to the cathode where reduces hydrogen cation to form hydrogen gas. Pitting corrosion is undetectable and unpredictable, so damage of a protective oxide layer leads to the formation of cavities with different diameters and holes of varying dimensions, which can rapidly perforate the wall thickness of a metal [24–27].

For prevent pitting corrosion, materials with known resistance to the service environment are used; pH (~7), chloride concentration, halogen concentration, oxygen content, and temperature can be controlled; and higher alloys (ASTM G48) with higher resistance to pitting corrosion are used (Figure 4.3) [18, 20].

4.1.2.4 Crevice Corrosion

Crevice corrosion is a type of localized attack on a metal surface that can be found within gap between two joining surfaces where a stagnant solution is present. Crevice corrosion occurs at narrow openings between two metals or a metal and nonmetallic material, and other surfaces are resistant to corrosion. A concentration cell forms with the crevice being depleted of oxygen. The damage caused by crevice corrosion is normally confined to one metal at the localized area within or close to the joining surfaces. This kind of corrosion is dependent on the difference in concentration of some chemical constituents, usually oxygen, which sets up an electrochemical concentration cell. Parameters affecting crevice corrosion are crevice type (metal-to-metal, metal-to-nonmetal), crevice geometry (gap size, depth, surface roughness), material (alloy composition (e.g., Cr, Mo), structure), and environment (pH, temperature, halide ions, oxygen) (Figure 4.4) [28, 29].

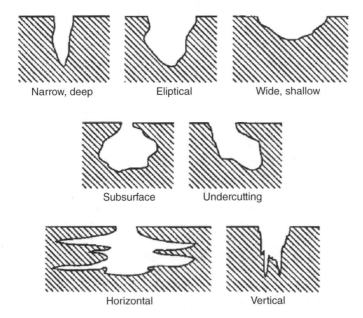

Figure 4.3 ASTM-G46 has a standard visual chart for rating of pitting corrosion.

Figure 4.4 Crevice corrosion of stainless steel tube due to the presence of crevice (gap) between the tube and tube sheet.

Figure 4.5 Intergranular corrosion or intergranular stress corrosion cracking.

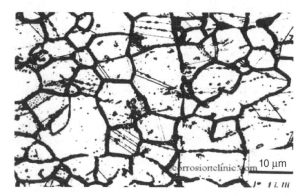

4.1.2.5 Intergranular Corrosion (IGC)

Intergranular corrosion or intercrystallite corrosion is caused by tensile stress, due to which cracking may occur along grain boundaries zone, but the bulk grain of metal or alloy is not attacked [30]. Grain boundaries are sites for precipitation and segregation, which make them physically and chemically different from the matrix. Differences in composition such as physical and chemical properties of matrix or accumulation of impurities result in this corrosion. Precipitation of chromium carbides in stainless steel leads to this type of corrosion and consumes the alloying element – chromium from a narrow band along the grain boundary and this makes the zone anodic to the unaffected grains (Figure 4.5) [28, 29].

4.1.2.6 Dealloying Corrosion

Dealloying corrosion, also called selective leaching, parting or selective attack corrosion, causes selective corrosion in one or more components

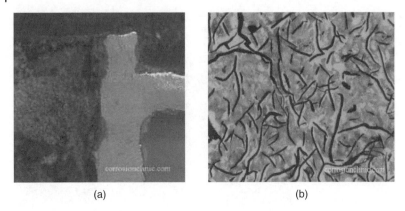

(a) (b)

Figure 4.6 Graphitic corrosion of a gray cast iron valve.

of a solid solution alloy such as decarburization, decobaltification, denickelification, dezincification, and graphitic corrosion.

Selective loss of carbon from the surface layer of a carbon-containing alloy due to reaction with one or more chemical substances in a medium that contacts the surface is called decarburization. Decobaltification is selective leaching of cobalt from cobalt-based alloys, such as Stellite, or from cemented carbides. Denickelification is selective leaching of nickel from nickel-containing alloys, which is found to occur in copper-nickel alloys after extended service in fresh water. Dezincification is selective leaching of zinc from zinc-containing alloys, which is encountered in copper–zinc alloys containing less than 85% copper after extended service in water containing dissolved oxygen [18, 21, 22].

Graphitic corrosion is deterioration of gray cast iron in which the metallic constituents are selectively leached or converted into corrosion products leaving the graphite intact. Graphitic corrosion should not be confused with another term graphitization, which is used to describe the formation of graphite in iron or steel, usually from decomposition of iron carbide at elevated temperatures [23, 24].

Different metals and alloys have different electrochemical potentials (or corrosion potentials) in the same electrolyte, which leads to preferential attack on the more "active" element in the alloy. For prevention of dealloying corrosion, metals/alloys containing inhibited brass and ductile iron can be used and environment must minimize the selective leaching. Another way is use of sacrificial anode cathodic protection or impressed current cathodic protection (Figure 4.6) [28, 29].

4.1.2.7 Stress Corrosion Cracking (SCC)

Simultaneous action of rust and sustained tensile stress leads to stress corrosion cracking, which prevents intercrystalline or trans-crystalline corrosion and corrosion-reduced sections that fail by fast fracture. Stress corrosion cracking may occur in combination with hydrogen and results from the conjoint action of a sensitive material, a specific chemical environment, and tensile stress, and it does not follow the specific mechanism for stress corrosion cracking in the literature. Copper and its alloys are susceptible to ammonium compounds, mild steels to alkalis, and stainless steels to chlorides [18, 21, 22].

This type of corrosion can be prevented by the following: controlling hardness and stress level (residual or load), controlling operating temperature and/or the electrochemical potential of the alloy, introducing compressive stress by shot-peening, using materials that do not crack in the specific environment [1–3, 18, 21, 22].

4.1.2.8 Erosion Corrosion

Erosion and corrosion combined in the presence of a moving corrosive fluid or a metal component moving through the fluid lead to accelerated loss of metal and is called erosion corrosion. Erosion corrosion usually takes place at high flow rates around tube blockages, tube inlet ends, or in pump impellers. Mechanical effect of flow or velocity of a fluid combined with the corrosive action of the fluid causes accelerated loss of metal. Erosion corrosion can be prevented through streamlining the piping to reduce turbulence, controlling fluid velocity, using more resistant materials, and using corrosion inhibitors or cathodic protection [31].

4.2 Corrosion Protection

Between metals and alloys, corrosion of aluminum and its alloys is more important. Aluminum has gained much more attention because of its high production and wide consumption and applications such as chemical, aerospace, food, electronics, and marine industries, lower price, lighter in its density, and higher strength compared to iron [2, 4, 6].

Metals are active and tend to react with oxygen and oxidized in nature. Some alloys such as aluminum form aluminum oxide protection layer on the surface, which is hard and self-renewing, thus protecting the inner aluminum from further corrosion [5], and is not always resistant to severe corrosive environment. Protection of metals and alloys is of great importance for the durability of products structures and prevents great economic loss. Various methods can be used for protection of metals, among which use of coatings or anticorrosive coatings containing inhibitors is the most popular and economical method.

4.2.1 Anticorrosion Methods

Three methods have been used for corrosion protection of metals: coating, inhibitors, and cathodic protection [32]. In cathodic protection, an impressed current is forced from inert anodes to the large area of applied structure or an active metal can be used as anodes and connected to the applied structure; and the active anode provides a cathodic protection current [11, 33]. In another protection method, inhibitors to inhibit corrosion such as organic amines can be adsorbed on the anode and cathode to reduce the corrosion current [34, 35].

Corrosion inhibitors can also directly influence or control the anodic or cathodic process or accelerate the formation of passive layers on metal surface and enhance the ability of corrosion control [36]. This method has many weaknesses, so their application is restricted. By using coating or paints on the surface of metals and alloys, these layers protect the surface and control corrosion by decreasing mass transfer rates of electrolyte of corrosion environments and prevent the occurrence of the cathode reaction [37].

4.2.2 Anticorrosion Coating

Anticorrosion coating prevents corrosion via an external layer of film onto a metal substrate, which prevents contact of the environment with surfaces, and must be flexible and must have resistance against impact, chemical resistance to the environment, resistance to permeation of moisture, good adhesion, and cohesion [38, 39]. Protective coatings minimize/control corrosion of the underlying substrate based on one or a combination of three main mechanisms: (i) barrier protection where the coating prevents/reduces ingress of corrosive agents to the metal/coating interface, (ii) cathodic protection where the coating acts as a sacrificial anode, and (iii) active protection as a result of inclusion of inhibitors, imparting anodic or/and cathodic protection to the coating [1, 2].

Coatings can be divided into three main groups: organic, inorganic, and hybrid coatings. Organic coatings are the most widely used anticorrosion method due to low cost factors. Paints, lacquers, and varnishes are the examples of organic coatings. Oxygen and silicon are the most important inorganic elements for organic coatings and provide high thermal resistance, but these coatings are brittle [16–18]. If a coating contains metallic and ceramics particles, corrosion protection is higher and called inorganic coating, such as electroplating, cladding, flame spraying, and vapor deposition and can provide resistance to heat and radiation, biological inertness, and electrical conductivity.

In another way, protective coating may consist of organic–inorganic resins that are combined for enhancing general coating properties and is called hybrid coating. Coatings are applied at low additional cost to the least expensive structural products and are economic livability of coating as a corrosion control solution, so are most effective for usage.

Factors such as surface preparation, selection of primer coat and top coat, and mechanical bonding affect coating performance and coating adhesion [16, 17].

Anticorrosive coatings are an efficient method for protection of metal structures against corrosion because of good adhesion, mechanical properties, durability, and various mechanical, chemical, or thermal impacts [40, 41].

These coatings are effective for large and industrially used steel structures such as offshore oil rigs, sea ships, whose repair is so arduous and costs for repair or replacement are so high. Anticorrosion coatings generally do not last as long as the operating lifetime of the material to be protected and lead to failure of coating systems [40, 41].

For these defects and to extend service life of anticorrosive coatings, scientists have designed self-repair coating and developed a class of polymeric materials called self-healing with extended lifetime, no aftercare, or a much smaller amount of repairing costs [42].

Coatings that can restore their structural integrity once a microdamage has happened and prevent propagation of the crack and reduce penetration of water, oxygen, and ions down to the substrate are called self-healing and show long-term durability and have gained high priority protecting the metal substrate from corrosion. Another meaning of self-healing is the ability of a material to heal (recover/repair) damages automatically and autonomously without any external intervention [1, 2, 41, 42].

4.3 An Introduction to Self-healing Coatings

Metals and alloys are used widely in engineering structures, cars, aircrafts, chemical factories and products, pipelines, storage tanks, automobiles, airplanes, ships and offshore installations, bridge and oil rig, household equipment. They need to be effectively protected against corrosion by coatings that provide a dense barrier against corrosive species. Corrosion processes involve the transfer of charges, so anticorrosion coatings on metallic surface must be resistant to a flow of charge; in the other words, the surface must be protected against an environment-induced corrosion attack and must be provided long-term protection [39–41].

The application of coatings is the most common and cost–effective method of improving the corrosion resistance and, consequently, the durability of metallic structures. The main role of a coating in corrosion protection is to provide a dense barrier against corrosive species.

However, defects appear in the protective films during operation of the coated structures, allowing a direct access of corrosive agents to the metallic surface. The corrosion processes develop faster after disruption of the protective barrier. Therefore, active "self-healing" of defects in coatings is necessary to provide long-term protection [41, 42].

Self-healing coatings are the most common and cost–effective method for improving the corrosion resistance and durability of metallic structures, which can act in two strategies: first, restore their mechanical or barrier properties and second act actively against corrosion of the underlying substrate by reducing the corrosion activity in the defect. Partial or total recovery of the main functionality of the material can also be considered as a self-healing or self-recovery ability. The classical understanding of self-healing is based on the complete recovery of the coating functionalities due to real healing of the defect, which restores the initial validity of the coating [39–41].

Self-healing materials offer a new route toward safer, longer lasting products and components [43]. According to the ways of healing, smart materials are classified into three categories: capsule based, vascular, and intrinsic [44, 45]. These methods differ by the mechanism that is used to detain the healing agent until triggered by damage [46, 47]. The damage volume that can be healed, repeatability of healing, and recovery rate for each approach are important parameters for self-healing property [47, 48].

4.3.1 Classification of Self-healing Approaches

4.3.1.1 Materials with Intrinsic Self-healing

Intrinsic and extrinsic self-healing strategies can be used to simplify the effects of local damage in order to (partially) restore a lost property or functionality and to avoid failure of the whole system [48–51]. Intrinsic self-healing polymers use an inherent reversibility of bonding of the matrix polymer such as thermally reversible reactions, hydrogen bonding, ionomeric coupling, a dispersed meltable thermoplastic phase, or molecular diffusion [51–54]. These systems do not have a reverse healing agent but possess a latent self-healing agent that is caused by damage or by external forces such as pressure, light, or heat that enable crack healing under certain stimulation [54–56].

This ability is largely dependent on entanglement processes and chain mobility, reversible polymerizations, melting of thermoplastic phases, hydrogen bonding, or ionic interactions to initiate self-healing within the polymer matrix [55–58]. This means that the interfaces created by the damage event virtually disappear when healing takes place by the formation of chain entanglements and chemical or physical cross-links as strong as the bulk material [59–61]. The ability of these self-healing polymers depends on the characteristics of association, overall flexibility of the molecule, and environment of the system [62, 63].

The most important advantage of intrinsic over extrinsic healing relies on the possibility of fully or partially repairing the initial properties multiple times. The intrinsic healing approach can be applied to both thermoplastic and thermoset polymers and elastomers [64, 65]. Intrinsic healing has also been pursued by chemical principles in many dedicated studies leading to multiple chemistries susceptible to be used for macroscale healing [66, 67]. Based on main molecular principle, the existing chemistries can be grouped in two broad categories: (i) reversible covalent bonds and (ii) supramolecular interactions [68, 69]. Two other categories of intrinsic healing systems combining physical and chemical approaches are shape memory polymers and polymer blends as recently highlighted in a review on self-healing composites [70, 71].

Researchers are interested in developing intrinsic polymer systems combining more than one chemical healing principle because of increasing mechanical properties of self-healing materials [72, 73]. Intrinsic self-healing materials based on reversible covalent bonds are systems in which a dynamic dissociation and reassociation of stress-bearing bonds allows rapid conformational changes that ensure the healing action as reaction to damage [74–76].

Supramolecular interactions impart the capability of self-assembly or self-organization using highly directional and reversible noncovalent interactions that dictate the overall mechanical properties of a material. These interactions, reversible by definition, are ideal to create self-healing intrinsic polymers [62].

Self-healing materials based on reversible reactions include components that can be reversibly transformed from the monomeric state to the highly cross-linked polymeric state through the addition of external energy with mechanical properties similar to those of epoxy and show healing properties by adding to thermoset systems and do not require solvents or catalysts to begin their work [75].

This coating work based on retro-Diels–Alder reaction in which the energy to break Diels-Alder adduct is much lower than to break covalent bonds of the matrix, which implies that the retro-DA reaction will be the main pathway for crack propagation (Figure 4.7) [41, 75–79]. Thermoset materials show self-healing ability by incorporating a melting thermoplastic additive into the crack plane, which fills the crack and interlocks with the surrounding matrix material via mechanical bonds [80, 81].

Ionomeric copolymers are noncovalent self-healing systems that are capable of conducting electric current because of reversible hydrogen bonding or ionic bonding and form clusters acting as reversible

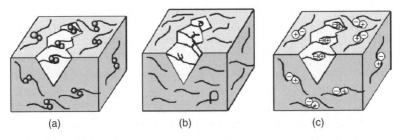

Figure 4.7 Intrinsic self-healing materials (a) reversible bonding schemes make use of the reversible nature of certain chemical reactions; (b) chain entanglement approaches use the mobility at crack faces to entangle chains that span the crack surfaces; and (c) noncovalent self-healing systems rely on reversible hydrogen bonding or ionic clustering, which manifests as reversible cross-links in polymers.

cross-link polymers [82–85]. Ionomers governed by ionic interactions within discrete regions of the polymer structure and monovalent and divalent cations lead to partially neutralized manner. In these systems, cations lead to strong Coulombic interactions between the ion pairs, yielding ionic aggregates, which act as multifunctional reversible electrostatic cross-links and called clusters affecting the viscoelastic behavior and absorption of solvents and exhibit self-healing property [86].

Clusters are stimuli-responsive because of reversible bonding and can be activated by external stimuli such as temperature or ultraviolet irradiation [87]. Sometimes, reversible linkages are not very strong, so the efficiency of self-healing polymer decreases and additional triggers are needed, such as elevated temperature or strong irradiation; in the absence of a trigger, healing does not happen.

4.3.1.2 Capsule-based Sealing Approach

One of the most promising healing systems is coating based on microcapsules filled with active healing agents. Microencapsulated agents are incorporated into the coating matrix and contain solid polymeric core–shell. When an internal stress or a physical damage causes a microcrack, microcapsules rupture, healing agent is released and propagates through the coating, dissolves the catalyst, which initiated the reaction in the fracture plane, leaching the film-forming components in the immediate binding of the damage, and fills any defects or cracks in the coating, renewing the protective barrier. Released healing agents then start to react with polymer network and fill the crack (Figure 4.8) [79].

Capsule-based self-healing materials detain the healing agent in discrete capsules until the damage occurs [88, 89]. When the capsules are cracked by damage, the healing agent is released and reacts by region of damage [90–92]. After release, leakage of the healing agent from

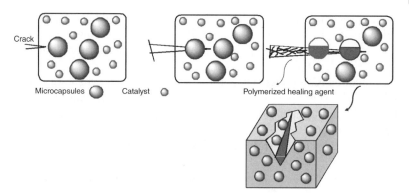

Figure 4.8 Capsule-based healing process.

the capsules into the crack may recover the damage part [93, 94]. In the capsule-based approach, spherical or cylindrical capsules may be used [95–97].

Various methods have been used for encapsulation of reactive materials such as interfacial, in situ, coacervation, meltable dispersion, or physical on the basis of the mechanism of wall formation [98–100]. These materials have been developed for some of the most commonly used synthetic polymers and elastomers using a variety of capsule-based systems, in which healing agent is stored in a capsule until damage triggers its release [89, 101].

The encapsulation strategy is also suggested for the preparation of a polymeric self-healing coating that contains organic or inorganic compounds. Epoxy-amine microcapsules containing $MgSO_4$ were prepared by interfacial polymerization in inverse emulsion and used on a steel substrate for cathodic protection conditions when magnesium sulfate was a healing agent. Magnesium sulfate can form insoluble $Mg(OH)_2$ precipitations at high pH, which can arise in paint defects due to cathodic processes [66].

Microencapsulation is based on the formation of polymeric core–shell, in which a healing agent is covered with a thin film of a shell material [46]. In chemical formation, chemical reactions of interfaces are the driving force to encapsulation such as in situ encapsulation. In this method, monomers, which are used for shell formation, is soluble in continuous, often water, phase; thus, the core material is immiscible with water and presents in the form of emulsion [46–48]. For this reason, an aqueous solution of one of the monomers is prepared and participates in the shell wall creation in the presence of a stabilizer and a substance adjusting pH to 3.5. Depending on the desired microcapsule size, the solution is agitated, and the core material is poured slowly into the system. The most common core–shell is urea–formaldehyde shell, in which nucleophilic addition starts first giving prepolymer methylol–urea product, which undergoes polycondensation reaction forming a linear polymer (Figure 4.9) [46–49].

Oligocondensate molecules deposit on the surface of the core material droplets where polycondensation continues. Finally, under heating, the highly cross-linked and water-insoluble polymer shell is produced (Figure 4.10) [66, 102].

Epoxide-based film-forming particles are incorporated in epoxy-amine coatings to protect steel surfaces under cathodic protection, which activate only under potential within a film defect and form barrier on the surface [67].

Self-healing coating based on poly(phenylenesulfide) by Sugama in which calcium alumina fillers were incorporated into coating and seal and repair microcracks generated during exposure to hydrothermal environment at 200 °C [103].

Guilbert et al. dispersed sealant-containing vesicles into fusion-bonded epoxy coating, which release an inhibitor or a sealant upon experiencing

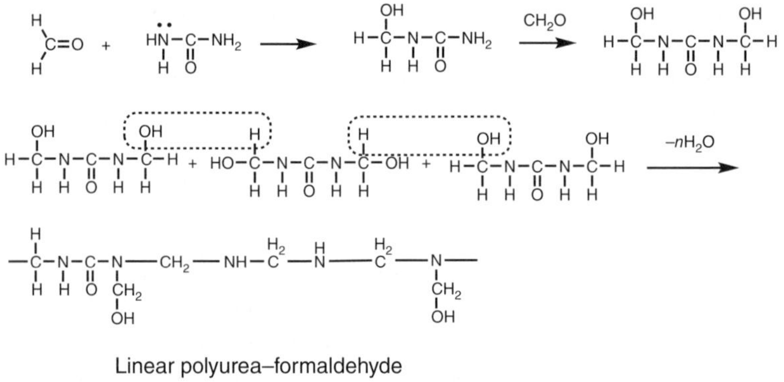

Figure 4.9 Chemical reaction of urea with formaldehyde prepolymer formation.

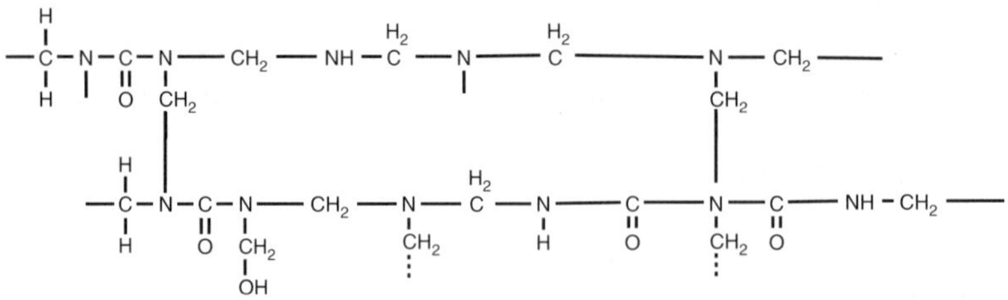

Figure 4.10 Chemical formula of the cross-linked polyurea–formaldehyde capsule shell.

a mechanical or chemical stress, and designed a more damage-tolerant fusion-bonded epoxy coating compared to traditional ones [104]. Cho *et al.* used multicomponent siloxane-based systems in epoxy and vinyl ester coatings on steel in the presence of polyurethane capsules containing healing agent [60].

Microcapsules should form strong bond with the healing agent to remain strong during usage and will rupture when the polymer is damaged. Also, microcapsules must be impervious to leakage and diffusion of the encapsulated (liquid) healing agent for considerable time to provide the capsules long shelf-life. Microencapsulation is limited by the amount of material that is incorporated inside the capsules and so leads to the situation that only relatively small damage volumes can be effectively healed. Also, volume of sealant is limited and only singular local healing is possible, which is overcome by elongated capsules or tubes, which results in sufficient increase of the healable damage volumes [101, 105, 106].

4.3.1.3 Vascular Self-healing Materials

Another self-healing coating is based on a network of interconnected hollow fibers that can carry a pressurized healing fluid, break by damage or stress, release healing agent during damage, flow into the damage part, and heal it [107].

Vascular self-healing materials hold the healing agent in capillaries or hollow channels until damage occurs [49, 108]. When the vasculature is damaged and the first release of healing agent occurs, the network may be recovered. The refilling action allows for multiple local healing events [109, 110]. Damage in vascular self-healing materials is repaired by the release of reactive healing agent that subsequently reacts with the damage part and repairs crack damage in coatings (Figure 4.11) [111, 112].

Vascular networks are similar to capsule-based healing systems, which must be chosen with regard to mechanical characterization, the triggering mechanism, and the healing performance of agent [69, 108–110].

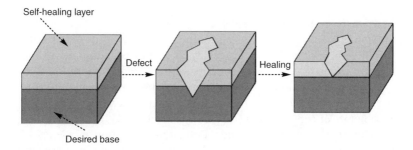

Figure 4.11 Mechanism of self-healing coating.

Depending on the fabrication method, the interactions between the matrix materials, the healing agents, and the network materials are different and play a critical role in the effective development of a self-healing system. For vascular systems, we can expect various performances and advantages by choosing various components. Any location in the network has multiple connection points, which lead to increased reliability on the network channel and the healing agent(s) and allow the easier refilling of the network after depletion [110–112].

Vascular self-healing materials have control over network shape and connectivity, but limit the choice of matrix to materials depending on surface wettability, chemical reactivity, and viscosity. These properties also affect vascular network design, especially the channel diameter, because viscosity and wettability affect the release and transport of the healing agent(s) [109–112].

After release of the healing agent, the network may be refilled by an external source for multiple local healing. Grid network of vascular systems shows a negligible effect on the mechanical properties of the material and illustrates complete recovery with epoxy resin as the healing agent [113].

Vascular self-healing networks employ hollow glass fibers (HGFs) as channels, which are filled with the healing agent and are compatible with many standard polymer matrices [114].

Multiple connection points between channels increase validity with regard to channel blockages and a larger accessible reservoir for the healing agent(s). Also refilling of the network after depletion is easier [18]. Once the system is stabilized, the scaffold is removed and hollow channel network also separated. This method has thus control over network shape and connectivity but restricts the choice of matrix to materials that can be formed around the fugitive scaffold. Surface wettability, chemical reactivity, and viscosity determine the type of healing agent for efficient filling of the network and chemical incompatibility, which provide long-term stability of the system [110–112, 115].

4.3.1.4 Active Anticorrosion Coatings

Corrosion of metals and alloys is one of the main failure processes resulting in huge economic losses, especially in aerospace, automotive, and petroleum industries. So using anticorrosion coatings is more interesting and applicable. Conducting polymers have been mainly investigated as barrier films for the protection of iron and copper, which reduce the corrosion rate of copper in neutral chloride solutions [116].

Graham *et al.* reported dependency of the protection efficiency of polyaniline in the redox state [117]. Coating based on conducting polymers show passivating effect on small defects; but in the presence of larger defects, the coating is no longer able to provide passivation and delamination occurs [118]. Also, conducting polymers show weak chemical stability and lack of UV resistance. If conducting polymers are used as particles or clusters, degradation can be slowed down, providing a long-term passivation effect. Coatings are usually applied on the metal surface and provide a barrier against corrosion propagation and its damages [68, 69].

However, corrosion processes can occur under the coating, and corrosive species can be extremely aggressive and can cause fast corrosion. Also when the barrier is damaged, corrosive agents penetrate the metal surface and the coating system cannot stop the corrosion process. Therefore, active corrosion protection should be used to deactivate corrosion species. The most effective active protection of metals is anticorrosion coatings containing chromate derivatives, which are environmentally unfriendly and cause diseases such as cancer. So the best choice for active anticorrosion is to introduce a nonchromate, environmentally friendly corrosion inhibitor directly into the coating, which releases the inhibitor just after crack formation and sufficiently slows down or terminate the corrosion propagation at already damaged corrosion defect sites [36–38]. In other words, inhibitor nanocontainers serve as a reservoir from which inhibitors may leach out during its service life [30, 119, 120].

Solubility of corrosion inhibitors must be moderate. Low solubility leads to a lack of the inhibitor at the substrate interface, thus leading to weak feedback activity, while high solubility causes inhibitor to leach out from the coating rapidly and the substrate will be protected for short time only. Recent achievements in surface science led to active anticorrosion coatings based on nanoscale containers (carriers) loaded with inhibitors or any other active compound into existing "passive" films. Anticorrosion coating is usually a complex system constituted by several layers; thus, corrosion inhibitors can be added to the different parts of the coating system [30, 119–123].

Recently, sensitive inhibitor nanocontainers to the external (mechanical damage) or internal (pH changes) corrosion trigger have been developed. When corrosion process is induced on the metal surface by changes in environment, nanocontainers release encapsulated active material (inhibitor) directly into the damaged area, preventing undesirable leakage of the inhibitor and reduction in the barrier properties of the film. In addition to inhibitor nanocontainers, mesoporous containers can be dispersed in a polymeric matrix material, which are filled with either organic or inorganic inhibitors or a combination of inhibitors and oil-loaded containers [124, 125].

4.4 Protective Coatings Containing Corrosion Inhibitors

Introducing a corrosion inhibitor to coating is so important and it affects efficiency of anticorrosion property. The most usual way is to mix corrosion inhibitor into the coating formulation. Solubility of corrosion inhibitors in corrosive environment is more important; if solubility is too high, leaching of inhibitor from the coating is rapid and protected corrosion is only for a short time, on the other side, low solubility of inhibitors leads to shortage of active healing agent and weak self-healing ability [119, 126].

If corrosion inhibitor formed chemical bonds with coating, weakening the barrier properties of coating and it may inhibitor caused coating degradation or deactivation of inhibiting activity. Therefore, direct dissolution of an inhibitor in the formulation of organic coatings is not used in practice. So, addition of corrosion inhibitor is a challenge leading to several attempts to produce self-healing hybrid films doped with organic and inorganic corrosion inhibitors [30, 119–122, 126].

There are several techniques for deposition of coatings on metals, including physical vapor deposition (PVD), chemical vapor deposition (CVD), electrochemical deposition, plasma spraying, and wet method [127, 128].

Among various coating technologies for synthesis of anticorrosion coatings, sol–gel method has the most attractive features such as low processing temperature, high chemical versatility, ease of application, strong bonding to a wide range of metallic substrates, and an environmentally friendly mode of deposition. Sol–gel coatings are glass like and can be used for preparation of organic–inorganic hybrid (OIH) films and coatings with a broad range of applications [30, 119–123].

Incorporation of inorganic or organic corrosion inhibitors into the hybrid films can significantly improve the corrosion protection properties and show the best results such as incorporation of cerium dopants into sol–gel coatings, which enhances the corrosion protection of aluminum alloys, magnesium alloys, galvanized steel, and stainless steel [30, 119–123].

4.5 An Introduction to Sol–Gel

Sol–gel process provides an appropriate route to combine inorganic and organic components as a homogeneous hybrid system in which organic and inorganic components can be chemically linked or just physically blended. An important factor that enhances compatibility in hybrid materials is the formation of covalent bond between organic polymers and inorganic components [129]. During the past 30 years, sol–gel hybrid materials prepared as organic–inorganic nanocomposites have become a fascinating new area of study in materials science. Increased efforts in this field of research have led to development of novel organic/inorganic hybrid multifunctional coating materials with interesting chemical and physical properties for promising applications [30, 119, 120].

The preparation of sol–gel materials can be controlled by varying the process parameters to achieve the desired properties of the final coatings, which allow the deposition of a thin oxide film at room temperature in contrast to common ceramic processing technologies. Sol–gel processes are mainly based on hydrolysis and condensation reactions of metal alkoxides. The resulting oxide materials present structures varying from nanoparticulate sols to continuous polymer gels depending on the rate of the involved reactions and subsequent drying and processing steps. The exact control of sol–gel reaction parameters leads to the design of new materials with interesting properties for many applications in the form of glasses, fibers, ceramic powders, and thin films. Silicon–organic based sol–gel materials are developed more in comparison with other metal–organic-based precursors [30, 119–123].

The main facet of the sol–gel chemistry and the application of sol–gel-derived materials are already described in detail in many reviews and books. Low reaction temperature and pressure allow introducing organic groups in the inorganic materials to obtain a novel class of materials composed of both inorganic and organic components. The inorganic portion imparts enhanced mechanical properties, while the organic one leads to better flexibility and functional compatibility with organic paint systems [129].

This review introduces the basic chemistry involved in sol–gel processes, the progress and development of sol–gel protective coatings on metal substrate, and various novel applications of sol–gel-derived coatings included protective coating on metal sheets, micropatterning on glass substrates, water repellant coating, colored coating on glass bottles for easy recycling, super hydrophobic and super hydrophilic coating on glass plates, and antireflective, antifog coating [30, 119–121].

4.5.1 Sol–Gel Chemistry

The main advantages of sol–gel techniques for preparation of hybrid materials, as mentioned earlier, are low temperature and pressure of processing, versatility, and flexibility. This technique has been widely employed in hybrid materials field [129].

The most important advantages of the sol–gel process for preparation of hybrid materials are as follows:

1) The temperatures required for all stages are low and close to room temperature. Thus, thermal decomposition of organic material itself and any entrapped species is minimized, and high purity and stoichiometry can be achieved.
2) As organometallic precursors for different metals are miscible, homogenous sol solutions are easily achieved.
3) Precursors such as metal alkoxides and mixed alkyl/alkoxides are easily purified (e.g., by distillation or sublimation) using common techniques, and high-purity products are obtained.
4) The chemical conditions are mild. Hydrolysis and condensation are catalyzed by acids and bases, at mild pH conditions.
5) Highly porous materials and nanocrystalline materials can be prepared by this process.
6) By chemical modification of the precursors, controlled rates of hydrolysis and condensation are achieved to control colloid particle size and pore size, porosity and pore wall chemistry of the final material.
7) Several components are introduced in a single or two steps.
8) Production of hybrid samples is possible in different physical forms.

Factors that affect the properties of final products prepared by sol–gel technique include hydrolysis ratio, acidity of the hydrolyzing agent, gelation condition, drying condition and procedures, and solvent properties [130–136].

4.5.2 General Procedures Involved in the Preparation of Sol–Gel Coatings

Sol–gel process can be described as the development of an oxide network by advanced condensation reactions of molecular precursors in a liquid medium. Basically, there are two ways to prepare sol–gel coatings: the inorganic method and the organic method. The inorganic method involves the growth of networks through the formation of a colloidal suspension (usually oxides) and gelation of the sol (colloidal suspension of very small particles, 1–100 nm) to form a network in continuous liquid phase [129–131]. Organic method consists of a reaction starts with a solution of monomeric metal or alkoxide precursors $M(OR)n$ in an alcohol or other low-molecular-weight organic solvent, where M can be Si, Ti, Zr, Al, Fe and R is typically an alkyl group (C_xH_{2x+1}).

Generally, the sol–gel formation occurs in four stages: (a) hydrolysis, (b) condensation and polymerization of monomers to form chains and particles, (c) growth of particles, (d) agglomeration of the polymer structures followed by the formation of networks that extend throughout the

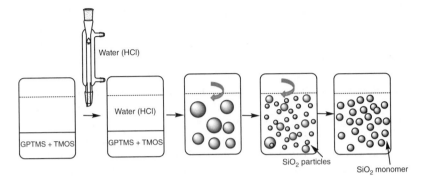

Figure 4.12 Schematic diagram for sol–gel process.

liquid medium, resulting in thickening, which forms a gel. Both hydrolysis and condensation reactions occur simultaneously once the hydrolysis reaction has been initiated. Both hydrolysis and condensation steps generate low-molecular weight by-products such as alcohol and water. Upon drying, these small molecules are driven off and the network shrinks as further condensation may occur (Figure 4.12) [130, 131].

4.5.3 Applications of Sol–Gel-Derived Coating

The sol–gel process is considered to be the most feasible method in recent years for preparing chemically homogeneous coatings and powders and will open promising applications in many areas such as optics, electronics, ionics, mechanics, energy, environment, biology, medicine, for example, as membranes and separation devices, fuel and solar cells, catalysts, sensors, and functional smart coatings. The emphasis of this work, however, is placed on recent developments and applications of the hybrid coatings, including self-cleaning, depolluting, scratch-resistant, anti-icing and anti-fogging, antimicrobial, UV protection, corrosion-resistant, and waterproofing [129].

4.5.3.1 Corrosion Protective Sol–Gel Coatings

Protection coatings from sol–gel process have shown excellent chemical stability, oxidation control, and modified corrosion resistance for metal substrates. Further, the sol–gel method is an environmentally friendly technique of surface protection and is proven to be an effective replacement for toxic pretreatments and coatings, which have commonly been used for increasing corrosion resistance of metals [129–131]. Applying protective films or coatings is a generic approach to improve corrosion resistance, and through modification of the chemical composition of the coating, such protective coatings can also permit the implant of other desired chemical and physical properties such as mechanical strength and hydrophobicity. An improvement in resistance of metals and alloys to corrosion is due to the formation of a protective layer keep apart the substrate from the surrounding oxidant atmosphere [129–132].

A variety of protective ceramic coatings, such as nitrides, carbides, silicates, or transition metal oxides, have been deposited on steel. We have illustrated that the hybrid coatings are promising surface treatment systems, which provide improved corrosion protection for aluminum alloys. Electrochemical experiments of these coatings on aluminum alloy substrates demonstrated enhanced corrosion protection by forming an efficient barrier to water and corrosive agents (e.g., chloride ion, oxygen), which effectively separated the anode from the cathode electrically [30, 119–123].

Corrosion resistance of the sols with lower silane content was better than that of the sols with high silane content. The hybrid coating also demonstrated good adhesion, which could be attributed to the formation of chemical bonding at the interface [129, 132].

4.5.3.2 Organic–Inorganic Hybrid (OIH) Sol–Gel Coatings

The combination of inorganic and organic components in a single-phase hybrid network provides a new route of tailoring the desired coating properties for diverse applications. The convenient synthetic pathway to prepare these kinds of materials in thin film coatings is sol–gel process, which leads to the formation of transparent hybrid coatings, with molecular-scale interactions between the constituents [129, 131, 132]. With this approach, it is possible to make dispersed materials through the growth of metal-oxo polymers in an appropriate solvent, and by the addition of organic components, the hybrid system can be formed. In sol–gel process, which is a wet chemical method for manufacturing thin film coatings with a thickness of 0.5–5 μm, a liquid mixture (sol) of the appropriate starting materials is prepared and applied on the surface of the substrate by different coating methods to form an amorphous network (gel) to modify the substrate surface [129, 131, 132].

The gel layer that forms acts as physical and chemical protection agents and provides a wide range of properties to the coated surface that find applications in many diverse areas such as decorative, functional and photoactive coatings, abrasion resistant and corrosion protection coatings, and antistatic films. Silica is preferred as the inorganic component in these coating applications because of its good process ability and the stability of the Si—C bond during the formation of a silica network [129].

Organic/inorganic hybrid materials prepared by different chemical routes have rapidly become an interesting new field of research in materials science. The increased activities in this area in recent years have made great contributions to the development of the new hybrid materials with outstanding properties for novel applications [129, 131, 132].

Material properties of hybrids strongly depend on the nature and strength of the interactions between the organic and inorganic components. The strength of the interactions gradually decreases from strong covalent bonding to weak van der Waals forces, and based on the type of interaction or the nature of chemical bonding between the organic and inorganic species, organic–inorganic hybrid materials can be classified into two broad families: in class I weak interactions exist between the two phases and class II shows strong chemical interactions between the two building units. Structural properties can also be used to distinguish between various hybrid materials [129–132].

Maya blue is a beautiful example of a man-made class I hybrid material, an ancient painting found in Mexico and characterized by bright blue colors that had been miraculously preserved. In this hybrid, molecules of the natural blue indigo are encapsulated within the channels of a clay mineral that withstood more than 12 centuries of harsh environment with synergic performance beyond those of a simple mixture of two components [129, 132].

Organically modified silicates (ormosils) are organic–inorganic hybrid materials, which can be applied in many branches of materials chemistry such as high reflective optical materials, hard coatings, colored glasses, abrasion resistant, antisoiling, and antifogging coatings on various substrates, corrosion protection coatings, porous materials for chromatography, and catalyst supports because of their simple process and design dependent on the molecular scale. Combination of rigidity of an inorganic phase and toughness of an organic phase makes such systems interesting also from the aspect of mechanical behavior. In hybrid systems, organic groups that are dispersed throughout the film apparently serve to increase the hydrophobicity of the coatings, repel water, and enhance the corrosion protection properties [30, 119–123, 130–133].

The final properties of sol–gel coatings depend on the starting materials and the processing conditions, for example, pH and temperature. For synthesis of anticorrosion coating by sol–gel process, organic inhibitors introduced into inorganic sol–gel networks cause crack-free hybrid coating, which slows down corrosion [137–139].

Sol–gel-derived coating can be hydrophobic or hydrophilic. Hydrophobic coating prolongs penetration of water and other electrolytes toward the metal/coating interface, acts as anticorrosion coating, and improves general passive protective properties [6, 140].

Nature of organic inhibitors and ratio of organic/inorganic components affect coating performance [137, 141].

4.6 Addition of Corrosion Inhibitors to Sol–Gel Coating Micro-/Nanoparticles

For improved mechanical properties, nanoparticles such as silica, ceria, zirconia, alumina, titania, zeolite increase thickness and lower crack sensitivity by the addition of controlled amount of the particles and enhance corrosion protection of the substrate [12, 16].

Sol–gel-derived coatings are employed for particle surface modification with desired functional groups, which cause necessary interactions between the particles and the coating matrix for good particulate composite formulations. Micro-/nanoparticles or microcapsules entrapping active species can be formed in situ in the organic–inorganic hybrid coatings or can be added to the system after coating formation via sol–gel and improve the protective properties of sol–gel coatings [137, 142].

4.6.1 Direct Addition of Inhibitor

Corrosion inhibitors or nanocapsules containing inhibitors can be mixed with the coating formulation directly, which is highly dependent on solubility of the inhibitor. High-soluble inhibitors lead to rapid leaching of the active agents from coating and limit extended healing effect and delamination processes; on the other side weak self-healing effect is due to low-solubility inhibitors and the low concentration of active agents at damaged site[137–139, 141, 142].

Organic corrosion inhibitors form hydrophobic layer on the metal surface, increase the anodic or cathodic polarization resistance of the corrosion cell, and slows down metal surface corrosion. Efficiency of anticorrosion coating is dependent on chemical composition and molecular structure of inhibitors and affinity for the metal surface. Corrosion processes cause change in local pH, so for smart and controlled release of corrosion inhibitors at damaged areas from coating matrix, capsules containing inhibitor are pH-sensitive, which lead to intelligent release of corrosion inhibitors by crack formation and improve corrosion protection of coating by inducing active protection [143]. Phosphonic acid, 2-mercaptobenzothiazole (MBT), 2-mercaptobenzimidazole (MBI), benzotriazole (BTA) are the most useful organic inhibitors [30, 119–123].

4.6.1.1 Inorganic Inhibitors

The most notable inorganic corrosion inhibitor is Cr(VI)-based inhibitors, which show good efficiency. But these compounds are toxic and cause various cancers, so use of these compounds is restricted. Cr(VI)-based inhibitors replaced by Ce, La, and Zr inhibitors improve electrochemical performance of organic–inorganic coating [28, 29, 31]. Incorporation of inorganic corrosion inhibitors into organic–inorganic coating simplifies formation of dense and defect-free coating, and also by migration of inhibitor ions to the damage site, coating's anticorrosion properties are restored [143].

Direct addition of corrosion inhibitor into organic–inorganic coating has unavoidable disadvantages, which causes indirect addition of inhibitor into coating and improves active corrosion protective coating. By direct addition, inhibitors may chemically interact with the coating matrix and lower activity of coating and barrier properties. First, it is quite difficult to control leaching of entrapped inhibitors especially when they are poorly soluble within the coating matrix. Also, encapsulation of corrosion inhibitor reduces the possible interactions with the matrix and control release of the inhibitor [30, 119–123].

Encapsulation of organic corrosion inhibitors into hybrid coatings has been achieved as a result of physical entrapment of the inhibitor within the coating material at the stage of film formation and cross-linking. Entrapped corrosion inhibitor becomes active in corrosive electrolyte and can slowly diffuse out of the host material. To ensure continuous delivery of the inhibitor to corrosion sites and long-term corrosion protection, a sustained release of the inhibitor is achieved by a reversible chemical equilibrium of the inhibitor with the coating material or through cyclodextrin-assisted molecular encapsulation [30, 119–123].

Several organic compounds, such as mercaptobenzothiazole, mercaptobenzimidazole, mercaptobenzimidazole sulfonate, and thiosalicylic acid, have been selected to evaluate the effectiveness of these two approaches. To improve corrosion protection properties of the coating when it is mechanically damaged, the incorporation of active corrosion inhibitors into the coating is needed. Organic corrosion inhibitors are promising candidates, as they appear to be easily compatible with hybrid coating material that can be loaded with inhibitors by adding the inhibitor into application solution prior to cross-linking and film formation [30, 119–123].

The supramolecular nanocontainers are useful in the formation of anticorrosion coating, which is prepared by uniform distribution of the nanocontainers into the sol–gel coating. Inclusion complex formation of corrosion inhibitor with cyclodextrin is the result of reversible hydrogen bonds formation, leading to anticorrosion property of coating, and it can be simply repaired by bringing together the fractured ends for as little as few minutes at ambient temperature. Hydrogen bonding units may react with the closest ones in their section, which is crucial to obtain sufficient self-healing property [30, 119–123].

4.7 Self-healing Coating Containing Corrosion Inhibitor Capsules

4.7.1 Self-healing Anticorrosion Coatings Based on Nano-/Microcontainers Loaded with Corrosion Inhibitors

In order to uniformly distribute active protection of the substrate, corrosion inhibitors should be introduced into a coating in the form of nano- and microcontainers loaded with active inhibiting compounds into existing conventional coatings. By crack formation or mechanical impact, coating is damaged and active inhibitor is released from nanocarriers [129]. Encapsulation of corrosion inhibitors acts as nanoreservoir and improves their applicability on self-healing anticorrosion coatings, which provide protection of metal/coating interface with controlled release of active substances of inhibitor and long-term protection. Entrapment of corrosion inhibitors is done by complexation of organic molecules by cyclodextrins [30, 119–123]. Cyclodextrins are promising candidates to serve this purpose, which are cyclic oligosaccharides consisting of several glucopyranose units and are often described as truncated cone-shaped structures with a hydrophilic exterior surface and a hydrophobic interior cavity. Cyclodextrins are effective complexing agents, which have an ability to form inclusion complexes with various organic guest molecules that fit the size of the cyclodextrin cavity [144–146].

Inclusion complexes of inhibitors and CDs are effective delivery systems of organic inhibitors in active corrosion protection applications. Amiri *et al.* designed novel self-repairing anticorrosion coating by using the encapsulated corrosion inhibitor in CDs (α, β, γ-CD) as smart corrosion inhibitor nanocontainers in different conditions (at room temperature and under sonic energy). The encapsulated particles were used as reservoirs for repairing agent and chemical initiator, which imparted anticorrosion ability to the protective system [30, 119–123].

Cyclodextrins act as host and form inclusion complexes with various organic guest molecules. Guest molecules are incorporated within the cyclodextrin cavities and immobilize them [49]. Khramov *et al.* investigated corrosion protection of AA2024 aluminum by adding organic compounds such as mercaptobenzothiazole, mercaptobenzimidazole-sulfonate, thiosalicylic acid, and mercaptobenzimidazole into hybrid sol–gel films. Nanocapsules based on β-CD/inhibitor slowed the release of the inhibitor and its continuing and slow delivery to the corrosion sites was followed by self-healing of corrosion defects [146].

Nanoparticles have small size nanoparticles and high surface area so can act as carrier and entrapped inhibitor, provides sufficient loading capacity for the inhibitor and enhances long-term corrosion protection compared to the case when the inhibitor is directly added to the sol–gel matrix. Addition of encapsulated inhibitor into hybrid sol–gel formulations improves the self-repairing properties of hybrid sol–gel coating, provides thick and low crack sensitivity coating and epress corrosion initiated at the coating defects[30, 119–123].

Zheludkevich *et al.* used Ce^{3+} ions as inhibitor capsulated by zirconia nanoparticles and synthesized hybrid sol–gel coatings containing oxide nanocontainers of cerium ions, which slow the release of inhibitor from the surface of oxide nanoparticles and achieve long-term corrosion protection [137]. Silica- and ceria-based nanocarriers obtained by this method provide additional active corrosion protection to an organosilane coating applied to galvanized steel [147].

Ion-exchange mechanism is another method to immobilize inorganic corrosion inhibitors onto the surface and the release of inhibitors is due to ion exchange with anions or cations (e.g. chlorides, sulfates, sodium ions) transported into the coating from the environment via water that penetrates through the coating. Drawback of this method is harmless ions in the surrounding environment, which causes the release of inhibitor without crack formation. Thus, the release of the inhibitor starts only by the corrosion process, which prevents undesirable leakage of the inhibitor from an intact coating during exploitation [148].

4.7.2 Preparation of Supramolecular Corrosion Inhibitor Nanocontainers for Self-protective Hybrid Nanocomposite Coatings

New generation of smart anticorrosion coatings composed of passive coatings and the smart nanocontainers, which are uniformly dispersed in the passive coatings, has been investigated. Encapsulated inhibitors in nanocontainers can quickly respond to local changes associated with corrosion processes and release [30, 119–123].

Sol–gel technology offers various ways to prepare functional hybrid coatings with unique chemically tailored properties. Addition of corrosion inhibitors into the hybrid sol can also improve the anticorrosion properties of hybrid sol–gel coating and design of functional nanostructured materials through the use of controlled hybrid organic–inorganic interface. However, slow and controllable release of inhibitors from coatings is important for anticorrosion properties, which strongly dependent upon the type of compound selected for capsulation of corrosion inhibitor [130–136].

Supramolecular nanocontainers containing corrosion inhibitors can be explained by the act of slow release of the inhibitor from the microcontainers and by the self-healing of corrosion defects. The effectiveness of the approach for long-term protection of high-strength aluminum alloys against atmospheric corrosion is discussed [129].

Supramolecular chemistry has made a dramatic progress in self-healing coating because it is based on reversible hydrogen intermolecular forces. Incorporation of encapsulated corrosion inhibitors in the hybrid sol–gel systems has been proven to enhance the corrosion protection properties, which lead to long-term corrosion protection with slow release of inhibitors. Anticorrosion nanocontainers can be formed by inclusion complex formation between corrosion inhibitors and cyclodextrins (Figure 4.13) [30, 119–123].

Molecular structure of cyclodextrins concept is shown in Figure 4.14. Organic aromatic and heterocyclic compounds are usually predominant candidates for the inclusion complexation reaction. Cyclodextrins (CD) are a family of cyclic oligosaccharides consisting of (α-1,4)-linked α-D-glucopyranose units with a slight lipophilic central cavity and a hydrophilic outer surface [144, 145] (Figure 4.13).

Truncated cone shape of cyclodextrins with secondary hydroxyl groups extending is due to the chair formation of the glucopyranose units. Cyclodextrins are widely used as "molecular cages" in the pharmaceutical, agrochemical, food, cosmetic industries, analytical sciences, separation processes (e.g., for environmental protection) and catalysis, as well as in cosmetic, textile, food, and packaging industries. Cyclodextrins have limited aqueous solubility due to the strong intermolecular hydrogen bonding in the crystal state, so substitution of the H—bond forming —OH group has improved their solubility [145, 146].

Amiri *et al.* focused on the synthesis and characterization of the corrosion inhibitor nanocontainers via inclusion complex formation of MBI and MBT with α-, β-, and γ-cyclodextrin at room temperature and under

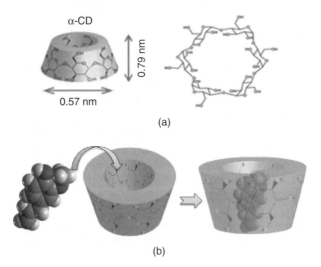

Figure 4.13 (a) Molecular structure of cyclodextrins and (b) formation of inclusion complex with guest molecule. *Source*: Amiri 2014 [119]. Reproduced with permission of Springer.

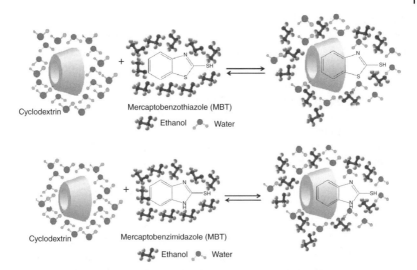

Figure 4.14 Schematic presentation of CD/mercaptobenzimidazole (MBI) and CD/mercaptobenzothiazole (MBT) inclusion complex formation.

sonic energy. The encapsulation of organic corrosion inhibitors in the form of their inclusion complexes with cyclodextrin is more bulky and expected to be more easily trapped within the cross-linked nanoporous coating material, making leaching of the inhibitor more difficult and thus prolonging the inhibition effect of the doping agent. This is the first observation in which α-, β-, and γ-cyclodextrin forms a complex with MBI and MBT at various conditions (without sonic energy at room temperature and under sonic energy) [30, 119–123].

The encapsulation of organic corrosion inhibitors into the coating host material as inclusion complexes with cyclodextrins is an effective delivery system of organic inhibitors in active corrosion protection applications. The slow release of organic corrosion inhibitor from the molecular cavity of cyclodextrins ensures the long-term delivery of corrosion inhibitor and thus the healing of a damaged coating. Additionally, the kinetics of self-healing process characterized by EIS measurement was parametrically analyzed in an equivalent circuit when the coating was exposed to salt solution [30, 119–123].

4.7.2.1 Formation of Cyclodextrin–Inhibitor Inclusion Complexes

Cyclodextrins are cyclic oligosaccharides with hydrophobic interior cavity and hydrophilic surface, which cause the formation of inclusion complexes between cyclodextrins and various guest molecules. Corrosion inhibitors can form inclusion complexes with CDs and are entrapped in CD cavities. 2-Mercaptobenzothiazole (MBT) and

2-mercaptobenzimidazole (MBI) were successfully loaded in CDs and exhibited slow and controlled release, which prolonged self-healing potential of the coating. Formation of micro-/nanocontainers immobilizes corrosion inhibitors and not only gives active corrosion properties to the organic–inorganic hybrid coatings but also can prevent particle agglomeration by stabilization of the particle surface charge. Nanocapsules of CD and corrosion inhibitor can effectively tune the release of the entrapped inhibitor depending on pH, temperature, and redox reaction inducing self-diagnosis and self-healing characteristics to the organic–inorganic hybrid sol–gel coatings [30, 119–123].

4.7.2.2 Characterization of Encapsulated Organic Corrosion Inhibitors

Different parameters are important in choosing organic corrosion inhibitors such as inhibitor activity for corrosion protection of high-strength aluminum alloys, capability to form inclusion complexes with CDs, and compatibility of the corrosion inhibitor with the coating material. To the best of our knowledge, none of the selected organic corrosion inhibitors has been reported in complexation reaction with CDs, and no data on their stability constants are available in the literature. The ability of MBT and MBI to form inclusion complexes with CDs has been assumed based on certain structural similarities with other reported aromatic and heterocyclic organic compounds [30, 119–123, 129].

In aqueous solutions, CDs are able to form inclusion complexes with many organic compounds by taking up the organic compounds or some lipophilic moiety of the molecule into the central cavity. No covalent bonds are formed or broken during complex formation, and the organic compounds in complex are in rapid equilibrium with free molecules in the solution. The driving forces for the complex formation include release of enthalpy-rich water molecules from the cavity, hydrogen bonding, van der Waals interaction, charge transfer interaction, etc. [30, 119–123, 129].

The physicochemical properties of free cyclodextrin molecules differ from those in complexes. Because of the special steric effect resulted from the occupation of CDs cavity space by MBT and MBI phenyl ring and with its thiazole ring protruding from the cavity, not the whole hydrophobic part of MBT and MBI, but its hydrocarbon chain could interact with CDs and had no choice but to enter CDs cavity along its narrow rim to form a ternary inclusion complex of stoichiometry. Other parts of MBT and MBI located outside the cavity and the phenyl ring covered at the narrow mouth of the cavity (Figure 4.14) [30, 119–123, 145, 146].

^{1}H NMR spectra of the CDs, MBI/CD, and MBT/CD inclusion complex at various conditions have been illustrated in Figures 4.15–4.17.

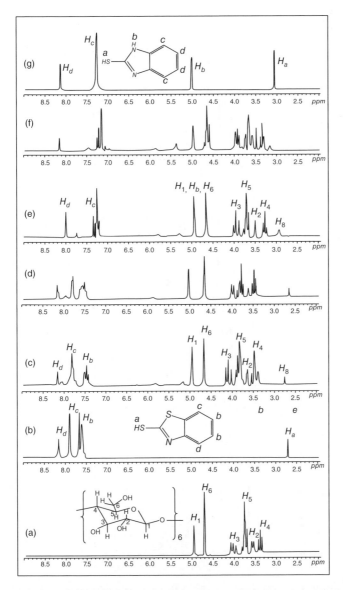

Figure 4.15 ^{1}H NMR of (a) α-CD, (b) 2-MBT, (c) α-CD/MBT at room temperature, (d) α-CD/MBT under sonic energy, (e) α-CD/MBI at room temperature, (f) α-CD/MBI under sonic energy, and (g) MBI.

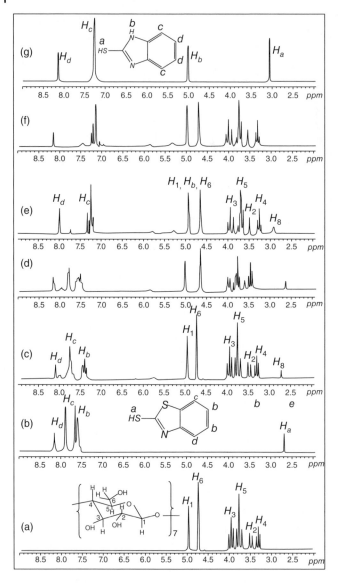

Figure 4.16 ^{1}H NMR of (a) β-CD, (b) MBT, (c) β-CD/MBT at room temperature, (d) β-CD/MBT under sonic energy, (e) β-CD/MBI at room temperature, (f) β-CD/MBI under sonic energy, and (g) MBI.

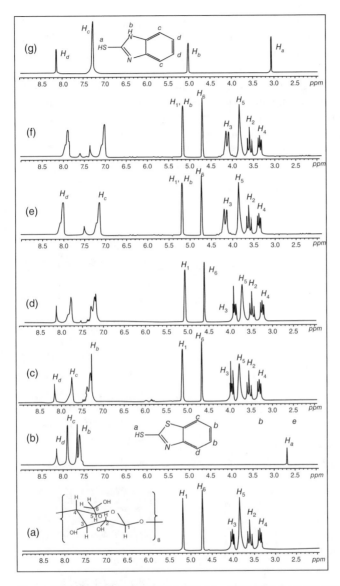

Figure 4.17 ^{1}H NMR of (a) γ-CD, (b) MBT, (c) γ-CD/MBT at room temperature, (d) γ-CD/MBT under sonic energy, (e) γ-CD/MBI at room temperature, (f) γ-CD/MBI under sonic energy, and (g) MBI.

All signals of the ¹H NMR spectra were assigned to their corresponding monomers, and it can be declared the synthesis of MBI/CD and MBT/CD inclusion complexes [30, 119–123, 145, 146]. Insertion of a guest molecule into the hydrophobic cavity of the CD will result in the chemical shift of guest and host molecules in the ¹H NMR spectra. In general, large chemical shifts will be observed at H_3 and H_5, which are located in the inner cavity of CDs due to the inclusion phenomenon. However, the chemical shifts for protons such as H_3 and H_5 (located

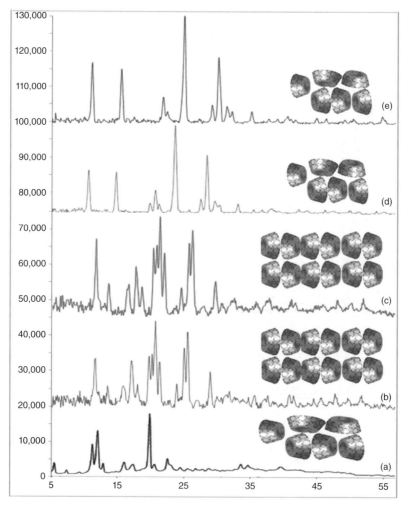

Figure 4.18 XRD spectra of (a) α-CD, (b) α-CD/MBT at room temperature, (c) α-CD/MBT under sonic energy, (d) α-CD/MBI at room temperature, and (e) α-CD/MBI under sonic energy.

inside the cavity of CD) are slightly higher compared to other protons, which are located at the exterior side of the cavity [30, 119–123].

Further evidence for the formation of the CD–inhibitor was obtained through the X-ray powder diffraction (XRD) as demonstrated in Figures 4.18–4.20. Basically, no peaks are clearly visible in the XRD of the pure CD, which is due to cage type structure of CD where each cavity is blocked off on both sides by adjacent CDs. The XRD results confirmed that the inclusion complexes formed are crystalline. Moreover, the appearance of a strong peak at $2\theta = 13$, 18, 20, and 25 in the CD/MBT inclusion complex suggests that the complexes adopt head-to-head channel-type structures in which the CD molecules are stacked along a MBT axis to form a cylinder where strong peaks at $2\theta = 8$, 12, and 23 in the CD/MBI inclusion complex suggest that the complexes adopt cage-type structures [30, 119–123, 145, 146].

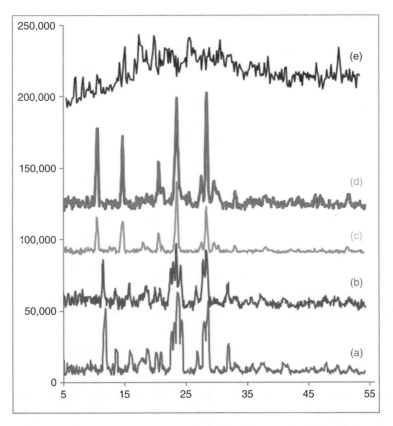

Figure 4.19 XRD spectra of (a) β-CD/MBT under sonic energy, (b) β-CD/MBT at room temperature, (c) β-CD/MBI under sonic energy, (d) β-CD/MBI at room temperature, and (e) β-CD.

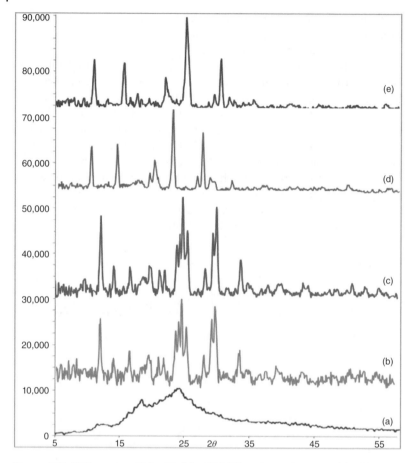

Figure 4.20 XRD spectra of (a) γ-CD, (b) γ-CD/MBT under sonic energy, (c) γ-CD/MBT at room temperature, (d) γ-CD/MBI under sonic energy, and (e) γ-CD/MBI at room temperature, and (e) γ-CD.

4.8 Morphology of the Smart Corrosion Inhibitor Nanocontainers

Corrosion inhibitor nanocontainers were synthesized via encapsulation of corrosion inhibitors in cyclodextrin core following an established method. The SEM images in Figures 4.21 and 4.22 clearly show the spherical shape of the resultant inhibitor nanocontainers. Self-healing performance of inhibitor nanocontainers is greatly dependent on the diameter of embedded capsules [16].

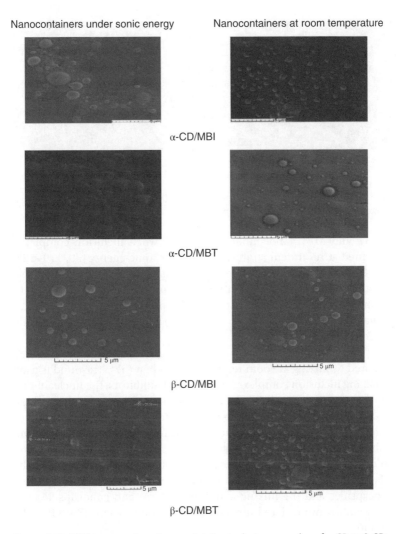

Figure 4.21 SEM images of coating containing inclusion complex of α-CD or β-CD and inhibitor nanocontainers at various conditions.

Results show that mixing, hydrolysis, and condensation play an important role in the final product of sol–gel coating. Mixing at room temperature during the synthesis of inhibitor nanocontainers causes a reduction in the initial size, so change the hydrolyzed state of the inhibitor nanocontainers and affecting their subsequent evolution into stable sols or gels by agglomeration. It is found that the average diameter

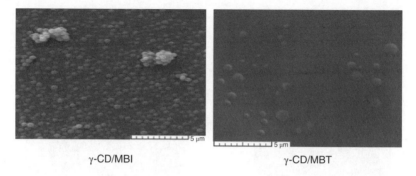

γ-CD/MBI   γ-CD/MBT

Figure 4.22 SEM images of coating containing inclusion complex of γ-CD and inhibitor nanacontainers at room temperature.

of MBT and MBI nanocontainers is lower when inclusion complexes are formed at room temperature than under sonic energy [30, 119–123, 145, 146]. It is therefore possible to conclude that mixing can be actively used and must be properly controlled in sol–gel processes in order to guarantee the final characteristics of the particulate product. Micro- and nanocapsules can self-assemble into ordered film on the substrate surface [129] (Figure 4.21).

The size of nanoparticles becomes bigger with ultrasonic energy compared to mixing at room temperature. When the ultrasound is used for making inclusion complexes of β-CD and inhibitors, the nucleant rate is further increased, so it is easy to get bigger nanoparticles due to high particle density, high impact probability, and the serious aggregation of particles. But at room temperature, nucleant rate is much slower and yields nanocontainers of smaller size. The coating thickness is 100 nm. The degree of orderliness of these microspheres self-assembled on the surface of AA2024 alloy depends on both dispersion of micro- and nanocapsules and the mixing condition. Micro- and nanocapsules can self-assemble into ordered films on the substrate surface [30, 119–123, 145, 146].

4.8.1 Microstructural Characterization

The microstructure of self-healing coatings deposited on AA2024 was studied by TEM in Figure 4.23. The estimated thickness of the sol–gel film is around 100 nm. The obtained TEM images do not show large particles or agglomerates, which were shown previously by SEM (Figure 4.23). Apparently, the nanocapsules were successfully dispersed in the sol–gel solution [30, 121].

Basics of Corrosion | 181

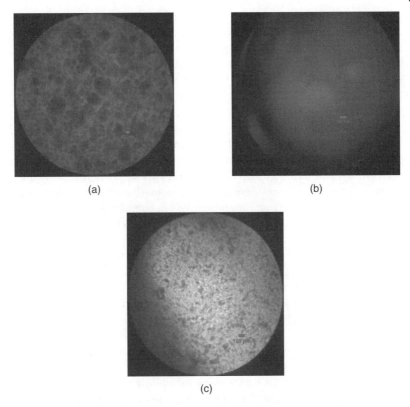

Figure 4.23 TEM images of coating containing CD/MBT nanocontainers at room temperature (a) α-CD, (b) β-CD, and (c) γ-CD.

4.8.2 Self-healing Mechanism of Corrosion Inhibitor Nanocontainers

Structural changes of atoms or molecules may cause cracks and decline in strength, so inverse reaction, that is, recombination of the broken molecules, should be one of the repairing strategies. Reversible self-healing smart materials are more applicable and are able to undergo multiple repair cycles even upon damage at the same site, which are based on covalent and noncovalent bonding. The need of large amount of energy to break and reform covalent bonds is a drawback for self-healing systems based on covalent bonding; in contrast, noncovalent bond systems need low energy for reversible self-healing materials [129].

Covalent bonds require a large amount of energy (heat or light) to break and reform covalent bonds, which indicate self-healing properties of supramolecular polymers with noncovalent bonding such as hydrogen, hydrophobic, metal coordination, and hydrogen bonds. Therefore, reversible self-healing materials requiring low energy have been designed successfully via encapsulation of organic compounds in cyclodextrin cavity by hydrogen bonding, which is reproducible, and the healed materials show reasonably conserved mechanical properties after several repetitions [30, 119–123].

In aqueous solutions, many organic and inorganic compounds are able to form inclusion complexes with cyclodextrins by taking up the organic compounds or some lipophilic moiety of the molecule, into the central cavity of cyclodextrin. Various driving forces are important in inclusion complex formation. No covalent bonds are formed or broken during complex formation, and the organic compounds in complex are in rapid equilibrium with free molecules in the solution. The driving forces for the complex formation include release of enthalpy-rich water molecules from the cavity, hydrogen bonding, van der Waals interaction, charge transfer interaction, and so on. [30, 119–123].

The physicochemical properties of free cyclodextrin molecules differ from those that are involved in inclusion complex with cyclodextrin. Because of the special steric effect resulted from the occupation of β-CDs cavity space by MBT and MBI phenyl ring and with its thiazole ring protruding from the cavity, not the whole hydrophobic part of MBT and MBI, but its hydrocarbon chain could interact with β-CD and had no choice but to enter β-CDs cavity along its narrow rim to form a ternary inclusion complex of stoichiometry [30, 119–123]. Encapsulation of corrosion inhibitor via inclusion complex formation of MBI or MBT with β-cyclodextrin is via reversible hydrogen bonds, and when the nanocontainer is broken or cut, it can be simply repaired by bringing together the fractured ends for as little as few minutes at ambient temperature, which is crucial to bring the ends of the material together as quick as possible to obtain sufficient self-healing as the hydrogen bonding units may react with the closest ones in their section [30, 121].

4.8.3 EIS Measurement of Coating Containing Inhibitor Nanocontainers

The self-healing performance of the coating based on corrosion inhibitor nanocontainers was further characterized by EIS measurement. As complementary experiments, potentiodynamic polarization curves were plotted (Figure 4.24 and 4.25) under extreme environmental conditions, consisting of an aqueous, air-exposed sodium chloride solution (5% NaCl). The potentiodynamic polarization curves in Figure 4.25 show that the corrosive behavior of the alloys with coating containing encapsulated

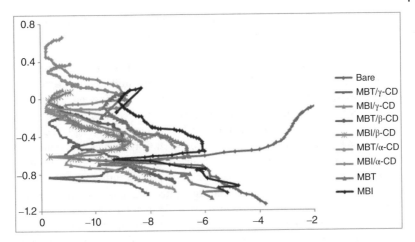

Figure 4.24 Potentiodynamic scans for coating containing encapsulated inhibitor at room temperature dilute 5% NaCl solution.

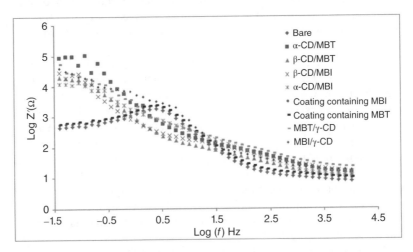

Figure 4.25 The electrochemical impedance spectra for scribed coatings with different inhibitor nanocontainers in 5% NaCl solution.

inhibitor coatings is very different from that observed for the alloy without protective coating (identified as "bare" in Figures 4.24 and 4.25) and coating containing MBT and MBI without CD [30, 119–123].

The protected alloys display very resistive behaviors with very low current density values (in the range from 10^{-12} to 5×10^{-10} A/cm^2), which are typical for nonconductive materials such as the ceramic coatings. Note that the polarization curves of the alloys with coating containing encapsulated inhibitors were carried out in a very wide range of potentials, between −1.4 and 1.0 V versus saturated calomel electrode (SCE) (2.4 V), which is higher than the commonly used range as observed for the bare alloy [30, 121].

A decrease in the corrosion current density of one order of magnitude is observed in the protected alloy, which is important for the corrosion protection of the alloy. Protective coatings containing encapsulated inhibitors prepared at room temperature are very efficient in the passivation and protection of the alloy surfaces, since they displayed very small current density values in the polarization curves. In contrast, the alloy with protective coating containing encapsulated inhibitors prepared under sonic energy presents a lower protection. Tafel extrapolations show that the SNAP coatings caused a positive shift in the corrosion potential compared with bare. The polarization curves for the coatings containing corrosion inhibitors demonstrate well-defined passivation regions that are indicative of retained inhibition activities of encapsulated compounds. The substantial reduction in corrosion current ranging from seven times for the coating with MBT to nine times for the MBI-doped coating represents a significant improvement in corrosion protection properties over the coatings without inhibitors [30, 119–123].

The impedance spectra of the SNAP coating with the MBI/α-CD and MBT/α-CD systems along with the spectra of the inhibitor-free sample are shown in Figure 4.26. Corrosion rates of uncoated samples and samples coated with inhibitor nanocontainers indicate the SNAP coating containing MBT provides better protection than MBI. A comparison of these results indicates that both coating systems exhibit similar low-frequency impedance modulus values at the initial immersion in corrosive electrolyte. However, the impedance values for the inhibitor-free coating quickly decrease, implying that coating is starting to delaminate, which is more likely to happen within the scribed area as a result of corrosion attack at the coating/substrate interface [30, 121, 129].

The coatings with the MBI/α-CD and MBT/α-CD inclusion complexes demonstrate steady impedance performance in the low-frequency region, suggesting that the coating/substrate interface is more resistant to corrosion damage, as a released inhibitor blocks corrosion sites. Additionally, the coatings with the inhibitor/cyclodextrin complexes have shown visually much less pitting corrosion within the area of the scribe compared to the inhibitor-free SNAP coatings (Figure 4.25). Overall, the results of the EIS analysis support our findings on the corrosion protection properties of inhibitor-doped SNAP coatings. The results confirm our previous conclusions of promising corrosion protection of the cyclodextrin-based coating formulations [30, 119–123].

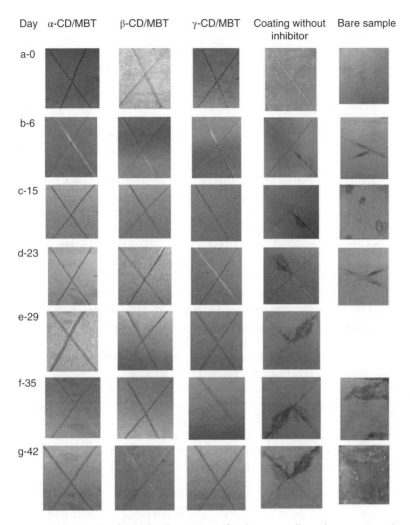

Figure 4.26 Results of 1000-h salt spray tests for aluminum alloy substrates coated with coating containing corrosion inhibitor nanocontainers.

4.8.4 Salt Spray Test for Investigation of Anticorrosive Performance of the Nanocapsules Incorporated Coating

For very long time, salt spray was used to investigate the anticorrosive performance of the nanocapsules incorporated coating. Figure 4.26 shows Al panel with coating without and with corrosion inhibitor incorporated into the coating via inclusion complex of MBT or MBI with α-, β-, and γ-CD after immersion in 10 wt.% NaCl solution for 1000 h. The influences of cavity size of cyclodextrin in self-healing property of coating and the thickness of final coating on the corrosion protection ability of the coatings were investigated in the salt spray test. Results have shown that the average diameter of the nanocapsules synthesized was highly dependent on cyclodextrin cavity size and mixing condition [30, 119–123].

The specimens coated with different nanocapsule-based coatings were scribed and exposed to salt fog for 1000 h in a monitored salt spray chamber. As shown in Figure 4.27, after exposure, bare samples (without coating) were seriously corroded along the scribes. And also, samples that were coated without nanocapsules showed corrosion after 6 days. Other coatings containing nanocapsules showed no corrosion after 1000 h, indicating an excellent corrosion protection performance in the coatings containing corrosion inhibitor nanocontainers. Coatings containing MBT/γ-CD and MBI/γ-CD could provide better corrosion protection than coating containing MBT/β-CD and MBI/β-CD and also coating containing MBT/α-CD and MBI/α-CD. After 15 h (as shown in Figure 4.27c), drastic corrosion was observed for bare samples with coating without corrosion inhibitor, indicating that protection of these coatings to steel substrates was quite poor. After 1000 h (as shown in Figure 4.27), no corrosion was observed for samples containing inhibitor nanocontainers, indicating protection of these coatings to steel substrates. The results revealed that the coatings based on corrosion inhibitor nanocontainers exhibit good corrosion protection to metal substrates if properly formulated [30, 119–123, 129].

The corrosion protection performance of the coatings based on corrosion inhibitor nanocontainers was influenced by the cavity size of cyclodextrins in the coating and the coating thickness. If the coating thickness were the same, the corrosion would be more severe when the cavity size of cyclodextrin is smaller. The results imply that higher

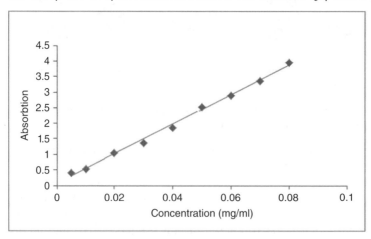

Figure 4.27 Calibration curve for nanocontainers at room temperature.

nanocapsule content in the coating and larger coating thickness are favorable for the self-healing coating based on corrosion inhibitor nanocontainers. Figure 4.27 demonstrates the process that corrosion developed on three corrosion-inhibitor-nanocontainers-based coatings and one control sample when the salt spray test proceeded. Corrosion started to appear after 6 days in bare samples, and more and more rust was observed when the exposure time of the specimens in salt fog increased, but samples with coatings based on nanocontainers did not rust after 1000 h [30, 121].

Actually, the comparison of other specimens also showed a similar trend, in which the coatings exhibited best corrosion resistance performance when the diameter of the incorporated nanocapsules was 100 nm, followed by the 80 nm specimens and the 60 nm specimens showing the worst corrosion protection. These results indicate that the coating based on corrosion inhibitor nanocontainers provided better corrosion protection toward Al substrate when the diameter of nanocapsules was bigger, and better self-healing properties were seen. This result is in good agreement with that reported in other publications [30, 121].

As discussed above, the salt-spray test reveals that the size of nanocapsules, weight fraction of nanocapsules in coating, and the coating thickness significantly influence the anticorrosion performance of the prepared coating. Generally, larger size of nanocapsules, more content of nanocapsules, and thicker coating will afford better corrosion protection. Some other factors influencing the test results include the quality of prepared nanocapsules, distribution of nanocapsules in coating, the position of specimen in the salt spray chamber and so on. Specifically, our study confirmed from the test that by optimizing the formulation, self-healing coating based on nanocapsules could provide excellent corrosion protection to Al substrates [30, 119–123].

4.8.5 Controlled Release of Inhibitors from Nanocontainers Obtained by Encapsulation of Inhibitor Corrosion in CDs

Nanocontainer solubility and solution stability are important properties to be considered when selecting the dissolution medium. In this study, 0.05 M acetate buffer pH 5 was used as dissolution medium. UV spectra of nanocontainers showed that the inhibitor absorbed appreciably at 237 and 239 nm for MBT and MBI, respectively. So these wavelengths were selected as the detection wavelength. The calibration curve was found to be linear (Figure 4.27) [30, 121].

The amount of inhibitor released was measured by UV–vis spectroscopy following the evolution of the absorption peaks of MBT (237 nm) and MBI (239 nm). UV spectra showed the controlled release of inhibitor from nanocontainers with time, which is based on reversible

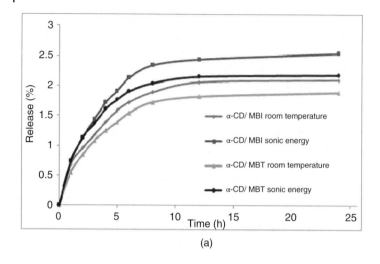

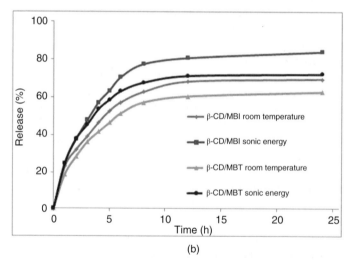

Figure 4.28 Release of coating containing nanocontainers of corrosion inhibitor and (a) α-CD and (b) β-CD at various conditions.

hydrogen bonds of inclusion complex formation of β-CD and inhibitors (Figure 4.28).

4.9 Concluding Remarks

Smart corrosion inhibitor nanocontainers were synthesized via incorporation of corrosion inhibitor into cyclodextrin cavities that are able to store an inhibitor and release it in the region of the damaged barrier coating, providing self-healing of the localized corrosion attack on the exposed metal. Reversible hydrogen bond formation via inclusion complex formation of CD and MBI or MBT leads to self-healing property of the coating and can simply repair by bringing together the fractured ends for as little as few minutes at ambient temperature.

From the salt-spray experiment and EIS measurements on anticorrosion coatings with nanocapsules containing inhibitor, it is found that the nanocapsules release healing material, which heals cracks efficiently with satisfactory anticorrosive properties during cracking. In general, this research provides appropriate nanocapsules for the self-healing composites, and the self-healing performance of the nanocapsule on the polymeric composites is satisfactory.

References

1 Davis, J.R. (1999) *Corrosion of Aluminum and Aluminum Alloys*, ASM International, Materials Park, OH.
2 Vasudevan, A.K. and Doherty, R.D. (2012) *Aluminum Alloys-Contemporary Research and Applications: Contemporary Research and Applications*, Elsevier, Amsterdam.
3 Davis, J.R. and Davis, J.R. (1993) *Aluminum and Aluminum Alloys*, ASM International.
4 Li, Y., Kalia, R.K., Nakano, A. and Vashishta, P. (2015) *Reactive Molecular Dynamics Study of Oxidation of Aggregated Aluminum Nanoparticles, MRS Proceedings*, Cambridge University Press, 1758:mrsf14-1758-vv1703-1703.
5 Doberdò, I., Löffler, N., Laszczynski, N. *et al.* (2014) Enabling aqueous binders for lithium battery cathodes−carbon coating of aluminum current collector. *J. Power Sources*, **248**, 1000–1006.
6 Marcus, P., Maurice, V. and Strehblow, H.H. (2008) Localized corrosion (pitting): A model of passivity breakdown including the role of the oxide layer nanostructure. *Corros. Sci.*, **50**, 2698–2704.
7 Perez, N. (2004) *Electrochemistry and Corrosion Science*, Springer Science & Business Media.
8 Strehblow, H.H. (1995) Mechanisms of pitting corrosion, in *Corrosion Mechanisms in Theory and Practice* (eds P. Marcus and J. Oudar), Marcel Dekker, New York, pp. 201–238.
9 Nguyen, T. and Foley, R. (1980) The chemical nature of aluminum corrosion III. The dissolution mechanism of aluminum oxide and aluminum powder in various electrolytes. *J. Electrochem. Soc.*, **127**, 2563–2566.

10 Pardo, A., Merino, M., Coy, A. et al. (2008) Corrosion behaviour of magnesium/aluminium alloys in 3.5 wt.% NaCl. *Corros. Sci.*, **50**, 823–834.
11 Paul, M.C. and John, P.B. (2011) *Cathodic Protection of Steel in Concrete and Masonry*, 2nd edn, CRC Press, Boca Raton.
12 Evans, U.R. (1937) *Metallic Corrosion and Protection*, Edward Arnold, London.
13 Wagner, C. (1974) Corrosion in aqueous solution and corrosion in gases at elevated temperature–analogies and disparities. *Werkst. Korros.*, **25**, 161–165.
14 Pourbaix, M. (1966) *Atlas of Electrochemical Equilibria in Aqueous Solutions*, Pergamon Press, Oxford.
15 Garreles, R.M. and Christ, C.L. (1965) *Solutions*, Minerals and Equilibria, Harper & Row, London, p. 172.
16 Jones, D.A. (1992) *Principles and Prevention of Corrosion*, 2nd edn, Macmillan Publishing Company, New York, p. 398.
17 Richards, J.W. (1896) *Aluminium, its History, Occurrence, Properties, Metallurgy and Application, including its alloys*, H.C. Baird & Co, Philadelphia.
18 Feliu, S., Barajas, R., Bastidas, J.M. and Morcillo, M. (1989) Mechanism of cathodic protection of zinc-rich paints by electrochemical impedance spectroscopy. *J. Coating Technol.*, **61**, 71–76.
19 Fragata, F.L., Mussoi, C.R.S., Moulin, C.F. et al. (1993) Influence of extender pigments on the performance of ethyl silicate zinc-rich paints. *J. Coating Technol.*, **65**, 103–109.
20 Armas, R.A., Gervasi, C.A., Sarli, A.D. et al. (1992) Zinc-rich paints on steel in artificial seawater by electrochemical impedance spectroscopy. *Corrosion*, **48**, 379–383.
21 Feliu, S., Barajas, R., Bastidas, J.M. et al. (1993) *Electrochemical Impedance: Analysis and Interpretation*, American Technical Publishers Ltd., Philadelphia.
22 Cheng, Y.L., Li, J.F., Zhang, J.Q. et al. (2003) Electrochemical features during pitting corrosion of ly12 aluminum alloy in different neutral solution. *Acta Metall. Sinica*(English Letter), **16**, 319–326.
23 Pourbaix, M. (1974) *Atlas of Electrochemical Equilibria in Aqueous Solutions*, NACE Cebelcor, Houston, p. 213.
24 Shimizu, K., Furneaux, R.C., Thompson, G.E. et al. (1991) On the nature of "easy paths" for the diffusion of oxygen in thermal oxide films on aluminum. *Oxid. Met.*, **35**, 427–439.
25 Davis, J.R. (1987) *Metals Handbook*, 9th edn, ASM International, Materials Park, OH, pp. 104–122 and 583–609.
26 Nisancioglu, K. (1992) Corrosion of aluminium alloys. *Proc. ICAA3*, Trondheim, NTH and SINTEF, **3**, 239.

27 Scamans, G.M., Hunter, J.A., and Holroyd, N.J.H. (1987) Corrosion of Aluminum – a New Approach, Proceedings of 8th International Light Metals Congress, Leoben-Wien, p. 699.
28 Bergmann, G., Waugh, L.T. and Pauling, L. (1957) The crystal structure of the metallic phase $Mg_{32}(Al,Zn)_{49}$. *Acta Crystallogr.*, **10**, 254.
29 Hatch, J.E. (1984) *Aluminium Properties and Physical Metallurgy*, vol. **242**, ASM, Material Park, OH.
30 Amiri, S. and Rahimi, A. (2014) Self-healing hybrid nanocomposite coatings containing encapsulated organic corrosion inhibitors nanocontainers. *J. Polym. Res.*, **22**, 624. doi: 10.1007/s10965-014-0624-z
31 Sato, N. (2010) Fundamental aspects of corrosion of metals and semiconductors, in *Electrocatalysis: Computational, Experimental, and Industrial Aspects, Surfactant Science Series*, vol. **149**, Chapter 22 (ed. C.F. Zinola), CRC Press, Taylor & Francis Book, Inc., pp. 531–588.
32 Shreir, L.L. (2013) *Corrosion: Corrosion Control*, Newnes.
33 Arman, S., Ramezanzadeh, B., Farghadani, S. *et al.* (2013) Application of the electrochemical noise to investigate the corrosion resistance of an epoxy zinc-rich coating loaded with lamellar aluminum and micaceous iron oxide particles. *Corros. Sci.*, **77**, 118–127.
34 Ali, A., Megahed, H., Elsayed, M. and El-Etre, A. (2014) Inhibition of acid corrosion of aluminum using *Salvadora persica*. *J. Basic Environ. Sci.*, **1**, 136–147.
35 Umoren, S., Eduok, U. and Solomon, M. (2014) Effect of polyvinylpyrrolidone–polyethylene glycol blends on the corrosion inhibition of aluminium in HCl solution. *Pigment Resin Technol.*, **43**, 299–313.
36 Sastri, V.S. (1998) *Corrosion Inhibitors: Principles and Applications*, Wiley, Chichester.
37 Carter, V. (2013) *Metallic Coatings for Corrosion Control: Corrosion Control Series*, Newnes.
38 Rau, S.R.S., Vengadaesvaran, B., Ramesh, K. and Arof, A.K. (2012) Studies on the adhesion and corrosion performance of an acrylic-epoxy hybrid coating. *J. Adhesion*, **88** (4–6), 282–293.
39 Wicks, J., Zeno, W., Jones, F.N. *et al.* (2007), *Organic Coatings: Science and Technology*, 3rd Edition, John Wiley & Sons.
40 Feng, W., Patel, S.H., Young, M.Y. *et al.* (2007) Smart polymeric coatings recent advances. *Adv. Polym. Technol.*, **26**, 1–13.
41 Murphy, E.B. and Wudl, F. (2010) The world of smart healable materials. *Prog. Polym. Sci.*, **35**, 223–251.

42 Wu, D.Y., Meure, S. and Solomon, D. (2008) Self-healing polymeric materials: a review of recent developments. *Prog. Polym. Sci.*, **33**, 479–522.

43 Williams, K.A., Dreyer, D.R. and Bielawski, C.W. (2008) The underlying chemistry of self-healing materials. *MRS Bull.*, **33**, 759–765.

44 White, S.R., Caruso, M.M. and Moore, J.S. (2008) Autonomic healing of polymers. *MRS Bull.*, **33**, 766–769.

45 Benita, S. (2006) *Microencapsulation: Methods and Industrial Applications*, CRC/Taylor& Francis, Boca Raton, FL.

46 Ghosh, S. (2006) *Functional Coatings: By Polymer Microencapsulation*, Wiley-VCH, Weinheim.

47 Blaiszik, B.J., Sottos, N.R. and White, S.R. (2008) Nanocapsules for self-healing materials. *Compos. Sci. Technol.*, **68**, 978–986.

48 Tittelboom, K.V. and Belie, N.D. (2013) Self-healing in cementitious materials—a review. *Materials*, **6**, 2182–2217.

49 Maiti, S., Shankar, C., Geubelle, P.H. and Kieffer, J. (2006) Continuum and molecular-level modeling of fatigue crack retardation in self-healing polymers. *J. Eng. Mater. Technol. Trans. ASME*, **128**, 595–602.

50 Wilson, G.O., Caruso, M.M., Reimer, N.T. *et al.* (2008) Evaluation of ruthenium catalysts for ring-opening metathesis polymerization-based self-healing applications. *Chem. Mater.*, **20**, 3288–3297.

51 Wilson, G.O., Porter, K.A., Weissman, H. *et al.* (2009) Stability of second generation Grubbs' alkylidenes to primary amines: formation of novel ruthenium–amine complexes. *Adv. Synth. Catal.*, **351**, 1817–1825.

52 Liu, X., Lee, J.K., Yoon, S.H. and Kessler, M.R. (2006) Characterization of diene monomers as healing agents for autonomic damage repair. *J. Appl. Polym. Sci.*, **101**, 1266–1272.

53 Kamphaus, J.M., Rule, J.D., Moore, J.S. *et al.* (2008) A new self-healing epoxy with tungsten(VI) chloride catalyst. *J. Royal Soc. Interface*, **5**, 95–103.

54 Rong, M.Z., Zhang, M.Q. and Zhang, W. (2007) A novel self-healing epoxy system withmicroencapsulated epoxy and imidazole curing agent. *Adv. Compos. Lett.*, **16**, 167–172.

55 Yin, T., Zhou, L., Rong, M.Z. and Zhang, M.Q. (2008) Self-healing woven glass fabric/epoxy composites with the healant consisting of microencapsulated epoxy and latent curing agent. *Smart Mater. Struct.*, **17**, 1–8.

56 Yin, T., Rong, M.Z., Zhang, M.Q. and Zhao, J.Q. (2009) Durability of self-healing woven glass fabric/epoxy composites. *Smart Mater. Struct.*, **18**, 4001–4011.

57 Yin, T., Rong, M.Z., Wu, J.S. et al. (2008) Healing of impact damage in woven glass fabric reinforced epoxy composites. *Composites A*, **39**, 1479–1487.
58 Yin, T., Rong, M.Z., Zhang, M.Q. and Yang, G.C. (2007) Self-healing epoxy composites: preparation and effect of the healant consisting of microencapsulated epoxy and latent curing agent. *Compos. Sci. Technol.*, **67**, 201–212.
59 Keller, M.W., White, S.R. and Sottos, N.R. (2008) Torsion fatigue response of self-healing poly (dimethylsiloxane) elastomers. *Polymer*, **49**, 3136–3145.
60 Cho, S.H., White, S.R. and Braun, P.V. (2009) Self-healing polymer coatings. *Adv. Mater.*, **21**, 645–649.
61 Beiermann, B.A., Keller, M.W. and Sottos, N.R. (2009) Self-healing flexible laminates for resealing of puncture damage. *Smart Mater. Struct.*, **18**, 5001–5011.
62 Caruso, M.M., Delafuente, D.A., Ho, V. et al. (2007) Solvent-promoted self-healing epoxy materials. *Macromolecules*, **40**, 8830–8832.
63 Caruso, M.M., Blaiszik, B.J., White, S.R. et al. (2008) Full recovery of fracture toughness using a nontoxic solvent-based self-healing system. *Adv. Funct. Mater.*, **18**, 1898–1904.
64 Zako, M. and Takano, N. (1999) Intelligent material systems using epoxy particles to repair microcracks and delamination damage in GFRP. *J. Intell. Mater. Syst. Struct.*, **10**, 836–841.
65 Suryanarayana, C., Rao, K.C. and Kumar, D. (2008) Preparation and characterization of microcapsules containing linseed oil and its use in self-healing coatings. *Prog. Org. Coat.*, **63**, 72–78.
66 Kumar, A., Stephenson, L. and Murray, J. (2006) Self-healing coatings for steel. *Prog. Org. Coat.*, **55**, 244–253.
67 Sauvant-Moynot, V., Gonzalez, S. and Kittel, J. (2008) Self-healing coatings: an alternative route for anticorrosion protection. *Prog. Org. Coat.*, **63**, 307–315.
68 Grigoriev, D.O., Kohler, K., Skorb, E. et al. (2009) Polyelectrolyte complexes as a "smart" depot for self-healing anticorrosion coatings. *Soft Matter.*, **5**, 1426–1432.
69 Huang, J., Kim, J., Agrawal, N. et al. (2009) Rapid fabrication of bio-inspired 3D microfluidic vascular networks. *Adv. Mater.*, **21**, 3567–3571.
70 Chatterjee, D. and Dasgupta, S. (2005) Visible light induced photocatalytic degradation of organic pollutants. *J. Photochem. Photobiol. C: Rev.*, **6**, 186–205.
71 Esther, M.V., Frederik, R., Daming, W. et al. (2013) Bioactivity in silica/poly(γ-glutamic acid) sol–gel hybrids through calcium chelation. *Acta Biomater.*, **9**, 7662–7671.

72 Song, X.F. and Gao, L. (2007) Fabrication of hollow hybrid microspheres coated with silica/titania via sol–gel process and enhanced photocatalytic activities. *J. Phys. Chem. C*, **23**, 8180–8187.

73 Lira-Cantu, M. and Gomez-Romero, P. (1998) Electrochemical and chemical syntheses of the hybrid organic–inorganic electroactive material formed by phosphomolybdate and polyaniline: application as cation-insertion electrodes. *Chem. Mater.*, **10**, 698–704.

74 Lim, H.S., Han, J.T., Kwak, D.H. *et al.* (2006) Photoreversibly switchable superhydrophobic surface with erasable and rewritable pattern. *J. Am. Chem. Soc.*, **128**, 14458–14459.

75 Bergman, S.D. and Wudl, F. (2008) Mendable polymers. *J. Mater. Chem.*, **18**, 41–62.

76 Murphy, E.B., Bolanos, E., Schaffner-Hamann, C. *et al.* (2008) Synthesis and characterization of a single-component thermally remendable polymer network: Staudinger and Stille revisited. *Macromolecules*, **41**, 5203–5209.

77 Park, J.S., Takahashi, K., Guo, Z. *et al.* (2008) Towards development of a self-healing composite using a mendable polymer and resistive heating. *J. Compos. Mater.*, **42**, 2869–2881.

78 Zhang, Y., Broekhuis, A.A. and Picchioni, F. (2009) Thermally self-healing polymeric materials: the next step to recycling thermoset polymers? *Macromolecules*, **42**, 1906–1912.

79 Davis, D.A., Hamilton, A., Yang, J. *et al.* (2009) Force-induced activation of covalent bonds in mechanoresponsive polymeric materials. *Nature*, **459**, 68–72.

80 Luo, X., Ou, R., Eberly, D.E. *et al.* (2009) A thermoplastic/thermoset blend exhibiting thermal mending and reversible adhesion. *ACS Appl. Mater. Interf.*, **1**, 612–620.

81 Hayes, S.A., Jones, F.R., Marshiya, K. and Zhang, W. (2007) A self-healing thermosetting composite material. *Compos. Part A-Appl. Sci. Manuf.*, **38**, 1116–1120.

82 Varley, R.J. and Van der Zwaag, S. (2008) Towards an understanding of thermally activated self-healing of an ionomer system during ballistic penetration. *Acta Mater.*, **56**, 5737–5750.

83 Kalista, S.J. (2007) Self-healing of poly(ethylene-*co*-methacrylic acid) copolymers following projectile puncture. *Mech. Adv. Mater. Struct.*, **14**, 391–397.

84 Kalista, S.J. and Ward, T.C. (2007) Thermal characteristics of the self-healing response in poly (ethylene-co-methacrylic acid) copolymers. *J. Royal Soc. Interf.*, **4**, 405–411.

85 Cordier, P., Tournilhac, F., Soulie-Ziakovic, C. and Leibler, L. (2008) Self-healing and thermoreversible rubber from supramolecular assembly. *Nature*, **451**, 977–980.

86 Eisenberg, A. and Rinaudo, M. (1990) Polyelectrolytes and ionomers. *Polym. Bull.*, **24**, 671.
87 Varley, R.J., Shen, S. and Van der Zwaag, S. (2010) The effect of cluster plasticisation on the self healing behaviour of ionomers. *Polymer*, **51**, 679–686.
88 Yuan, L., Liang, G.Z., Xie, J.Q. *et al.* (2006) Preparation and characterization of poly(ureaformaldehyde) microcapsules filled with epoxy resins. *Polymer*, **47**, 5338–5349.
89 White, S.S., Sottos, N.R., Geubelle, P.H. *et al.* (2001) Autonomous healing of polymer composites. *Nature*, **409**, 794–797.
90 Yuan, Y.C., Rong, M.Z., Zhang, M.Q. *et al.* (2008) Self-healing polymeric materials using epoxy/mercaptan as the healant. *Macromolecules*, **41**, 5197–5202.
91 Yuan, Y.C., Rong, M.Z. and Zhang, M.Q. (2008) Preparation and characterization of microencapsulated polythiol. *Polymer*, **49**, 2531–2541.
92 Yuan, Y.C., Rong, M.Z. and Zhang, M.Q. (2008) Preparation and characterization of poly (melamineformaldehyde) walled microcapsules containing epoxy. *Acta Polym. Sin.*, **5**, 472–480.
93 Liu, X., Sheng, X., Lee, J.K. and Kessler, M.R. (2009) Synthesis and characterization of melamine-ureaformaldehyde microcapsules containing ENB-based self-healing agents. *Macromol. Mater. Eng.*, **294**, 389–395.
94 Cho, S.H., Andersson, H.M., White, S.R. *et al.* (2006) Polydimethylsiloxane-based selfhealing materials. *Adv. Mater.*, **18**, 997–1000.
95 Xiao, D.S., Yuan, Y.C., Rong, M.Z. and Zhang, M.Q. (2009) Hollow polymeric microcapsules: preparation, characterization and application in holding boron trifluoride diethyl etherate. *Polymer*, **50**, 560–568.
96 Rule, J.D., Brown, E.N., Sottos, N.R. *et al.* (2005) Wax-protected catalyst microspheres for efficient self-healing materials. *Adv. Mater.*, **17**, 205–208.
97 Yeom, C.K., Oh, S.B., Rhim, J.W. and Lee, J.M. (2000) Microencapsulation of water-soluble herbicide by interfacial reaction. I. Characterization of microencapsulation. *J. Appl. Polym. Sci.*, **78**, 1645–1655.
98 Mookhoek, S.D., Blaiszik, B.J., Fischer, H.R. *et al.* (2008) Peripherally decorated binary microcapsules containing two liquids. *J. Mater. Chem.*, **18**, 5390–5394.
99 Voorn, D.J., Ming, W. and van Herk, A.M. (2006) Polymer-clay nanocomposite latex particles by inverse pickering emulsion polymerization stabilized with hydrophobic montmorillonite platelets. *Macromolecules*, **39**, 2137–2143.

100 Abate, A.R. and Weitz, D.A. (2009) High-order multiple emulsions formed in poly(dimethylsiloxane) microfluidics. *Small*, **5**, 2030–2032.
101 Brown, E.N., Sottos, N.R. and White, S.R. (2002) Fracture testing of a self-healing polymer composite. *Exp. Mech.*, **42**, 372–379.
102 Brown, E.N., Kessler, M.R., Sottos, N.R. and White, S.R. (2003) In situ poly(urea–formaldehyde) microencapsulation of dicyclopentadiene. *J. Microencapsul.*, **20**, 719–730.
103 Sugama, T. and Gawlik, K. (2003) Self-repairing poly(phenylenesulfide) coatings in hydrothermal environments at 200C. *Mater. Lett.*, **57**, 4282–4290.
104 Enos, D.G., Kehr, J.A., and Guilbert, C.R. (1999) A high-performance damage-tolerant fusion bonded epoxy coating, Proceedings of the Pipeline Protection Conference, 13.
105 Shukla, P.G. (2006) Microencapsulation of liquid active agents, in *Functional Coatings* (ed. S.K. Gosh), Wiley-VCH Verlag GmbH & Co. KGaA, Weinheim, pp. 153–186.
106 Brown, E.N., White, S.R. and Sottos, N.R. (2004) Microcapsule induced toughening in a self-healing polymer composite. *J. Mater. Sci.*, **39**, 1703–1710.
107 Toohey, K.S., Sottos, N.R., Lewis, J.A. *et al.* (2007) Self-healing materials with microvascular networks. *Nat. Mater.*, **6**, 581–585.
108 Maiti, S. and Geubelle, P.H. (2006) Cohesive modeling of fatigue crack retardation in polymers: crack closure effect. *Eng. Fract. Mech.*, **73**, 22–41.
109 Williams, H.R., Trask, R.S., Weaver, P.M. and Bond, I.P. (2008) Minimum mass vascular networks in multifunctional materials. *J. Royal Soc. Interface*, **5**, 55–65.
110 Williams, H.R., Trask, R.S., Knights, A.C. *et al.* (2008) Biomimetic reliability strategies for self-healing vascular networks in engineering materials. *J. Royal Soc. Interface*, **5**, 735–47.
111 Kim, S., Lorente, S. and Bejan, A. (2006) Vascularized materials: tree-shaped flow architectures matched canopy to canopy. *J. Appl. Phys.*, **100**, 63525.
112 Zhang, H.L., Lorente, S. and Bejan, A. (2007) Vascularization with trees that alternate with upside-down trees. *J. Appl. Phys.*, **101**, 094904.
113 Williams, B. and Boydston, A.J. (2007) Towards electrically conductive, self-healing materials. *J. Royal Soc. Interface*, **4**, 359–362.
114 Toohey, K.S., Hansen, C.J., Lewis, J.A. *et al.* (2009) Delivery of two-part self-healing chemistry via microvascular networks. *Adv. Funct. Mater.*, **19**, 1399–1405.

115 Blaiszik, B.J., Kramer, S.L.B., Olugebefola, S.C. et al. (2010) Self-healing polymers and composites, in *Annual Review of Materials Research*, vol. **40** (ed. D.R.R.M.Z.F. Clarke), pp. 179–211.
116 Fenelon, A.M. and Breslin, C.B. (2002) The electrochemical synthesis of polypyrrole at a copper electrode: corrosion protection properties. *Electrochim. Acta*, **47**, 4467–4476.
117 Brusic, V., Angelopoulos, M. and Graham, T. (1997) Use of polyaniline and its derivatives in corrosion protection of copper and silver. *J. Electrochem. Soc.*, **144**, 436–442.
118 Michalik, A. and Rohwerder, M. (2005) Conducting polymers for corrosion protection: a critical view. *Phys. Chem.*, **219**, 1547–1559.
119 Amiri, S. and Rahimi, A. (2014) Preparation of supramolecular corrosion inhibitor nanocontainers for self-protective hybrid nanocomposite coatings. *J. Polym. Res.*, **21** (556), 2014. doi: 10.1007/s10965-014-0566-5
120 Amiri, S. and Rahimi, A. (2015) Synthesis and characterization of supramolecular corrosion inhibitor nanocontainers for anticorrosion hybrid nanocomposite coatings. *J. Polym. Res.*, **22**, 66. doi: 10.1007/s10965-015-0699-1
121 Amiri, S. and Rahimi, A. (2015) Anti-corrosion hybrid nanocomposite coatings with encapsulated organic corrosion inhibitors. *J. Coating Technol. Res.*, **12**, 587–593.
122 Amiri, S. and Rahimi, A. (2016) Self-healing anticorrosion coating containing 2-mercaptobenzothiazole and 2-mercaptobenzimidazole nanocapsules. *J. Polym. Res.*, **23**, 83. doi: 10.1007/s11998-014-9652-1
123 Semsarzadeh, M.A. and Amiri, S. (2013) Silicone macroinitiator in atom transfer radical polymerization of styrene and vinyl acetate: synthesis and characterization of novel thermoreversible block copolymers, in *Progress in Silicones and Silicone-Modified Materials*, ACS Symposium Series, vol. **1154**, Chapter 7, ACS Publications, pp. 87–101.
124 Schmidt, C. (2002) Anti-corrosive coating including a filler with a hollow cellular structure. US Patent 6,383,271.
125 Lammerschop, O. and Roth, M. (2002) Silikatische Partikel. German Patent DE 10064638 A1.
126 Palanivel, V., Huang, Y. and Ooij, W. (2005) Effects of addition of corrosion inhibitors to silane films on the performance of AA2024-T3 in a 0.5 M NaCl solution. *Prog. Org. Coat.*, **53**, 153–168.
127 Matejka, L., Dukh, O. and Kolarik, J. (2000) Reinforcement of crosslinked rubbery epoxies by in-situ formed silica. *Polymer*, **41**, 1449–1459.

128 Liu, Y.L., Wu, C.S., Chiu, Y.S. and Ho, W.H. (2003) Preparation, thermal properties, and flame retardance of epoxy-silica: hybrid resins. *J. Polym. Sci.*, **41**, 2354–2367.

129 Amiri, S. and Rahimi, A. (2016) Sol–gel technology and various hybrid nanocomposite coatings applications. *Iranian Polym. J.*, **25**, 559–577.

130 Rahimi, A. (2004) Inorganic and organometallic polymers: a review. *Iranian Polym. J.*, **13**, 149–164.

131 Ershad Langroudi, A. and Rahimi, A. (2003) DGEBA TEOS ormosil coatings for metal surfaces, Proc. 6th Iran. Semi. on Polym. Sci. and Technol., (ISPST 2003) Tehran, Iran, 331, 12–15 May.

132 Zandi-zand, R., Ershad-langroudi, A. and Rahimi, A. (2005) Silica based organic–inorganic hybrid nanocomposite coatings for corrosion protection. *Prog. Org. Coating*, **53**, 286–291.

133 Zandi-zand, R., Ershad-langroudi, A. and Rahimi, A. (2005) Organic–inorganic hybrid coatings for corrosion protection of 1050 aluminum alloy. *J. Non-Cryst. Solids*, **351**, 1307–1311.

134 Farhadyar, N., Rahimi, A., and Ershad Langroudi, A. (2004) Synthesis and study of inorganic–organic hybrid produced from tetraethoxysilane and epoxy-aromatic amine, Proceedings of International union of pure and applied chemistry world Polymers congress, Paris, France, July 4–9.

135 Zandi-zand, R., Ershad-langroudi, A. and Rahimi, A. (2005) Improvement of corrosion resistance of organic –inorganic hybrid coating based on based on epoxy-silica via aromatic diol curing agent. *Iranian J. Polym. Sci. Technol.*, **6**, 350–367.

136 Zandi-zand, R., Ershad-langroudi, A. and Rahimi, A. (2005) Synthesis and characterization of nanocomposite hybrid coatings based on 3-glycidoxypropyl-trimethoxysilane and bisphenol A. *Iranian Polymer Journal*, **14**, 371–377.

137 Zheludkevich, M.L., Salvado, I.M. and Ferreira, M.G.S. (2005) Sol–gel coatings for corrosion protection of metals. *J. Mater. Chem.*, **15**, 5099–5111.

138 Figueira, R.B., Silva, C.J.R. and Pereira, E.V. (2015) Organic–inorganic hybrid sol–gel coatings for metal corrosion protection: a review of recent progress. *J. Coating Technol. Res.*, **12**, 1–35.

139 Brinker, C.J. and Scherer, G.W. (1990) *Sol–Gel Science: The Physics and Chemistry of Sol Gel Processing*, Academic Press, Inc.

140 Jensen, W.B. (1980) *The Lewis Acid–base Concepts*, John Wiley & Sons, Inc., New York, pp. 112–336.

141 Ooij, W.J., Zhu, D., Stacy, M. *et al.* (2005) Corrosion protection properties of organofunctional silanes-an overview. *Tsinghua Sci. Technol.*, **10**, 639–664.

142 Ghosh, S.K. (2009) *Self-healing Materials: Fundamentals, Design Strategies, and Applications*, Wiley-VCH.

143 Abdolah Zadeh, M., van der Zwaag, S. and Garcia, S.J. (2013) Routes to extrinsic and intrinsic self-healing corrosion protective sol–gel coatings: a review. *Self-Healing Mater.*, **1**, 1–18.

144 Semsarzadeh, M.A. and Amiri, S. (2013) Preparation and properties of polyrotaxane from α-cyclodextrin and poly (ethylene glycol) with poly (vinyl alcohol). *Bull. Mater. Sci.*, **36**, 989–996.

145 Semsarzadeh, M.A. and Amiri, S. (2012) Preparation and characterization of inclusion complexes of poly(dimethylsiloxane)s with gamma-cyclodextrin without sonic energy. *Silicon*, **4**, 151–156.

146 Khramov, A.N., Voevodin, N.N., Balbyshev, V.N. and Donley, M.S. (2004) Hybrid organo-ceramic corrosion protection coatings with encapsulated organic corrosion inhibitors. *Thin Solid Films*, **447**, 549–557.

147 Montemor, M.F. and Ferreira, M.G.S. (2007) Cerium salt activated nanoparticles as fillers for silane films: evaluation of the corrosion inhibition performance on galvanised steel substrates. *Electrochim. Acta*, **52**, 6976–6987.

148 Pippard, D.A. (1983) Corrosion inhibitors, method of producing them and protective coatings containing them, US Patent 4405493, 20 September.

149 Wefers, K. and Misra, C. (1987) Oxides and Hydroxides of Aluminium, Alcoa, Technical, Paper No. 19 Revised, Aluminum Company of America, Pittsburgh, PA, p. 92.

150 Xiao, D.S., Yuan, Y.C., Rong, M.Z. and Zhang, M.Q. (2009) Self-healing epoxy based on cationic chain polymerization. *Polymer*, **50**, 2967–2975.

151 White, D.M., Sottos, N.R., Geubelle, P.H., Moore, J.S., Sriram, R., Kessler, M.R., and Brown, E.N. (2003) Multifunctional Autonomically Healing Composite Material, US Patent 6,518,330 B2.

152 Zandi-zand, R., Ershad-langroudi, A. and Rahimi, A. (2005) Improvement of corrosion resistance of organic–inorganic hybrid coatings based on epoxy-silica via aromatic diol curing agent. *Iranian J. Polym. Sci. Technol.*, **17**, 359–367.

153 Wright, J.D. and Sommerdijk, N.A.J. (2001) *Sol–Gel Materials Chemistry and Applications*, CRC Press, OPA Overseas Publishers Association.

154 Amiri, S. and Rahimi, A. (2016) Anticorrosion behavior of cyclodextrins/inhibitor nanocapsule-based self-healing coatings. *J. Coating Technol. Res.*, **13**, 1095–1102.

5

Phytochemicals

Phytochemicals are a large group of chemicals present naturally in plants. They provide health benefits for humans that are biologically active and are used as preventive medicine for diseases such as asthma, arthritis, cancer, and so on without side effects [1]. Phytochemicals are being widely examined for their ability to provide health improvement and disease prevention. They develop safety and consumer acceptability in foods, fruits, legumes, vegetables, whole grains, fungi, herbs and spices, seeds, and nuts [2].

Phytochemicals are found in different parts of the plants, such as the roots, stems, leaves, flowers, fruits, or seeds [3], or concentrated in the outer layers of the various plant tissues such as pigment molecules [4] or in supplementary forms, but evidence is lacking toward the health benefits as dietary phytochemicals [5].

Phytochemicals act as neutralize free radicals by blocking or suppressing active carcinogens or tumor promoters from reaching the target tissue, reduce the risk of coronary heart disease by preventing the oxidation of low-density lipoprotein (LDL) cholesterol, reducing the synthesis or absorption of cholesterol, normalizing blood pressure and clotting, and improving arterial elasticity [1, 6]. They are used for the prevention and treatment of diabetes; high blood pressure; inflammation; microbial, viral, and parasitic infections; psychotic diseases; spasmodic conditions; and ulcers and act as antioxidant and antibacterial agents [5].

Phytochemicals also can act as ligands that agonize or antagonize cell surface or intracellular receptors and scavengers of reactive or toxic chemicals, and they enhance the absorption and/or stability of essential nutrients. On the basis of their biosynthetic origins, phytochemicals are classified as carotenoids, alkaloids, nitrogen-containing compounds, organ sulfur, and phenolic compounds (Figure 5.1).

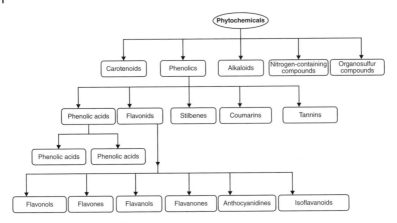

Figure 5.1 Classification of phytochemicals.

5.1 Phenolic Acids

Phenolic chemicals are the largest group of phytochemicals that are water soluble and are synthesized by plants during normal development in response to stress conditions such as infection, stabbing, UV irradiation, herbivores, and reactive oxygen species [7, 8], which form intermolecular complexes in the form of condensed proanthocyanidins, galloyl and hexahydroxydiphenoyl esters and derivatives, or phlorotannins [1].

Phenolic acids provide health benefits by several mechanisms, including the elimination of free radicals, the protection and regeneration of other dietary antioxidants (e.g., vitamin E), and the chelation of pro-oxidant metals. Hydroxybenzoic acid and hydroxycinnamic are single-ring phenolic acids. Gallic acid and ellagic acid are subgroups of hydroxybezoic acid that is found in onions, black radish, red fruits, and various kinds of tea [9]. These compounds are stable and powerful types of dietary antioxidants such as vitamins and carotenoids [10, 11] and have antimicrobial, anti-inflammatory, antiallergenic, antiatherogenic, and antithrombotic effects because of free radicals [1, 10–13].

Polyphenols historically are considered as antinutrients by nutritionists because some, e.g., tannins, have adverse effects such as decreasing the activities of digestive enzymes; energy, protein, and amino acid availabilities; and mineral uptake and have other toxic effects. These compounds have an aromatic ring with a hydroxyl substituent and a functional residue which these rings to another as phenolic acids, flavonoids, lignans, and stilbenes (Figure 5.2) and reveal their importance under oxidative stress [13].

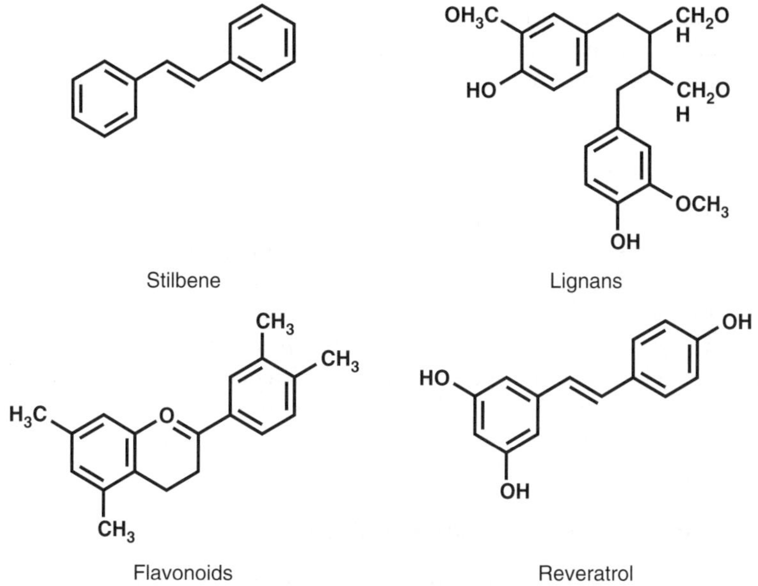

Figure 5.2 Chemical structures of some phenolic compounds.

Scavenging of various oxidizing species, superoxide anion, hydroxyl radical, and peroxyl radicals causes antioxidant activity of phenolic acids [1]. Free radicals in the environment that are produced by oxidant gases such as ozone, nitrogen dioxide, and sulfur dioxide attack and damage body cells, nucleic acids, lipids, and proteins and cause various chronic diseases such as coronary heart diseases, cancer, and atherosclerosis, and also aging. For this reason, using antioxidant compounds can help to protect from various diseases and prevent or decelerate harmful oxidation reactions [14–16] and also can be used as food additives to enhance the quality of foods. Phenolic antioxidants can be synthesized by various methods and classified as primary and secondary (deactivation of metals, inhibition of breakdown of lipid hydroperoxides, singlet oxygen quenching, and regeneration of "primary" antioxidants) antioxidants [17].

Primary antioxidants are synthesized via chain breaking as free radical scavengers such as natural and synthetic tocopherols, butylated hydroxyanisole, butylated hydroxytoluene, tertiary butyl hydroquinone, and alkyl gallate [18, 19]. Secondary antioxidants are synthesized via deactivation of metals, inhibition of breakdown of lipid hydroperoxides, singlet oxygen quenching, and regeneration of primary antioxidants such as ethylenediaminetetraacetic acid, citric acid, phosphoric acid, ascorbic acid, ascorbyl palmitate, and erythrobic acid [20]. Plants and various foods showed good antioxidants activities [21].

The important factor in using synthetic antioxidant is their concentration, which can be harmful to humans when present in high doses; high dose of *tert*-butylhydroquinone can damage DNA and lead to stomach tumors [22].

5.1.1 Polyphenolic Antioxidant Property

Natural antioxidant sources, such as phenolic compounds, are plant extracts and essential oils, which can be used as food supplements and in pharmaceutical applications (Figure 5.3). Phenolic compounds generally decrease the harmful and injurious effects of substances with high oxidative potential due to their redox properties which act as reducing agents, hydrogen donors, and singlet oxygen quenchers [23, 24].

Antioxidants show functions as terminators of free radical's chain or chelators of redox active transition metal ions that are capable of catalyzing lipid peroxidation [25, 26].

5.1.2 Phenolic Compounds and Free Radicals

Antioxidants prevent injuries and oxidative deterioration of cell structures by reducing free radicals and reactive species. Where active species area accumulated, antioxidants cannot overcome the demand due to the imbalance between the production of oxidative species and the protection system by antioxidants, cells will be damaged [27, 28].

Phenolic compounds have many phenolic hydroxyl groups that are inclined to donate a hydrogen atom or an electron to a free radical,

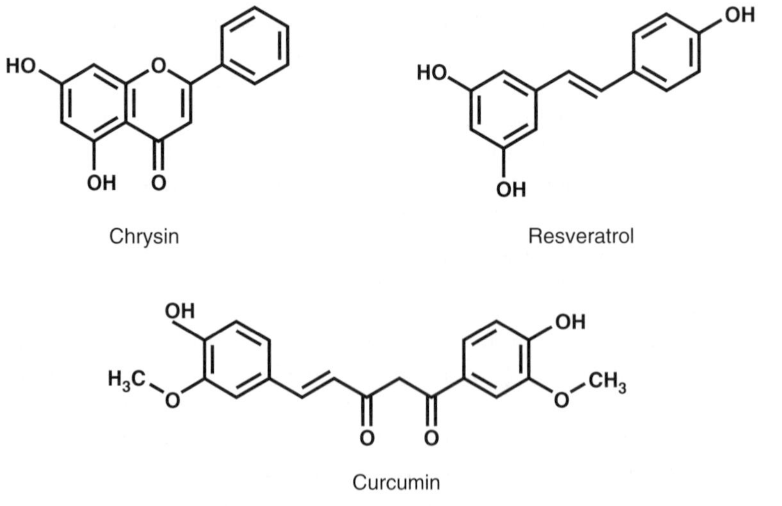

Figure 5.3 Some polyphenol structures.

although they are ideal for free radical scavenging activities and show antioxidant activity [29].

Free radicals improve their stability by obtaining electrons from fast reaction with the nearest molecules and capturing one of their electrons, and the kinetics is controlled by the diffusion rate of reactants. When molecules lose one electron, they become a free radical and may initiate a radical chain reaction, resulting in the damage and even disruption of cell membranes or DNA [30].

Metabolic reactions produce free radicals, so enzymatic antioxidant defense systems are active, involving oxydo-reductases, such as superoxide dismutase (SOD), peroxidases (POD), catalase (CAT), and glutathione peroxidase (GPx), and also nonenzymatic antioxidant systems, such as reduced glutathione (GSH), ascorbic acid, α-tocopherol, β-carotene, and polyphenols are available within the organisms [31]. Polyphenols inhibit platelet aggregation and protect low density lipoproteins (LDL) from oxidation [23]. Oxidants and antioxidants are generated in an oxidation–reduction cycle, where oxidations lead to electron loss and reduction electron gain, and this process is called "redox imbalance" [32–34].

5.1.3 Extraction of Plant Polyphenols

Owing to the high usage of phenolic products in food, pharmaceutical, nutraceutical, and cosmetic industries, extraction of phenolic compound from plants is very important. Extraction of bioactive material from plants or animal tissues is a critical step to the recovery of the phytochemicals that are commonly used as pharmaceutical food ingredients and cosmetic products by utilizing selective solvent and can be defined as separation of bioactive material from plants or animal tissues with various methods [1–4].

Pressurized liquid extraction, microwave, and ultrasound assisted extractions are some examples of different extraction types. Ethanol, methanol, acetone, water, or their mixtures can be used as solvent; nevertheless, special care is needed when foods are processed. It must not have any toxicity or should not leave any dregs after utilization with respect to health and safety concerns [35]. Selective solvent for extracts of phenolic compounds affects extraction yield, bioactivity of extracts, and antioxidant potential of compounds with different polarity [36]. Extraction is done by separation of bioactive portions of plant from the inactive component via selective solvents [37].

During extraction, solid particles are absorbed by selective solvents via osmotic forces and capillarity., phenolic compounds concentration increases by time and a concentration gradient is created, swelled and then diffuse from the plant matrix into the medium [38].

The extraction efficiency and yield depend on the type and polarity of solvents, extraction time and temperature, solvent viscosity, adjusting pH, temperature, solvent to solid ratio and sample particle size as well as on their chemical composition and physical characteristics [39]. The end step is to remove those unwanted components such as non-phenolic substances, that is, carbohydrates, protein, organic acids, and lipids. Solvents must be nontoxic with high capacity, high distribution coefficient, easily recoverable with best selective manner, environmentally safe, inexpensive, nonflammable, and nonexplosive, such as water, ethanol, methanol, acetone, and solvent mixtures with different proportions of water [39].

Among these solvents, the most efficient is ethanol for polyphenol extraction because it is safe for human health. Time and temperature have two-way effect, increasing time and temperature of extraction, increased solubility and mass transfer rate increased, and higher yield of phenolic compounds obtained, although long extraction times and high temperature increase cause oxidation of phenolics and decrease the yield of phenolics in the extracts; thus, the optimize time and temperature must be the choice [37, 39].

5.2 Flavanoids

Flavanoids are the largest group of plant phytochemicals, which include several hydroxyl groups with (C6—C3—C6) carbon structure consisting of two benzene rings linked by an oxygen containing heterocycle (Figure 5.4) [40]. Flavonoids are low–molecular-weight polyphenolic phytochemicals that are derived from secondary metabolism of plants and play an important role in various biological processes; thus flavonoids are responsible for 25% of the observed difference in mortality rates in the various countries studied [9, 41].

Flavanoids can be divided into two main group; first, anthocyanin such as color pigments (red, blue, or purple) and second, anthoxanthins that are colorless, white, or yellowish (Manach *et al.* [9]) and also subgrouped as flavones, flavonols, flavanols, and isoflavanols (Figure 5.5) [9].

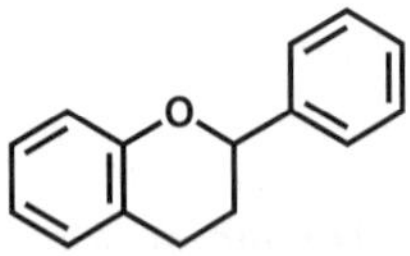

Figure 5.4 Flavanoid structure.

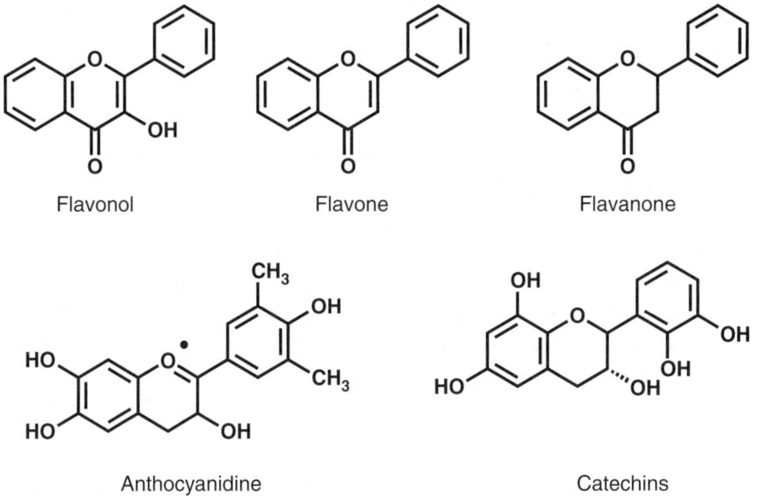

Figure 5.5 Chemical structures of flavonoids.

5.2.1 Flavones

Flavones are the phenolic groups containing one carboxyl group and can be synthesized via various methods that show several therapeutic activities like anti-inflammatory, antioestrogenic, antimicrobial, antiallergic, antioxidant, antitumor, and cytotoxic activities and are effective for diseases related to oxidative stress, such as atherosclerosis, diabetes, cancer, and Alzheimer's disease [42–44]. Flavones mainly consist of apigenin and luteolin. Olives, parsley, and celery are the most important edible sources. The most common type of flavonoids is flavonols and includes quercetin, kaempferol, and myricetin which are present in onion, kale, apple, red wine, and tea. Flavanols can be found in both the monomer form as catechins and the polymer form as proanthocyanidins [9].

5.2.2 Catechins

Catechin is a type of natural phenol and antioxidant which is synthesized via plant secondary metabolite and belongs to the group of flavan-3-ols (or simply flavanols) and is abound in grapes, berries, cocoa, green tea, red wine, broad beans, black grapes, apricots, and strawberries. Catechins include catechin, epicatechin, epigallocatechin, epicatechin gallate, and epigallocatechingallate, which are the main flavanols in fruits. Epicatechin concentrations are high in apples, blackberries, broad beans, cherries, blackgrapes, pears, raspberries, and chocolate. Black and green tea have high concentrations of gallic acid esters such as epigallocatechin, epicatechin gallate, and epigallocatechin gallate [45, 46].

5.2.3 Isoflavonoids

Isoflavonoids, including genistein and daidzein, are another subclass of the phenolic phytonutrients, which prevent and treat cancer and osteoporosis. Soybeans which are full of isoflavones can bind to estrogen receptor class of compounds and act as phytoestrogens. They inhibit the growth of most hormone-dependent and -independent cancer cells in vitro, including colonic cancer cells. Because of structural similarity of isoflavonoids to estrogens and estrogenic activity, they are called as estrogenic flavonoids [47, 48].

5.2.4 Tannins

Tannins are another group of phenolic polymers and are divided into hydrolizable tannins and condensed tannins. Hydrolizable tannins include a central core of polyhydric alcohol (such as glucose and hydroxyl groups). They are partially or completely esterified by gallic acid (gallotannins) or ellagic acid (ellagitannins). Condensed tannins are oligomeric derivatives of flovonols such as catechin and epicatechin and can be found in fruits, vegetables, cocoa, red wine, and legume family [9].

5.2.5 Anthocyanidins

Anthocyanidins, which are called anthocyanins, are water-soluble flavonoids that are aglycones of anthocyanins; are resistant to light, pH, and oxidation conditions; and have antioxidative and antimutagenic properties in vivo. Haslam1 cited the naturally occurring anthocyanidins as pelargonidin, cyanidin, paeonidin, delphinidin, petunidin, and malvidin. These compounds are among the principal pigments in fruits and flowers. The color of these pigments is influenced by pH and metal ion complexes. Anthocyanidins can form complex with other flavonoids that cause stabilization of these compounds [49]. Anthocyanins can form complexes with flavones and metal ions, namely, iron and magnesium in flowers. As the fruit nears its ripening stage, the anthocyanin content usually increases [50, 51].

5.2.6 Lignans and Stilbenes

Lignans are a type of phenylpropanoid, which consist of secoisolariciresinol and low quantities of matairesinol. The main sources of lignans are flax seeds, cereals, grains, some vegetables, and fruits. The edible plants comprise only less quantities of stilbenes [49].

5.2.7 Alkaloids and Other Nitrogen-containing Metabolites

Alkaloids are the other group of phytochemicals that contain mostly basic nitrogen atoms and show neutral and even weakly acidic properties. Alkaloids contain carbon, hydrogen, and nitrogen and also oxygen and sulfur and, more rarely, other elements such as chlorine, bromine, and phosphorus. They are found in cruciferous vegetables, are activators of liver detoxification enzymes, and provide protection against carcinogenesis, mutagenesis, and other forms of toxicity of electrophiles and reactive forms of oxygen [52].

5.3 Phytochemical Importance

Phytochemicals have a vital role in medicinal uses. Phytonutrients appear as potentially effective products and cure health problems. Various groups of phytochemicals can be used for prevention of chronic degenerative diseases and used widely in food industry and dietary products to enhance their nutritional value and taste. Other aspects of determining the role of phytochemicals in functional foods include consumer attitudes, any competitive advantage for manufacturers producing functional foods, and identification of those areas of research needed to produce foods with the desired health effects [53–55].

5.3.1 Oxidative Stress and Phenolic Compounds in Foods

Phenolic compounds act as neutralize free radicals by blocking or suppressing active carcinogens. They prevent oxidation of lipids (lipoxidation), glucose (glycation), and proteins (carbonylation), which are important for diabetes mellitus II and inflammation, and could correspond to important activators of inflammation in various tissues [56, 57].

Using phytochemicals in food industry prevent glucose oxidation by reactive species and formation of carbonyl toxic products, but it is possible that they attack the amino group of the amino acids that are expressed in some organs, such as kidneys, blood vessels, and adipose tissue, forming glycation end products that are very reactive; hence they increase the inflammatory process [58–60].

Polyphenols scavenge free radicals produced by oxidative stress and also slow down the production of reactive oxygen and nitrogen species and control the effects of oxidants, thus increasing the cell lifespan and reducing the risk of severe metabolic and cardiovascular diseases. Polyphenols such as resveratrol from red wine [61, 62], epigallocathechin-3-gallate from green tea [63, 64], curcumin from

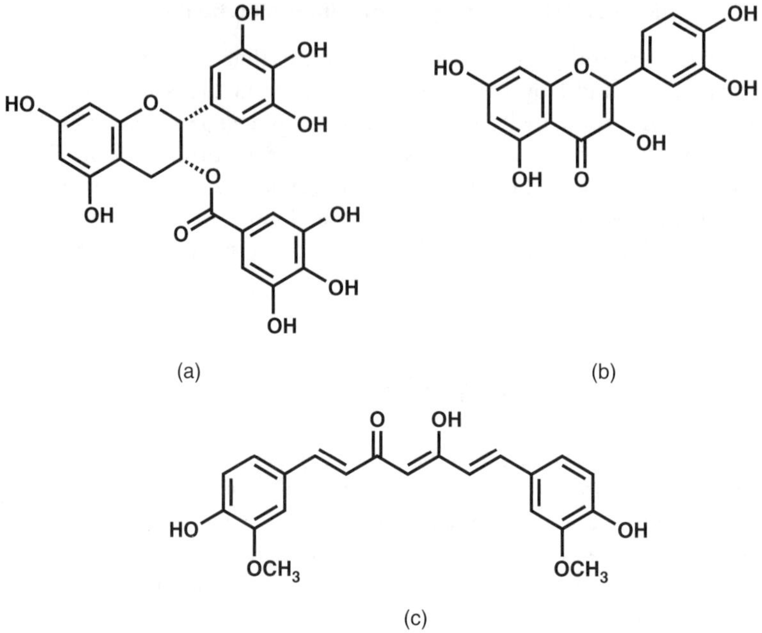

Figure 5.6 Schematic images of (a) epigallocatechin-3-gallate, (b) quercetin, and (c) curcumin.

turmeric [65], and quercetin (Figure 5.6) from different sources have all been studied as potential therapeutic agents [66].

5.3.2 Important Parameters for Phenolic Efficiency

The important parameters are the chemical structure, cultivation methods such as organic or conventional, concentration of polyphenols, stress inducers, light intensities, extreme temperatures, ionizing radiations, wounding and meteorological conditions determinant effectiveness, and antioxidant action of bioactive compounds [40, 67, 68].

Storage time and temperature control increase antioxidant activity and phenolic concentration in fruits, e.g., apples, where cold storage increases the polyphenol content [69]. Thus, the control of biotic and abiotic stresses, such as temperature, light intensity, herbivory, and microbial attack, may be used to modulate plant defense mechanisms, triggering many complex biochemical processes [70]. Due to high quantity of phenolic compounds present in fruits and vegetables, they are not eaten as raw, and special treatments such as heating (jelly) or freezing are done, which tend to change the content of polyphenols.

Heating may induce a decrease in flavonoid content vegetables such as spinach, potato, beans, carrots, tomatoes, eggplant, and broccoli [71, 72].

On the other hand, freezing affects the content of extracted polyphenols, preserving antioxidant properties of fruits, which induces an irreversible physical destruction of cell walls and protoplasts, damage to fruit texture, and therefore a loss of quality [73, 74].

Bioavailability and solubility of phenolic compounds determine the absorption efficiency of these compounds in the body. Isoflavones and gallic acid show better absorption efficiency than other polyphenols, followed by flavonones (catechins and quercetin glycosides) and then proanthocyanidins, anthocyanins, and catechins gallate, but with different kinetics [75]. However, the consumption of foods with antioxidant potential has received much attention and probably should have more benefits than harm. Otherwise, studies regarding the use of these substances have already adopted another direction, to prevent the intake of polyphenols and increase the life time of polyphenols [75].

Unfortunately, applicable properties of polyphenolic compounds are also responsible for a lack in long-term stability; thus they are very sensitive to light and heat. In addition, polyphenols show poor water solubility and biocompatibility. Furthermore, polyphenol molecules possess a very bitter taste, which must be masked before their incorporation in foodstuffs or oral medicines. Therefore, a finished protecting is required for improved polyphenol consumption or the administration, to mask its taste, increase water solubility and bioavailability, and extend it precisely toward a physiological target. Among the existing stabilization methods, encapsulation is an interesting means, which is a promising approach for prevalence of these problems and stabilizingd them [75, 76].

5.4 Encapsulation

Encapsulation in general involves packaging of active ingredients and protection of its contents from the environment that can allow small molecules to pass in and out of the membrane, such as dyes, proteins, vitamins, flavors, and living cells inside a capsule or the pores of a gel or another porous media. Encapsulation improves solubility and biocompatibility of phytochemicals, the retention time of the nutrient in food, and allows controlled release at specific times. Encapsulation involves the coating or entrapment of a pure material or a mixture into another material and protects the active ingredient from its surrounding environment for a specific period of time or triggering events including rupture, dissolution, and changing pH and temperature [77–80].

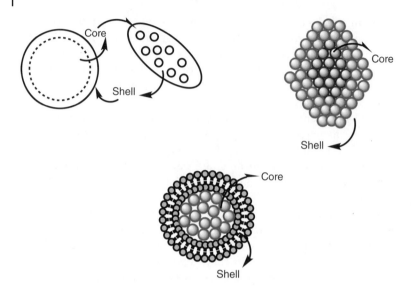

Figure 5.7 Various types of core–shell structures.

Core materials called actives fill the internal phase or payload and can be liquid, gas, or solid, while the coating material can also be defined as capsule, wall material, membrane, carrier, or shell. Microencapsulation can be used in preventing chemical reactions, masking undesirable flavor, enhancing the stability of formulations, improving solubility, controlled release for drugs, vitamins, agrochemicals, live cells, active enzymes, flavors, pharmaceuticals, and inks.

The coating can be a wall or a membrane of spherical or irregular shape and multiwall structure (Figure 5.7). Size, sensitive, permeability, mechanical strength, and compatibility of capsules can be easily controlled [77].

5.4.1 Polyphenol Encapsulation

Bioactivity and antioxidant properties of polyphenols depend on the stability of polyphenols and chemical, physical, and biological conditions [81, 82]. Encapsulation technology on natural compounds has gained great interest and is necessary to apply delivery or carrier systems in order to maximize the potential medical benefits of antioxidants. Encapsulation protects plant extracts from devastating environment effects such as undesirable effects of light, chemical reactions, moisture and oxygen, oxidation and dehydration which reduce the shelf life of compounds and to improve processing step, delivery of bioactive and stability of phytochemicals (Figure 5.8) [77].

Figure 5.8 Most common morphologies for encapsulation of polyphenols (a) monomer capsule and (b) aggregate.

(a)   (b)

Figure 5.8 shows the most common morphologies for encapsulation of polyphenols, which are called core–shell like and matrix [77, 83].

These coating materials may include polymers of natural or synthetic origin or lipids instability. There are many techniques for encapsulation of polyphenols for food, pharmaceutical, cosmetic industries, personal care, agricultural products, veterinary medicine, industrial chemicals, biotechnology, biomedical, and sensor industries. Encapsulation may protect a fragile or unstable compound from its surrounding environment or side effects, modify the density of a liquid trap a compound, and control the release of the encapsulated compound. Encapsulation is done based on physical phenomena or formation of capsule shell based on chemical reactions, or both physical and chemical phenomena can be combined together to prepare microcapsules (Table 5.1) [82, 83].

5.4.2 Physical Methods

In physical methods, encapsulation involves using mechanical phenomena to produce microcapsules loaded with designed core material, such as centrifugal force, extrusion, co-extrusion, and formation of sprays. This type of encapsulation process predates chemical encapsulation process as it was developed in the 1930s [83, 84].

Table 5.1 Encapsulation methods applied to polyphenols.

Physical methods	Physicochemical methods	Chemical methods
Spray-drying	Spray-cooling	Interfacial polycondensation
Fluid bed coating	Hot melt coating	In situ polymerization
Extrusion–spheronization	Ionic gelation	Interfacial polymerization
Centrifugal extrusion	Solvent evaporation–extraction	Interfacial cross-linking
Supercritical fluids	Coacervation	

5.4.2.1 Spray-drying

The spray drying technique involves the emulsification or dispersion of core material in a concentrated solution of shell material and from a dispersion of active compound in a solution of coating agent. The core materials can be water-immiscible liquids, and the shell material is usually a soluble polymer. The emulsion droplets are then passed into the heated chamber of a spray drier, where they rapidly dehydrate and then produce dry capsules. The capsules are harvested in the bottom of the spray-drying chamber [77, 78].

A liquid formulation contains a coating agent and the active ingredient droplets via either a nozzle or spinning wheel using compressed gas to atomize the liquid feed or a rotary atomizer using a wheel rotating at high speed. Then, the solvent is evaporated by contacting with hot air or gas and the particles collected after sedimentation [85–87].

Surface area is very large in this method so spray drying is a very rapid drying method in a single step, so is widely used in large-scale for production of microspheres or microcapsules encapsulated substances, such as antibiotics, medical ingredients, additives, vitamins, and polyphenols, among others. This method is relatively a low process cost and equipment, flexible, wide choice of carrier solids, good retention of volatiles, good stability of the finished product and large-scale production in continuous mode, so leads to the production of high quality and stable particles and making this technique the most used in the food industry [85–89].

Particle size, morphology, and the density of the powder can be controlled, which are vital for several delivery systems. The most common coating materials are proteins (sodium caseinate and gelatin), hydrocolloids (gum arabic), hydrolyzed starch (starch, lactose, and maltodextrin), sodium caseinate–maltodextrin conjugates, or mix of maltodextrin and gum arabic [83, 84].

5.4.2.2 Fluid Bed Coating

Fluid bed coating is used for drugs coating to provide sustained or controlled release, taste masking, improve stability or aesthetics with uniform quality and morphology, and is applicable for hot-melt coatings such as hydrogenated vegetable oil, stearines, fatty acids, emulsifiers, waxes, solvent-based coatings such as starches, gums, and maltodextrins [90, 91].

In fluid-bed coating, a natural or synthetic polymer coating (shell, wall, or membrane materials) is applied onto the powder particles in a batch processor or a continuous setup, which can be employed to improve functional ingredients and additives life time such as processing aids (leavening agents and enzymes), preservatives (acids and salts), vitamins and minerals, ferrous sulfate, ferrous fumarate, sodium ascorbate, potassium chloride, flavors (natural and synthetic), spices, and a variety of vitamin/mineral premixes [7]. The coating material may be an aqueous solution of cellulose or starch derivatives, proteins, and gums; thus, the powder particles are suspended by an air stream at a specific temperature and sprayed with an atomized, coating material [92, 93].

These capsules release their contents at controlled rates via various methods after mechanically rupturing the coating by the act of chewing (physical release), coating melt (thermal release), dissolve coating in solvents, or change pH [94–96]. Solvent evaporation and solids concentration increase the viscosity of the droplet and inhibit spreading and coalescence upon contact with the core material also distance that the droplets travel through the fluidization air before impinging on the core [93–95]. In a common fluidized bed, the particles are accelerated from the product container, past the nozzle which sprays the coating liquid counter currently onto the randomly fluidized particles [94–96].

5.4.2.3 Extrusion–Spheronization Technique

The extrusion–spheronization involves forcing a core material in a molten carbohydrate mass through a series of dies into a bath of dehydrating liquid and completely surrounds the core material with wall material. This process involves moistening the powder mixture and cylinder-shaped agglomerate formed through extrusion, breaking and rounding the extrudate to round pellets, intensify contacting the liquids and an encapsulating matrix formed which entrap the core material. In extruder method, core material is completely surrounded by the coating so stability against oxidation improved and increases the shelf life of core material. Efficiency of this method is significantly dependent on the ratio of liquid to solid material, the size of the extruder holes, and the drying process. The drawback of this method is the cost, low flavor loading, low solubility in cold water, and high process temperature [97, 98].

5.4.2.4 Centrifugal Extrusion

Centrifugal extrusion is used for encapsulation of vitamin A acetate, which forms capsules of a larger size, from 250µm up to a few millimeters in diameter. In this method, the core and the shell materials should be immiscible with one another and pumped separately to nozzles. On the basis of the composition and properties of the coating material, droplet walls are solidified by cooling or gelling bath. Centrifugal extrusion is used for encapsulation of food components such as flavorings, seasonings, vitamins, and many others [83, 84, 99].

5.4.2.5 Supercritical Fluids

Supercritical fluids show intermediate manner between a liquid and a gas, which is easily changed with variations in pressure and temperature, and hence can be used for encapsulation process. The most widely used supercritical fluid is carbon dioxide, which has low critical temperature, is nontoxic, nonflammable, inexpensive, and highly suitable for processing heat-sensitive materials. Supercritical fluid can be used as antisolvent for very weakly soluble solute in the supercritical fluid and injected into a pressurized container containing the solution. The precipitation cell is partially filled with the solution, and the supercritical antisolvent dissolves in the phase and decreases the density and the solvation power of the organic solvent. After the solvent evaporates in the supercritical phase, the solution is saturated, the solute is precipitated, and the particles are collected in the reactor after depressurization [83, 84, 88, 99].

5.4.3 Physicochemical Methods

5.4.3.1 Spray-cooling/Chilling

Spray-cooling or chilling is a fat-based system in which a molten lipid carrier for encapsulation of organic and inorganic salts is used in the pharmaceutical and food field [17]. This method improves heat stability, controlled release, and stabilizes liquid hydrophilic ingredient into free owing powders. The core and wall mixtures are atomized into the cooled or chilled air and heat exchange between the molten material and cold air solidifies the wall around the core. The core can be hydrophilic or hydrophobic, in liquid or solid forms, and the wall can be vegetable oils, fats, waxes, or their blends with oils and materials with a melting point of 12–45 °C. In this method, due to lipid coating, capsules are insoluble in water and need special handling and storage conditions [83, 85, 88, 89, 99].

5.4.3.2 Encapsulation by Emulsions

When the core component is water soluble, encapsulation is done by emulsion technology in which the active compound is dissolved or dispersed in a second immiscible liquid or melted wall under continuous phase heated. Emulsion-based techniques are essential for fabrication of functional microparticles and nanoparticles used for encapsulation and controlled delivery [82–84].

Emulsion hydrogels are used for drug delivery and biomedical applications including cell therapy, anticancer treatment, and tissue engineering, with hydrogels with imbedded oil droplets. Encapsulation occurs after the removal of organic solvent based on evaporation, extraction or heating, and/or under vacuum where the solvent must be miscible with water in all proportions [83, 84]. The polymer matrix is first dissolved in an organic solvent which is low miscible with water and active compound is dissolved in the polymer solution. This process is not used

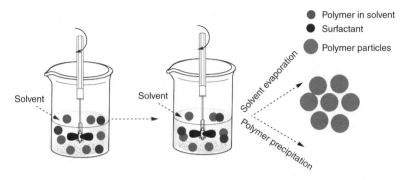

Figure 5.9 Encapsulation by emulsion/extraction and emulsion/evaporation method.

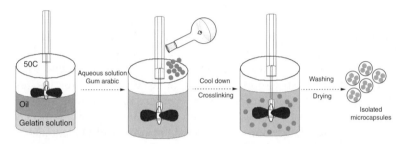

Figure 5.10 Schematic image of encapsulation by coacervation method.

for low-molecular-weight and volatile compounds or molecules with higher affinity for the continuous phase. By trapping active ingredient via solvent and then removal of the solvent, nanoparticles are precipitated, washed, collected by filtration or centrifugation, and then dry and porous microspheres formed (Figure 5.9) [83, 84, 100].

5.4.3.3 Coacervation

Coacervation is a modified emulsification technology, which is based on the interaction among different polymers with opposite charges and used to entrap hydrophobic substances. This method forms insoluble complexes around a hydrophobic core creating a barrier, and encapsulation is done. Phase separation is due to the electrostatic interactions through that process [83] (Figure 5.10).

5.4.3.4 Ultrasonication

Ultrasonics can accelerate or form functional materials in various fields including food, imaging, energy production, therapeutic/diagnostic medicine, and making emulsions. Ultrasound has broad active frequency region that generates microsize bubbles, and with pressure difference explosion of bubbles makes high levels of turbulence so intensity and number of cavitation evens, particle size, surface roughness, and structure are controlled [83, 101].

By increased sonication time, input energy is increased and smaller emulsion droplets are formed, which is related to increasing disruption with the droplet deformation. Ultrasound energy can be used for encapsulation of various compounds and increased emulsion size by decreased powder size and so protect and stabilize the internalized material from environmental deterioration and enables pharmaceuticals and/or nutrients to be delivered with enhanced efficacy in biological systems [83, 101]. Microencapsulation by ultrasonic atomizer was proposed in which lysozymes encapsulated for over 50 days with no loss of functional integrity and releasing were found to fit zero-order kinetics [83].

On the other hand, the most important problem for preparing functional food products is their low water solubility such as oils and vitamins and dispersion of an oil within water enables effective loading of oil soluble nutrients into aqueous food media so must be encapsulated as core–shell structures and reach target by respond to changes in pH, temperature, or other external stimuli [102]. Entrapped drug material are released by external stimuli under specific conditions and prolonging core efficacy and caused bio-compatible and biodegradable properties and have been extensively studied for pharmaceutical and food applications.

5.4.4 Chemical Methods

5.4.4.1 Micelles

Polymeric micelles attracted much attention because of small size and high structural stability for encapsulation of various compounds such as drugs, oil, and food ingredients. Micelles are usually within a range 5–100 nm and consist of mostly a hydrophilic head-group and a hydrophobic tail and hence named amphiphilic compound that is formed based on electrostatic interactions. In an aqueous media, amphiphilic micelles are able to self-organize into supramolecular arrangements possessing and can be used in various medical applications and drug delivery systems [102, 103]. Driving force for micellization are formation of van der Waals and hydrogen bonding and removal of hydrophobic core from the aqueous environment between hydrophobic blocks in the core of formed micelles. After formation of micelles, to improve the stability and reduce the polydispersity of system, a crosslinking agent may be used in the core or in the hydrophilic part. Polymer micelles are used as drug delivery systems such as anticancer agents, neurodegenerative

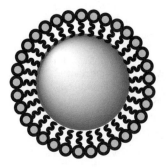 

Figure 5.11 Cross section of polymeric micelles.

drugs, antifungal agents, and nanocarriers for drug and gene delivery and ocular drug delivery (Figure 5.11) [83, 104].

Phytochemicals, such as tea polyphenol epigallocatechin-3-gallate or curcumin, have been encapsulated on micelles to enhance their solubility, absorption, bioavailability, therapeutic activity, and delivery[64]. Similarly, curcumin is absorbed in the intestine but is rapidly metabolized in both intestinal epithelial tissue and liver. Thus, very little curcumin reaches the circulation intact for subsequent bioaccumulation at the target organ [105, 106]. Addition of piperine along with curcumin treatment increases the circulating curcumin plasma levels by inhibiting the glucuronidation of curcumin and thereby inhibiting its elimination from the circulation [64, 105, 106].

In recent years, curcumin is the most widely polyphenols, which is used alone or in complex with other compounds in medical applications, but its poor solubility in aqueous environments limits its use. Micelles of sodium dodecyl sulfate (SDS) are used to improve solubilization and protection of polyphenols, such as α-tocopherol and curcumin, against oxidation phenomena. Micelle-based encapsulated curcumin is easily absorbed into the bloodstream and increases plasma and tissue concentrations of the active agent and causes controlled release of the compound directed at the target tissue [107, 108].

5.4.4.2 Liposomes

Liposomes are artificial spherical vesicles formed by one or more phospholipid bilayers separated by water compartments and are extensively used in target delivery such as drug, gen, and polyphenol compounds (Figure 5.12) which protect core agents from degradation, oxidation, deliver it at target site [109]. Polyphenols are located in the interior part of liposomes as core and used for target, protect, release, immobilize

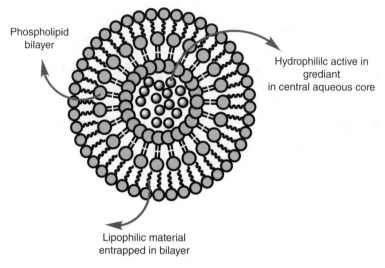

Figure 5.12 Phospholipid bilayers for protection of drug, gen, and polyphenol compounds.

hydrophilic, lipophilic, or amphiphilic substances. Liposomes are unstable in biological fluids and show low payload of the active ingredient, low control over ingredient release, low reproducibility, and may be stored for short time. Interaction between functional head groups, lipid chains, and linker groups in membrane components helps controlled release of encapsulated material under specific conditions from liposome. Polyphenols can be encapsulated via liposomes by reverse-phase evaporation method, which improves their bioactivity, bioavailability solubility, and stability increase [109, 110].

One of the most important applications of liposome-based encapsulation is curcumin encapsulation, and its bioavailability and absorption increased can be used for anti-HIV, anticancer, antioxidant, and anti-inflammatory applications [109].

5.4.4.3 In Situ Polymerization

In situ polymerization is a chemical microencapsulation process that takes place in oil-in-water emulsions and usually is used for synthesis of nanocomposites, smooth, spherical, and reservoir-type microcapsules with transparent polymeric pressure-sensitive microcapsule walls (Figure 5.13) and is similar to interfacial polymerization. In situ-based encapsulation consists of emulsifier, monomer mostly vinylic and acrylic compounds such as styrene or methyl methacrylate in an aqueous phase added with an appropriate surfactant, and water-insoluble polymer gives

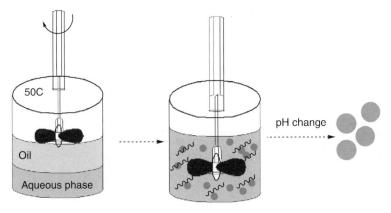

Figure 5.13 In situ polymerization is a chemical microencapsulation process.

microspheres resulted. Coating materials mostly are aminoplast resins and hardening of the wall material by dispersion of encapsulated oil droplets (Figure 5.13) [111].

5.4.4.4 Interfacial Polymerization
Interfacial polymerization or polycondensation is based on a chemical reaction involved in interfacial polycondensation polymerization in which a membrane made of polymers is created around emulsion droplets [112]. In this method, reaction takes place at the interface between the continuous and dispersed phases and usually used to produce synthetic fibers such as polyester, nylon, and polyurethane. Active compound and water soluble monomer (A) is dissolved in distilled water, aqueous phase emulsified in an organic external phase and formed oil-in-water (W/O) emulsion, and then an organo-soluble monomer B is added to the organic phase, and the interfacial polycondensation reaction done (Figure 5.14) [113].

5.4.4.5 Freeze-drying
Freeze-drying is based on the dehydration of heat-sensitive sample and stabilized thermosensitive and unstable molecules. This method is based on freezing the material, then reducing the surrounding pressure and adding sufficient heat; thus water in the material is frozen and purified directly from the solid phase to the gas phase [20], leading to encapsulation. Freeze drying is a powerful barrier against oxidation of core material and protects the particles against undesired physical and chemical changes and hence freeze-dried polyphenols are stable against the oxidation phenomenon, show long time storage, and antioxidant activity remains identical [114].

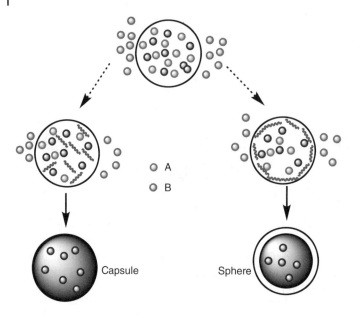

Figure 5.14 Principle of the microcapsulation by interfacial polymerization: (a) the oligomer is soluble in the droplet and (b) the oligomer is insoluble in the droplet.

5.5 Encapsulation of Phenolic Compounds Via Cyclodextrin

Cyclodextrins are cyclic, water-soluble, nonreducing, and nontoxic macrocycle oligosaccharides consisting of glucopyranose subunits bound through α-(1,4) links, which result from starch degradation by *Bacillus macerans*. CDs have a hydrophobic inner cavity and hydrophilic outer surface and so are able to form inclusion complexes with a wide range of organic and inorganic compounds. CDs have a truncated cone structure where the polar sugar hydroxyl groups are oriented to the cone exterior, and, consequently, the external faces of CDs are hydrophilic. Hydrophobic guests form inclusion complex with CDs and their solubility increases, which is attractive for drug delivery systems. CDs are torus shaped, and their molecular dimension allows total or partial inclusion complex of a range of aroma compounds.

CDs can form inclusion complex with guest molecules in aqueous solution or blend solid cyclodextrin with guest molecule or knead the flavor substances with cyclodextrin-water paste. Complex formation may change the physiochemical properties of the guest molecule such as stabilization, improvement of solubility, protection against degradation, and controlled guest released from complex. Polyphenols are potent antioxidants, antimicrobial and anticaries agents, but some of them are low soluble in water and have bitter taste, so their inclusion complex with cyclodextrins has been investigated [115–117].

Encapsulation immobilizes polyphenol molecules in CD cavity and improves thermo-oxidative stability, chemical stability, bioavailability, photo-stability, and water soluble compound. CDs have an empty cavity that can entrap guest molecules with appropriate geometry and polarity and encapsulation done. The core material can be protected against light, heat, and oxidation. The most important reason for encapsulation of polyphenols is controlled released from the molecular-inclusion complex on contact with water, exchange with another guest, temperature, moisture or pH, pressure or shear; and the addition of surfactants [118].

5.5.1 Cyclodextrin Inclusion Complexes Formation

The driving force for inclusion complex formation are noncovalent bonds such as van der waals and hydrogen bonding which are dependent on the guest and the CD. Polar–apolar interactions between the CD cavity and water molecules in the cavity and between water and an apolar guest compound are energetically unfavorable, but apolar–apolar interactions between the guest and cavity and polar–polar interactions between bulk water molecules and the released cavity–water molecules are energetically favorable [115–117].

Geometric and size compatibility between the CD cavity and guest species determine efficiency of inclusion complex formation. If guest molecule was larger than the cavity dimensions of CDs, certain groups or side chains actually penetrate the host cavity. The aqueous medium also plays an important role in the formation of inclusion complex and influence the inclusion equilibrium process [119].

5.5.2 Polyphenol Encapsulation in Cyclodextrins

Polyphenols are photosensitive, pH sensitive, and have reduced aqueous solubility, which lead to poor reabsorption and poor bioavailability in biological system, and their applications in oral absorption are restricted and are unstable in gastrointestinal tract [120–122]. By encapsulating polyphenols, their unwanted taste and odor are trapped and their shelf life and usage areas are extende. In addition to cyclodextrin, modified cyclodextrins such as 2-hydroxypropyl-α-cyclodextrin, 2-hydroxypropyl-β-cyclodextrin, 2-hydroxypropyl-γ-cyclodextrin, glycosyl-β-cyclodextrin, methylated derivatives of β-cyclodextrin, sulfobutylated β-cyclodextrin, and so on can be used for encapsulation of

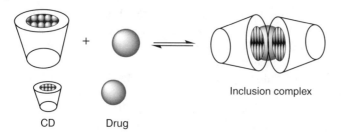

Figure 5.15 Inclusion complex formation between CD and the guest molecule.

polyphenols, which improves their solubility, stabilized light sensitive substances, increase shelf life, and catalytic activity of cyclodextrin with guest molecules which is shown in Figure 5.15. Cyclodextrin act as host molecules and polyphenols act as guest molecules, and by complexation, the polyphenol is solubilized in water [112, 119, 123].

Inclusion complex formation of curcumin with various hydroxypropyl derivatives of cyclodextrin such as 2-hydroxypropyl-α-cyclodextrin, 2-hydroxypropyl-β-cyclodextrin, and 2-hydroxypropyl-γ-cyclodextrin has been previously reported [112, 123]. In another study, inclusion complex formation of curcumin with of γ-cyclodextrin (γ-CD), methyl β-cyclodextrin (MβCD), hydroxypropyl β-cyclodextrin (HPβCD), and β-cyclodextrin (β-CD) via kneading method was reported by Yadav *et al.* in which solubility of curcumin is increased by formation of inclusion complex [124].

5.5.3 Solubilization and Stabilization of Polyphenols

Inclusion complex formation with cyclodextrins enhances the solubility of polyphenols in the aqueous solutions, bioavailability, stability or release controlled and masking undesirable tastes which allowing for rapid and quantitative delivery of sparingly soluble compound for intravenous and intramuscular dosing, decreasing irritation at the site of administration of parenterally administered, stabilization of unstable compounds in an aqueous environment [125].

5.6 Why Encapsulation by Cyclodextrin?

The most important property of CDs is formation of inclusion complexes with a wide range of bioactive molecules, organic, and inorganic compounds. Hydrophobic bioactive molecules are interesting for food, cosmetic, and pharmaceutical applications but are not available for release and delivery to target area. Polyphenols show antioxidant activity, antibacterial, and fungicidal actions, but their poor solubility in water and their bioavailability restricted their application [124–126].

To solve these problems, cyclodextrins are used, which show a relatively unique capacity of improving solubility of bioactive polyphenolic agents in aqueous systems; protect them from elevated temperatures, pH values, light, oxidation, or the moisture induced degradations phenomena which serve to increase their bioavailability and is a promising pathway for the development of pharmaceutical products friendlier to the user. This method is useful for polyphenolics compounds usage for food industry, cosmetic, and health product [77, 83, 126].

5.6.1 Cyclodextrins and Flavonoids

Flavonoids such as catechin, epicatechin, and epigallocatechin-gallate are the most important groups of polyphenols, and various studies were done in the area of formation of inclusion complex with cyclodextrins. Encapsulation by CDs improves the water solubility and stability, antidiabetic and antiobesity properties, and the antioxidant efficiency of the flavonoids and protect them against oxidation, light-induced reactions, sublimation, decomposition and loss of volatility or reduction of undesired taste, odors, microbial contaminations, hygroscopicity, and other undesired components [120, 127]. Another flavone that encapsulates with CDs is rutin, which is a plant pigment found in apple, buckwheat, Japanese pagoda tree, and *Eucalyptus macrorhyncha* and can strengthen blood vessels and can be used for varicose veins, internal bleeding, hemorrhoids, and to prevent side effect of cancer treatment and strokes due to broken veins or arteries (hemorrhagic strokes). Poor solubility of rutin restricts its application and so its inclusion complex with α-CD, β-CD, HPβ-CD, and DMβ-CD was described by different authors in order to improve rutin solubility, and, consequently, pharmacological activity [121, 122, 128, 129].

The formation of IC improves rutin solubility, antioxidant activity, bioavailability, and antioxidant protection of cells against oxidative stress, and hence increases its oral availability. Formation of inclusion complex increases rutin stability and protects rutin from thermal and UV degradation and also increases the phenolic antioxidant capacity [121, 122].

Another subgroup of flavonoids that is widely used is chrysin, which is most well known for being a testosterone boosting plant compound, sensitizes the testicles to produce more testosterone, and inhibits the conversion of testosterone to estrogen and are present naturally in plants such as the passionflower, silver linden, and some geranium species and in honey and bee propolis (glue). Chrysin is antioxidant, anti-inflammatory, and antihypertensive and used for body building, treating anxiety, inflammation, gout, HIV/AIDS, erectile dysfunction (ED), baldness, and preventing cancer. Although Chrysin appears to be less soluble in water and is poorly absorbed, it is readily metabolized, resulting in insufficient levels in the blood and testes to exert these beneficial effects [130, 131].

Quercetin is an antioxidant, anti-inflammatory, and antihistamine flavonoid present in plant pigments, which is used for heart and blood vessels such as hardening of the arteries, heart disease, circulation problems, diabetes, cataracts, hay fever, peptic ulcer, inflammation, asthma, gout, viral infections, preventing cancer, and for treating chronic infections of the prostate. Quercetin has low solubility in water, and formation of inclusion complex with CDs improve solubility and photostability of quercetin and protect quercetin against rapid oxidation, by free radicals [132]. In another study, Carlotti et al. reported that photodegradation ratio of the quercetin is reduced by complexation with MβCD, and the amount of light capable of reach this flavonol was lower, since it had to cross the CD molecule [129, 132]. Kaempferol is an antioxidant, anti-inflammatory, and antiangiogenic polyphenol, which is found in fruits and vegetables and reduces the risk of chronic diseases, especially cancer, and increases the body's antioxidant defense against free radicals and inhibits cancer cell growth and angiognesis and induces cancer cell apoptosis. The important parameter is its low solubility in water and low cell viability, so its inclusion complex with CDs and their derivatives was investigated [43, 133].

Taxifolin is a subgroup compound of quercetin that can act as a potential chemo preventive agent and inhibit the ovarian cancer cell growth and used as antiproliferative of human breast cancer cells and enhance the efficacy of conventional antibiotics. Yang et al. reported IC formation of CDs with taxifolin for enhanced water solubility and thermal stability, and CDs act as dilators of bloods vessels, enhance microcirculation and cerebral blood flow, and prevent platelet aggregation [134].

In another study, encapsulation of genistein with β-CD, γ-CD, HPβ-CD, and RMβ-CD and its solubility, anti-inflammatory effects, platelet aggregation inhibition, biological effects, and capacity to cross biological membranes were investigated [135].

Nonflavonoids had great importance on the pharmaceutical and cosmetic industry due to their biological properties that are limited because of degradation by oxidation and environmental factors such as light, temperature, and pH [106].

Thus, their inclusion complex with CDs is investigated. The most important nonflavonoids polyphenol is ferulic acid, which is extracted from rice bran oil and is one of the most powerful natural antioxidants used for the prevention of UV-light-induced skin tumor, but it has low stability under thermal and physical stress and may prevent the propagation of the free radical chain. Its inclusion complex with CDs increased the ferulic acid resistance to the degradation by UVB and also decreases it rate release [136]. They showed that IC formation prevents the formation of the less-active cis-isomer of ferulic acid and its degradation by UV light and protects the skin against UV damages and remains at the skin surface [137–140].

The IC obtained had lower stability (K 166.3 M-1) and the same stoichiometry of the other ICs described earlier. Nevertheless, the solubility and protection against decomposition caused by irradiation with UV light was enhanced by the complexation of the ferulic acid with this CD [135, 136]. Caffeic acid is found in plants and foods such as apples, artichoke, berries, pears, coffee, and wine. Caffeic acid shows antibacterial, biological, and antioxidant activities but is sensible to oxidation and has low solubility [132]. Caffeic acid is used in supplements for boosting athletic performance, HIV/AIDS, weight loss, and cancer.

5.7 Concluding Remarks

Phytochemical compounds, particularly polyphenols, are the most active compounds that are naturally present in plants and vegetable oils and show a unique combination of chemical, biological, and physiological, antioxidant, antibacterial, anti-inflammation, and antiangiogenic activities. Polyphenol application, stability, and solubility limit their application, which is caused by poor bioavailability and water solubility. Therefore, various methods are used for encapsulation of these compounds and potentialization of their activity of which spray-drying is the most common technique used to encapsulate polyphenols to protect them from oxidation, temperatures, air, metal ions. Encapsulation also may change composition, particle size and density, release mechanism and kinetics, degradation mechanism and kinetics, and finally the physical form. Formation of inclusion complex is another encapsulation method that increases solubility of core material and increases the shelf life, antioxidant, and antibacterial activities, and anti-inflammation activity and bioavailability also mask an unwanted flavor, smell, and controlled release of core material.

References

1 Mathai, K. (2000) Nutrition in the adult years, in *Krause's Food, Nutrition, and Diet Therapy*, 10th edn, vol. 271 (eds L.K. Mahan and S. Escott-Stump), W.B. Saunders Co., Philadelphia, pp. 274–275.

2 Bellik, Y., Boukraâ, L., Alzahrani, H.A. et al. (2012) Molecular mechanism underlying anti-inflammatory and anti-allergic activities of phytochemicals: an update. *Molecules*, **18**, 322–353.

3 Costa, M.A., Zia, Z.Q., Davin, L.B. and Lewis, N.G. (1999) Chapter four: toward engineering the metabolic pathways of cancer-preventing lignans in cereal grains and other crops. *Recent Adv. Phytochem.*, **33**, 67–87. New York: Kluwer Academic/Plenum Publishers.

4 King, A. and Young, G. (1999) Characteristics and occurrence of phenolic phytochemicals. *J. Am. Dietetic Assoc.*, **99**, 213–218.

5 American Cancer Society. Phytochemicals. Available at: http://www.cancer.org/eprise/main/docroot/ETO/content/ETO_5_3X_Phytochemicals/ (accessed July 22, 2002).

6 Hahn, N.I. (1998) Are phytoestrogens nature's cure for what ails us? A look at the research. *J. Am. Dietetic Assoc.*, **98**, 974–976.

7 Beckman, C.H. (2000) Phenolic-storing cells: keys to programmed cell death and periderm formation in wilt disease resistance and in general defence responses in plants. *Physiol. Molec. Plant Pathol.*, **57**, 101–110.

8 Dixon, R.A. and Paiva, N.L. (1995) Stress-induced phenylpropanoid metabolism. *Plant Cell*, **7**, 1085–1097.

9 Manach, C., Scalbert, A., Morand, C. et al. (2008) Polyphenols: food sources and bioavailability. *Am. J. Clin. Nutr.*, **79**, 727–747.

10 Gardner, P.T., White, T.A.C., McPhail, D.B. and Duthie, G.G. (2000) The relative contributions of vitamin, C, carotenoids and phenolics to the antioxidant potential of fruit juices. *Food Chem.*, **68**, 471–474.

11 Lee, K.W., Kim, Y.J., Kim, D.O. et al. (2003) Major phenolics in apple and their contribution to the total antioxidant capacity. *J. Agric. Food Chem.*, **51**, 6516–6520.

12 Puupponen-Pimiä, R., Nohynek, L., Meier, C. et al. (2001) Antimicrobial properties of phenolic compounds from berries. *J. Appl. Microbiol.*, **90**, 494–507.

13 Hercberg, S., Preziosi, P., Galan, P. et al. (1999) The su.vi.max study: a primary prevention trial using nutritional doses of antioxidant vitamins and minerals in cardiovascular diseases and cancers. *Food Chem. Toxicol.*, **37**, 925–930.

14 Chung, Y.C., Chien, C.T., Teng, K.Y. and Chou, S.T. (2006) Antioxidative and mutagenic properties of Zanthoxylum ailanthoides Sieb & zucc. *Food Chem.*, **97**, 418–425.

15 Kirkham, P. and Rahman, I. (2006) Oxidative stress in asthma and COPD: antioxidants as a therapeutic strategy. *Pharmacol. Therap.*, **111**, 476–494.

16 Wong, C.C., Li, H.B., Cheng, K.W. and Chen, F. (2006) A systematic survey of antioxidant activity of 30 Chinese medicinal plants using the ferric reducing antioxidant power assay. *Food Chem.*, **97**, 705–711.
17 Gordon, M.H. (1990) The mechanism of antioxidant action in vitro, in *Food Antioxidants* (ed. B.J.F. Hudson), Elsevier, London, pp. 1–18.
18 Valentão, P., Fernandes, E., Carvalho, F. *et al.* (2002) Antioxidative properties of cardoon (*Cynara cardunculus* L.) infusion against superoxide radical, hydroxyl radical, and hypochlorous acid. *J. Agric. Food Chem.*, **50**, 4989–4993.
19 Velioglu, Y.S., Mazza, G., Gao, L. and Oomah, B.D. (1998) Antioxidant activity and total phenolics in selected fruits, vegetables, and grain products. *J. Agric. Food Chem.*, **46**, 4113–4117.
20 Madhavi, D.L., Deshpande, S.S. and Salunkhe, D.K. (1995) *Food Antioxidants: Technological: Toxicological and Health Perspectives*, Marcel Dekker, Inc., New York, pp. 65–159.
21 Yen, G.C. and Duh, P.D. (1994) Scavenging effect of methanolic extract of peanut hulls on free-radical and active oxygen species. *J. Agric. Food Chem.*, **42**, 629–632.
22 Okubo, T., Yokoyama, Y., Kano, K. and Cell, I. (2003) Death induced by the phenolic antioxidant *tert*-butylhydroquinone and its metabolite *tert*-butylquinone in human monocytic leukemia U937 cells. *Food Chem. Toxicol.*, **41**, 679–688.
23 Pereira, C.A.M. and Maia, J.F. (2007) Study of the antioxidant activity and essential oil from wild basil (*Ocimumgratissimum* L.) leaf. *Ciênc. Tecnol. Aliment.*, **27**, 624–632.
24 Rice-Evans, C.A., Miller, N.J. and Paganga, G. (1997) Antioxidant properties of phenolic compounds. *Trends Plant Sci.*, **2**, 152–159.
25 Al-Mustafa, A.H. and Al-Thunibat, O.Y. (2008) Antioxidant activity of some Jordanian medicinal plants used traditionally for treatment of diabetes. *Pak. J. Biol. Sci.*, **11**, 351–358.
26 Wellwood, C.R., Cole, L. and Rosemary, A. (2004) Relevance of carnosic acid concentrations to the selection of rosemary, *Rosmarinus officinalis* (L.), accessions for optimization of antioxidant yield. *J. Agric. Food Chem.*, **52**, 6101–6107.
27 Linares, E., Mortara, R.A., Santos, C.X. *et al.* (2001) Role of peroxynitrite in macrophagemicrobicidal mechanisms in vivo revealed by protein nitration and hydroxylation. *Free Radic. Biol. Med.*, **30**, 1234–1242.
28 Cerqueira, F.M., de Medeiros, M.H.G. and Augusto, O. (2007) Antioxidantes dietéticos: Controvérsias e perspectivas. *Química Nova*, **30**, 441–449.

29 Petko, N., Kratchanov, C.G., Milan, C. et al. (2012) Bioavailability and antioxidant activity of black chokeberry (*Aronia melanocarpa*) polyphenols: in vitro and in vivo evidences and possible mechanisms of action: a review. *Compr. Rev. Food Sci.Food Safety*, **11**, 471–489.
30 Halliwell, B., Aeschbach, R., Löliger, J. and Aruoma, O.I. (1995) The characterization on antioxidants. *Food Chem. Toxicol.*, **33**, 601–617.
31 Mallick, N. and Mohn, F.H. (2000) Reactive oxygen species: response of algal cells. *J. Plant Physiol.*, **157**, 183–193.
32 Grant, C.M. (2008) Metabolic reconfiguration is a regulated response to oxidative stress. *J. Biol.*, **7**, 1.
33 Ralser, M., Wamelink, M.M., Kowald, A. et al. (2007) Dynamic rerouting of the carbohydrate flux is key to counteracting oxidative stress. *J. Biol.*, **6**, 10.
34 Poli, G., Schaur, R.J., Siems, W.G. and Leonarduzzi, G. (2008) 4-Hydroxynonenal: a membrane lipid oxidation product of medicinal interest. *Med. Res. Rev.*, **28**, 569–631.
35 Hasbay-Adil, İ., Yener, M.E. and Bayındırlı, A. (2008) Extraction of total phenolics of sour cherry pomace by high pressure solvent and subcritical fluid and determination of the antioxidant activities of the extracts. *Sep. Sci. Technol.*, **43**, 1091–1110.
36 Moure, A., Cruz, J.M., Franco, D. et al. (2001) Natural antioxidants from residual sources. *Food Chem.*, **72**, 145–171.
37 Handa, S.S., Khanuja, S.P.S., Longo, G., and Rakesh, D.D. (2008) Extraction technologies for medicinal and aromatic plants: earth, environmental and marine sciences and technologies.
38 Saltmarsh, M., Santos-Buelga, C. and Williamson, G. (2003) *Methods for Polyphenol Analysis*, Royal Society of Chemistry, UK, pp. 1–2.
39 Dai, J. and Mumper, R.J. (2010) Plant phenolics: extraction, analysis and their antioxidant and anticancer properties. *Molecules*, **15**, 7313–7352.
40 Pietta, P.G. (2000) Flavonoids as antioxidants. *J. Nat. Prod.*, **63**, 1035–1042.
41 Lakhanpal, P. and Rai, D.K. (2007) Quercetin: a versatile flavonoid. *IJMU*, **2**, 141–148.
42 Cushnie, T.P.T. and Lamb, A.J. (2005) Antimicrobial activity of flavonoids. *Int. J. Antimicrob.Agents*, **26**, 343–356.
43 Havsteen, B. (1983) Flavonoids, a class of natural products of high pharmacological potency. *Biochem. Pharmacol.*, **32**, 1141–1148.
44 Middleton, E. and Chithan, K. (1993) The impact of plant flavonoids on mammalian biology: implications for immunity, inflammation and cancer, in *The Flavonoids: Advances in Research Since 1986* (ed. J.B. Harborne), Chapman and Hall, London, pp. 145–166.

45 Hanasaki, Y., Ogawa, S. and Fukui, S. (1994) The correlation between active oxygens scavenging and antioxidative effects of flavonoids. *Free Radic. Biol. Med.*, **16**, 845–850.

46 Moroney, M.A., Alcaraz, M.J., Forder, R.A. *et al.* (1988) Selectivity of neutrophil 5-lipoxygenase inhibition by an antiinflammatory flavonoid glycoside and related aglycone flavonoids. *J. Pharm. Pharmacol.*, **40**, 787–792.

47 Messina, M.J. (1999) Legumes, soybeans: overview of their nutritional profiles and health effects. *Am. J. Clin. Nutr.*, **70** (Suppl 3), 439S–450S.

48 Zhou, J.R., Gugger, E.T., Tanaka, T. *et al.* (1999) Soybean phytochemicals inhibit the growth oftransplantable human prostate carcinoma and tumor angiogenesis in mice. *J. Nutr.*, **129**, 1628–1635.

49 Manach, C., Scalbert, A., Morand, C. *et al.* (2004) Polyphenols: food sources and bioavailability. *Am. J. Clin. Nutr.*, **79**, 727–747.

50 Peterson, J. and Dwyer, J. (1998) Flavonoids: dietary occurrence and biochemical activity. *Nutr. Res.*, **18**, 1995–2018.

51 Rice-Evans, C.A., Miller, N.J. and Paganga, G. (1996) Structure-antioxidantactivity relationships of flavonoids and phenolic acids. *Free Radic. Biol. Med.*, **20**, 933–956.

52 Fahey, J.W., Zhang, Y., Talalay, P. and Broccoli, T. (1997) Sprouts: an exceptionally rich source of inducers of enzymes that protectagainst chemical carcinogens. *Proc. Natl. Acad. Sci. USA*, **16**, 10367–10372.

53 Dillard, C.J. and German, J.B. (2000) Review phytochemicals: nutraceuticals and human health. *J. Sci. Food Agric.*, **80**, 1744–1756.

54 Sharma, G., Singh, R.P., Chan, D.C.F. and Agarwal, R. (2003) Silibinin induces growth inhibition and apoptotic cell death in human lung carcinoma cells. *Anticancer Res.*, **23**, 2649–2655.

55 Singh, R.P., Sharma, G., Sivanandhan, D. *et al.* (2003) Suppression of advanced human prostate tumor growth in athymic mice by silibinin feeding is associated with reduced cell proliferation, increased apoptosis and inhibition of angiogenesis. *Cancer Epidemiol. Biomark. Prev.*, **12**, 933–939.

56 Brownlee, M. (2001) Review article biochemistry and molecular cell biology of diabetic complications. *Nature*, **414**, 813–820.

57 Brownlee, M. (2005) The pathobiology of diabetic complications: a unifying mechanism. *Diabetes*, **54**, 1615–1625.

58 Harcourt, B.E., Sourris, K.C., Coughlan, M.T. *et al.* (2011) Targeted reduction of advanced glycation improves renal function in obesity. *Kidney Int.*, **80**, 190–198.

59 Yamagishi, A., Kunisawa, T., Nagashima, M. *et al.* (2009) Clinical usefulness of continuous cardiac output measurement: PulseCO masui. *Jpn. J. Anesthesiol.*, **58**, 422–425.

60 Rodiño-Janeiro, B.K., Salgado-Somoza, A., Tejeira-Fernandez, E. et al. (2011) Receptor for advanced glycation end-products expression in subcutaneous adipose tissue is related to coronaryartery disease. *Eur. J. Endocrinol.*, **164**, 529–537.

61 Ahn, J., Cho, I., Kim, S. et al. (2008) Dietary resveratrol alters lipid metabolism-related gene expressionof mice on an atherogenic diet. *J. Hepatol.*, **49**, 1019–1028.

62 Szkudelska, K., Nogowski, L. and Szkudelski, T. (2009) Resveratrol, a naturally occurring diphenolic compound, affects lipogenesis, lipolysis and the antilipolytic action of insulin in isolated rat adipocytes. *J. Steroid Biochem. Molec. Biol.*, **113**, 17–24.

63 Li, R.W., Douglas, T.D., Maiyoh, G.K. et al. (2006) Green tea leaf extract improves lipidand glucose homeostasis in a fructose-fed insulin-resistant hamster model. *J. Ethnopharmacol.*, **104**, 24–31.

64 Bose, M., Lambert, J.D., Ju, J. et al. (2008) The major green tea polyphenol, (−)-epigallocatechin-3-gallate, inhibits obesity, metabolic syndrome, and fatty liver disease in high-fat-fed mice. *J. Nutr.*, **138**, 1677–1683.

65 Ejaz, A., Wu, D., Kwan, P. and Meydani, M. (2009) Curcumin inhibits adipogenesis in 3T3-L1 adipocytes and angiogenesis and obesity in C57/BL Mice1-3. *J. Nutr. Dis.*, **139**, 919–925.

66 Egert, S., Bosy-Westphal, A., Seiberl, J. et al. (2009) Quercetin reduces systolic blood pressure and plasma oxidised low-density lipoprotein concentrations in overweight subjects with a high-cardiovascular disease risk phenotype: a double-blinded, placebo-controlled cross-over study. *Brit. J. Nutr.*, **102**, 1065–1074.

67 Art, I.C.W. and Hollman, P.C.H. (2005) Polyphenols and disease risk in epidemiological studies. *Am. Soc. Clin. Nutr.*, **81**, 317S–325S.

68 Asami, D.K., Hong, Y.J., Barret, D.M. and Mitchell, A.E. (2003) Comparison of the total phenolic and ascorbic acid content of freeze-dried and air-dried marionberry, strawberry, and corn grown using conventional, organic, and sustainable agricultural practices. *J. Agric. Food Chem.*, **51**, 1237–1241.

69 Matthes, M. and Schmitz-Eiberger, M. (2009) Polyphenol content and antioxidant capacity of apple fruit: effect of cultivar and storage conditions. *J. Appl. Botany Food Qual.*, **82**, 152–157.

70 Holopainen, J.K. and Gershenzon, J. (2010) Multiple stress factors and the emission of plant VOCs. *Trends Plant Sci.*, **15**, 176–184.

71 Rocha-Guzmán, N.E., González-Laredo, R.F., Ibarra-Pérez, F.J. et al. (2007) Effect of pressure cooking on the antioxidant activity of extracts from three common bean (*Phaseolus vulgaris* L.) cultivars. *Food Chem.*, **100**, 31–35.

72 Gitanjali, D.P.Y. and Shivaprakash, M. (2004) Effect of shallow frying on total phenolic content and antioxidantactivity in selected vegetables. *J. Food Sci. Technol.*, **41**, 666–668.
73 de Ancos, B., González, E.M. and Cano, M.P. (2000) Ellagic acid, vitamin c, and total phenolic contents and radical scavenging capacity affected by freezing and frozen storage in raspberry fruit. *J. Agric. Food Chem.*, **48**, 4565–4570.
74 Rodrigo, K.A., Rawal, Y., Renner, R.J. *et al.* (2006) Suppressionof the tumorigenic phenotype in human oral squamous cell carcinoma cells by an ethanol extract derived from freeze-dried black raspberries. *Nutr. Cancer*, **54**, 58–68.
75 Medina, I., Gllardo, J.M., Gonzalez, M.J. *et al.* (2007) Effect of molecular structure of phenolic families as hydroxycinnamic acid and catechins on their antioxidant effectiveness in minced fish muscle. *J. Agric. Food Chem.*, **55**, 3889–3895.
76 He, Y. and Shahidi, F. (1997) Antioxidant activity of green tea and its catechins in fish meat model system. *J. Agric. Food Chem.*, **45**, 4262–4266.
77 Munin, A. and Edwards-Lévy, F. (2011) Encapsulation of natural polyphenolic compounds; a review. *Pharmaceutics*, **3**, 793–829.
78 Nedovic, V., Kalusevica, A., Manojlovicb, V. *et al.* (2011) An overview of encapsulation technologies for food applications. *Procedia Food Sci.*, **1**, 1806–1815.
79 Seok, J.S., Kim, J.S. and Kwak, H.S. (2003) Microencapsulation of water-soluble isoflavone and physico-chemical property in milk. *Arch. Pharm. Res.*, **26**, 426–431.
80 Schrooyen, P.M.M., Meer, R. and Kruif, C.G. (2001) Microencapsulation: its application in nutrition. *Proc. Nutr. Soc.*, **60**, 475–479.
81 Li, Y., Lim, L.T. and Kakuda, Y. (2009) Electrospun zein fibers as carriers to stabilize (−) epigallocatechin gallate. *J. Food Sci.*, **74**, C233–C240.
82 Luo, Y., Wang, T.T., Teng, Z. *et al.* (2013) Encapsulation of indole-3-carbinol and 3,3′-diindolylmethane in zein/carboxymethyl chitosan nanoparticles with controlled release property and improved stability. *Food Chem.*, **139**, 224–230.
83 Fang, Z. and Bhandari, B. (2010) Encapsulation of polyphenols – a review. *Trends Food Sci. Technol.*, **21**, 510–523.
84 Shahidi, F. and Han, X.Q. (1993) Encapsulation of food ingredients. *Crit. Rev. Food Sci. Nutr.*, **33**, 501–547.
85 Zuidam, N.J. and Nedovic, V. (2010) *Encapsulation Technologies for Active Food Ingredients and Food Processing*, 1st edn, Springer Science+Business Media, LLC.

86 (a) Vidhyalakshmi, R., Bhakyaraj, R. and Subhasree, R.S. (2009) Encapsulation "the future of probiotics" – a review. *Adv. Biol. Res.*, **3**, 96–103; (b) Burgain, J.B.J., Gaiani, C., Linder, M. and Scher, J. (2011) Encapsulation of probiotic living cells: from laboratory scale to industrial applications. *J. Food Eng.*, **104**, 467–483.

87 Hamad, S.A., Paunov, V.N. and Stoyanov, S.D. (2012) Triggered cell release from shellac–cell composite microcapsules. *Soft Matter*, **8**, 5069–5077.

88 Madene, A., Jacquot, M., Scher, J. and Desobry, S. (2006) Flavour encapsulation and controlled release – a review. *Int. J. Food Sci. Technol.*, **41**, 1–21.

89 Vehring, R. (2008) Pharmaceutical particle engineering via spray drying. *Pharm. Res.*, **25**, 999–1022.

90 (a) Hall, H.S. and Pondell, R.E. (1999) The Wurster process' in controlled release technologies: methods, theory, in *Controlled Release Technologies: Methods, Theory and Applications* (ed. A.F. Kydonieus), CRC Press, New York, pp. 133–155; (b) Dewettinck, K. (1999) in *Trends in Food Science & Technology*, vol. 10 (ed. A. Huyghebaert), Science, New York, pp. 163–168.

91 Mehta, A.M., Valazza, M.J. and Abele, S.E. (1986) Evaluation of fluid-bed processes for enteric coating systems. *Pharm. Technol.*, **10**, 46–56.

92 Arshady, R. (1993) Microcapsules for food. *J. Micro-encapsul.*, **10**, 413–435.

93 Todd, R.D. (1970) Microencapsulation and the flavour industry. *Flavor Industry*, **1**, 768–771.

94 Sparks, R.E. (1981) Microencapsulation. *Encycl. Chem. Technol.*, **15**, 470–493, John Wiley & Sons, New York.

95 Karel, M. and Langer, R. (1988) Controlled release of foodingredients, in *Flavor Encapsulation* (eds G.A. Reineccius and S.J. Risch), American Chemical Society, Washington, DC, pp. 177–191.

96 Reineccius, G.A. (1995) Controlled release techniques in the food industry, in *Encapsulation and Controlled Release of Food Ingredients* (eds S.J. Risch and G.A. Reineccius), American Chemical Society, Washington DC, pp. 8–25.

97 Variable Impact Tester Model 304, Instrument Manual, Erichsen GmbH & Co, Hemer (2005).

98 Liu, X., Sheng, X., Lee, J.K. and Kessler, M.R. (2009) Synthesis, characterization of melamine–urea–formaldehyde microcapsules containing ENB-based self-healing agents. *Macromol. Mater. Eng.*, **294**, 389–395.

99 Giunchedi, P. and Conte, U. (1995) Spray-drying as a preparation method of microparticulate drug delivery systems: an overview. *S.T.P. Pharma Pratiques*, **5**, 276–290.
100 Yoksan, R., Jirawutthiwongchai, J. and Arpo, K. (2010) Encapsulation of ascorbyl palmitate in chitosan nanoparticles by oil-in-water emulsion and ionic gelation processes. *Colloid Surf. B: Biointerf.*, **76**, 292–297.
101 Yeo, Y. and Park, K. (2004) A new microencapsulation method using an ultrasonic atomizer based on interfacial solvent exchange. *J. Control. Release*, **100**, 379–388.
102 Yamamoto, T., Yokoyama, M., Opanasopit, P. *et al.* (2007) What are determining factors for stable drug incorporation into polymeric micelle carriers? Consideration on physical and chemical characters of the micelle inner core. *J. Control. Release*, **123**, 11–8.
103 Yamamoto, Y., Nagasaki, Y., Kato, Y. *et al.* (2001) Long-circulating poly(ethylene glycol)–poly(D,L-lactide) block copolymer micelles with modulated surface charge. *J. Control. Release*, **77**, 27–38.
104 Contal, E. (2010) Synthèse de nano-objets lipidiques et leur fonctionnalisation interne par reaction "click". Ph.D. Thesis, Strasbourg University, Strasbourg, France.
105 Ravindranath, V. and Chandrasekhara, N. (1981) Metabolism of curcumin-studies with [3*H*] curcumin. *Toxicology*, **22**, 337–344.
106 Pan, M.H., Huang, T.M. and Lin, J.K. (1999) Biotransformation of curcumin through reduction and glucuronidation in mice. *Drug Metab. Dispos.*, **27**, 486–494.
107 Gabizon, A. and Martin, F. (1997) Polyethylene glycol coated (pegylated) liposomal doxorubicin – rationale for use in solid tumours. *Drugs*, **54**, 15–21.
108 Gabizon, A.A. (2001) Pegylated liposomal doxorubicin: metamorphosis of an old drug into a new form of chemotherapy. *Cancer Invest.*, **19**, 424–436.
109 Colletier, J.P., Chaize, B., Winterhalter, M. and Fournier, D. (2002) Protein encapsulation in liposomes: efficiency depends on interactions between protein and phospholipidbilayer. *BMC Biotechnol.*, **2**, 1–8.
110 Ma, O.H., Kuang, Y.Z., Hao, X.Z. and Gu, N. (2009) Preparation and characterization of tea polyphenols and vitamin E loaded nanoscale complex liposome. *J. Nanosci. Nanotechnol.*, **9**, 1379–1383.
111 Vandamme, T.F., Poncelet, D. and Subra-Paternault, P. (2007) *Microencapsulation: des sciences aux technologies*, Paris, France, Lavoisier Tec & Doc.
112 Dandawate, P.R., Vyas, A., Ahmad, A. *et al.* (2012) Inclusion complex of novel curcumin analogue CDF and β-cyclodextrin (1:2) and

its enhanced in vivo anticancer activity against pancreatic cancer. *Pharm. Res.*, **29**, 1775–1786.
113 Janssen, L.J.J.M. and Nijenhuis, K. (1992) Encapsulation by interfacial polycondensation. I. The capsuleproduction and a model for wall growth. *J. Membr. Sci.*, **65**, 59–68.
114 Sanchez, V., Baeza, R., Galmarini, M. et al. (2013) Freeze-drying encapsulation of red wine polyphenols in an amorphous matrix of maltodextrin. *Food Bioprocess Technol.*, **6**, 1350–1354.
115 Semsarzadeh, M.A. and Amiri, S. (2012) Preparation and characterization of inclusion complexes of poly(dimethylsiloxane)s with gamma-cyclodextrin without sonic energy. *Silicon*, **4**, 151–156.
116 Semsarzadeh, M.A. and Amiri, S. (2013) Preparation and properties of polyrotaxane from α-cyclodextrin and Poly (ethylene glycol) with poly (vinyl alcohol). *Bull. Mater. Sci.*, **36**, 989–996.
117 Semsarzadeh, M.A. and Amiri, S. (2013) Polyrotaxane macroinitiator in Atom Transfer Radical Polymerization of Styrene and vinyl acetate: synthesis and characterization of PVAc-*b*-PSt-*b*-(PDMS/cyclodextrin)-*b*-PSt-*b*-Pvac pentablock copolymers. *J. Incl. Phenom. Macrocycl. Chem.*, **77**, 489–499.
118 Chandrasekar, V. (2010) Optimizing the microwave-assisted extraction of phenolic antioxidants from apple pomace and microencapsulation in cyclodextrins. MSc. Thesis, Purdue University Graduate School.
119 Szejtli, J. and Osa, T. (1996) *Comprehensive Supramolecular Chemistry, Vol 3, Cyclodextrins*, Permagon, Oxford.
120 Krishnaswamy, K., Orsat, V. and Thangavel, K. (2012) Synthesis and characterization of nano-encapsulated catechin by molecular inclusion with beta – cyclodextrin. *J. Food Eng.*, **111**, 255–264.
121 Shuang, S., Pan, J., Guo, S. et al. (1997) Fluorescence study on the inclusion complexes of rutin with β-cyclodextrin, hydroxypropyl-β-cyclodextrin and γ-cyclodextrin. *Anal. Lett.*, **30**, 2261–2270.
122 Sri, K.V., Kondaiah, A., Ratna, J.V. and Annapurna, A. (2007) Preparation and characterization of quercetin and rutin cyclodextrin inclusion complexes. *Drug Dev. Ind. Pharm.*, **33**, 245–253.
123 Tobar, E.L., Blanch, G.P., Castillo, M.L.R. and Cortes, S.S. (2012) Encapsulation and isomerization of curcumin with cyclodextrins characterized by electronic and vibrational spectroscopy. *Vib. spectrosc.*, **62**, 77–84.
124 Yadav, V.R., Suresh, S., Devi, K. and Yadav, S. (2009) Effect of cyclodextrin complexation of curcumin on its solubility and antiangiogenic and anti-inflammatory activity in rat colitis model. *PharmSciTech.*, **10**, 752–762.

125 Awad, A.B., Williams, H. and Fink, C.S. (2003) Effect of phytosterols on cholesterol metabolism and MAP kinase in MDA-MB-231 human breast cancer cells. *J. Nutr. Biochem.*, **14**, 111–119.
126 Marques, H.M.C. (2010) A review on cyclodextrin encapsulation of essential oils and volatiles. *Flavour Fragr. J.*, **25**, 313–326.
127 Haidong, L., Fang, Y., Zhihong, T. and Changle, R. (2011) Study on preparation of β-cyclodextrin encapsulation tea extract. *Int. J. Biol. Macromol.*, **49**, 561–6.
128 Haiyun, D., Jianbin, C., Shuang, Z.G. and Jinhao, P. (2003) Preparation and spectral investigation on inclusion complex of β-cyclodextrin with rutin. *Spectrochim. Acta Part A Mol. Biomol. Spectrosc.*, **59**, 3421–3429.
129 Yu, Z., Cui, M., Yan, C. et al. (2007) Investigation of heptakis(2,6-di-O-methyl)-beta-cyclodextrin inclusion complexes with flavonoid glycosides by electrospray ionization mass spectrometry. *Rapid Commun. Mass Spectrom.*, **21**, 683–690.
130 Chakraborty, S., Basu, S., Lahiri, A. and Basak, S. (2010) Inclusion of chrysin in β-cyclodextrin nanocavity and its effect on antioxidant potential of chrysin: a spectroscopic and molecular modeling approach. *J. Mol. Struct.*, **977**, 180–188.
131 Kim, H., Kim, H. and Jung, S. (2008) Aqueous solubility enhancement of some flavones by complexation with cyclodextrins. *Bull. Korean Chem. Soc.*, **29**, 590–594.
132 Jullian, C., Moyano, L., Yañez, C. and Olea-Azar, C. (2007) Complexation of quercetin with three kinds of cyclodextrins: an antioxidant study. *Spectrochim. Acta A Mol. Biomol. Spectrosc.*, **67**, 230–234.
133 Kim, H., Choi, J. and Jung, S. (2009) Inclusion complexes of modified cyclodextrins with some flavonols. *J. Incl. Phenom. Macrocycl. Chem.*, **64**, 43–47.
134 Yang, L., Chen, W., Ma, S.X. et al. (2011) Host–guest system of taxifolin and native cyclodextrin or its derivative: Preparation, characterization, inclusion mode, and solubilization. *Carbohydr. Polym.*, **85**, 629–637.
135 Yatsu, F.K.J., Koester, L.S., Lula, I. et al. (2013) Multiple complexation of cyclodextrin with soy isoflavones present in an enriched fraction. *Carbohydr. Polym.*, **98**, 726–735.
136 Del, R.D., Costa, L.G., Lean, M.E.J. and Crozier, A. (2010) Polyphenols and health: what compounds are involved? *Nutr. Metab. Cardiovasc. Dis.*, **20**, 1–6.
137 Liu, R.H. (2003) Health benefits of fruits and vegetables are from additive and synergistic combination of phytochemicals. *Am. J. Clin. Nutr.*, **78**, 517S–520S.

138 Aherne, S.A. and O'Brien, N.M. (2002) Dietary flavonols: chemistry, food content, and metabolism. *Nutrition*, **18**, 75–81.

139 Ré, M.I. (1998) Microencapsulation by spray drying. *Drying Technol.*, **16**, 1195–1236.

140 Felnerova, D., Viret, J.F., Gluck, R. and Moser, C. (2004) Liposomes and virosomes as delivery systems for antigens, nucleic acids and drugs. *Curr. Opin. Biotechnol.*, **15**, 518–529.

6

Cyclodextrins Application as Macroinitiator

Well-defined polymers and copolymers can be synthesized with a combination of the core-first method and various polymerization systems such as controlled radical polymerization. CD-based macroinitiators can be used for synthesize of star and block copolymers. These polymers are able to undergo a conformational change or phase transition as a reply to an external stimulus, resulting in the formation of core–shell nanoparticles, which further tend to aggregate.

6.1 Cyclodextrins Application as Macroinitiator in Polyrotaxane Synthesis Via ATRP

Cyclodextrins can form inclusion complexes with silicon-containing polymers and new organic–inorganic hybrids with exact stoichiometric relationships. Cyclodextrins are cyclic oligosaccharides, the most common consisting of 6, 7, and 8 glucopyranose units linked by R-1,4 glucosidic bonds, which are called α-cyclodextrin (α-CD), β-cyclodextrin (β-CD), and γ-cyclodextrin (γ-CD), respectively [1–3]. CDs' specific truncated cone shape with a hydrophobic core can incorporate nonpolar compounds when two hydrophilic rims are composed of —OH groups. The inclusion complex (IC) formation depends on the internal parameters such as the nature of the CD and polymer and solvent media and also the external parameters such as temperature and pressure. Accordingly, new strategies were developed to fabricate novel supramolecular hydrogels via several routes. Recently, attention has been paid on ICs formed by CDs and inorganic polymers that offer different sites of binding and may be selectively threaded by CDs [4–8].

Recent advances in controlled radical polymerization techniques have led to facile synthesis of the well-defined block copolymers having a wide range of functional monomers. Atom transfer radical polymerization (ATRP), reversible addition fragmentation chain transfer

Cyclodextrins: Properties and Industrial Applications, First Edition. Sahar Amiri and Sanam Amiri.
© 2017 John Wiley & Sons Ltd. Published 2017 by John Wiley & Sons Ltd.

(RAFT) polymerization, and nitroxide-mediated radical polymerization techniques have been utilized to develop well-defined functional polymers [9–12]. However, block and graft copolymers were prepared through ATRP. For example, PDMS containing internal were used for the synthesis of segmented multiblock copolymers with various vinyl or styrene monomers which opens the way to various industrial applications such as thermoplastic elastomers, compatibilizers, and surfactants in polymer blends [13–15]. Such inclusion complexes may form interesting structures with good and new structures, and CD segments thread onto one of the blocks [14].

Supramolecular self-assembly structures based on CDs may show stimuli-sensitive properties and are capable of forming a multitude of sophisticated suprastructures. These novel complexes can be used in various areas such as drug delivery, sensors, separations, and membranes [15–19]. Pentablock polyrotaxane was synthesized based on γ-CD threaded onto a PDMS as macroinitiator in the presence of PVAc and PSt monomers by ATRP mechanism [16]. The presence of styrene as the outer block copolymer can enhance the rigidity of the chain, and it can thus be used in the macromolecular design of silicone-based copolymers [17–19].

The findings suggest that the reaction condition changes the crystalline structure. The results indicate that light and mixing are necessary for the formation of an inclusion complex between cyclodextrins and PDMS and may change the structure of polyrotaxane and a number of cyclodextrins that capped in the main polyrotaxane. Results of this study demonstrated that pentablock copolymers containing Br–PDMS–Br/γ-CD have narrow molecular weight distribution. Our group focused on the preparation of triblock and pentablock copolymers using Br–PDMS–Br/γ-CD macroinitiator and compared the effect of various monomers on the synthesized block copolymers [19–21]. It seems that no other report can be found in the literature on the application of Br–PDMS–Br/γ-CD macroinitiators for making pentablock copolymers. Temperature-dependent complexation with γ-CD has been extensively studied earlier due to their ability to produce supramolecular structures of desired architectures [19–21]. Raising the temperatures caused significantly increased rate of formation of inclusion complex polymer complexation with CD [22–25].

This chapter reports supramolecular polymeric host–guest system consisting of γ-CD core (as the host polymer) and bis(boromoalkyl)-terminated PDMS (as the guest polymer) in an aqueous solution. These interesting host–guest systems exhibited a dual thermoresponsive behavior due to two existing kinds of thermoresponsive segments, PDMS and cyclodextrin. In other words, this polymeric host–guest system was able to form ABA and CABAC-type supramolecular triblock and pentablock copolymers via inclusion complexation in aqueous solution [26–28]. Adjusting temperature and concentration of the components and modifying host/guest ratio and the length of PDMS blocks in the guest polymer affect the size of the resultant noncovalently connected micelle and can be tuned simply [19–21, 24, 25].

These pentablock copolymers revealed a thermosensitive micelle formation behavior in the solid state. As temperature increased, the PDMS/CD chains were dehydrated, while the color of the complex was changed from white to green through hydrophobic–hydrophobic interactions. The complex is colorless; however, the presence of a catalyst, such as CuCl-ligand, turns the block copolymer green. The catalyst impurities of the prepared rotaxane were removed by alumina column. Further increase of the temperature caused the phase transition-related PDMS to occur on the corona of noncovalently connected micelles. Therefore, the micelles were destabilized, which resulted in micelle aggregation and precipitation. The supramolecular system presented here is rather different from the reported systems. Our work has formed dual thermoresponsive tri and pentablock copolymers in order to propose a novel supramolecular approach in designing and constructing the noncovalently connected polymeric micelles with modifiable properties.

6.2 Inclusion Complexes of PDMS and γ-CD Without Utilizing Sonic Energy

γ-Cyclodextrin (γ-CD) formed crystalline inclusion complexes with poly(dimethylsiloxane)s (PDMS) under sonic energy. An inclusion complex between PDMS and γ-CD was synthesized at room temperature both in the presence and absence of light and mixing. By addition of PDMS (liquid) aqueous solutions of γ-CD and mixing at room temperature for 7 days, the heterogeneous solution became turbid, and the complexes were formed as a crystalline precipitate. This is the first observation in which γ-CD forms a complex with inorganic polymers at room temperature without sonic energy. Table 6.1 shows the results of the complex formation and conversion between γ-CD and PDMS under various conditions. γ-CD forms a complex as soon as PDMS is added, and the reaction yield becomes higher after 6 days at room temperature without sonic energy. The cavity of the γ-CD is large enough to incorporate PDMS [7, 12–14]. At the presence of sonic energy for 15 min, the yield of the complex between γ-CD and PDMS is lower at room temperature (Table 6.1) [13, 29].

Table 6.1 Conversion of γ-CD/PDMS for various conditions.

Condition	Conversion (%)
Room temperature (7 days)	71
Sonic energy (15 min)	56
Without light and mixing (7 days)	34
Without light (7 days)	49
At room temperature (7 days) PDMS–Br	67

It is well known that ultrasonication is frequently used for complexation of axle and ring molecules. Usually, the sonication promotes a reaction to make a homogeneous system. Complex formation between CDs and PDMS was also promoted under sonication condition; however, it was too fast to form large particles, and actually turbid solution was observed. The complex particles could not be caught by filtration and precipitation by centrifugation. When PDMS (liquid) was added to aqueous solutions of γ-CD under sonic energy, the heterogeneous solution became turbid, and the complex formed as crystalline precipitate [30, 31]. The complex formation of γ-CD with PDMS was studied quantitatively. The amount of the complex formed increased with an increase in the amount of PDMS added to the aqueous solution of γ-CD. The amount of the complex showed similar values even if excess amounts of PDMS were used; this indicates the stoichiometric complexation. The continuous variation plot for the formation of the complex between γ-CD and PDMS is at maximum level in 0.40, that is, 2:3 (monomer unit: γ-CD) stoichiometry (Figure 6.1).

The results of Figure 6.1 revealed that 1.5 units of PDMS were bound in each γ-CD cavity. The stoichiometry was confirmed by the use of ^{1}H NMR spectroscopy. The length of the 1.5 monomer units corresponds to the depth of the γ-CD cavity. The complexes were isolated by centrifugation, washed, and dried. The inclusion complexes were thermally stable and insoluble in water, even under boiling conditions [29, 30]. The FTIR spectra of the PDMS–γ-CD are shown in Figure 6.2. The spectra showed strong Si—O—Si stretching absorptions at 400–800 cm^{-1}, which is characteristic of a siloxane backbone. The complete list of FTIR structure assignments is given in Table 6.2.

For investigation of inclusion complex between γ-CD and PDMS at various conditions, X-ray diffraction pattern was used, which shows that the complexes are crystalline (Figure 6.3). The complexes are soluble in DMSO, DMF, and pyridine. The powder XRD shows that all of the complexes are crystalline, although linear PDMS is a liquid at room

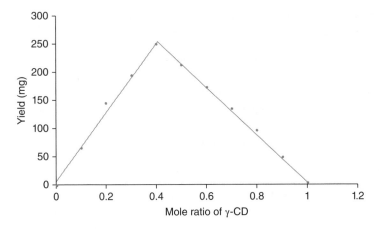

Figure 6.1 Continuous variation plot for complex formation between γ-CD and PDMS under sonic energy.

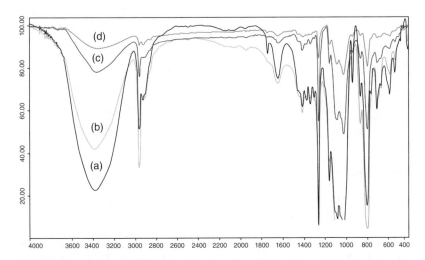

Figure 6.2 FTIR spectra of γ-CD/PDMS complexes at various conditions: (a) PDMS–γ-CD, 7 days at room temperature; (b) PDMS–γ-CD under sonic energy at room temperature; (c) PDMS–Br, 7 days at room temperature; and (d) PDMS–γ-CD without mixing or light, 7 days at room temperature.

temperature. Harada reported that the crystal structures of CD complexes are classified mainly into three types: channel, cage, and layer [16].

The γ-CD/PDMS and γ-CD/PDMS–Br complexes were found to have a cage-type structure at room temperature (Figure 6.3a,b). However, the γ-CD/PDMS complexes in other conditions adopt a head-to-head channel-type structure in which γ-CD molecules are stacked along

Table 6.2 Assignments of FTIR of PDMS/γ-CD.

Wavenumber (cm^{-1})	Assignment
2960	v(C—H) in CH$_3$
1259.90	δ(C—H) in Si—CH$_3$
791	v_a(Si—O—Si) in Si—O—Si
723	v_s(Si—O—Si) in Si—O—Si
601	P(C—H) in Si—CH$_3$
3369	v (O—H) in γ-CD

v = stretching mode, v_a = asymmetric stretching,
v_s = symmetric stretching, δ = in-plane bending or scissoring,
ρ = in-plane bending or rocking.

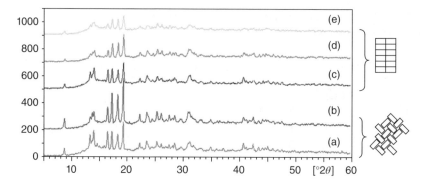

Figure 6.3 Powder X-ray diffraction patterns (solid state) of γ-CD/PDMS complexes at various conditions: (a) PDMS–γ-CD, 7 days at room temperature; (b) PDMS–Br, 7 days at room temperature; (c) PDMS–γ-CD under sonic energy at room temperature; (d) PDMS–γ-CD without mixing at room temperature; and (e) PDMS–γ-CD, 7 days at room temperature without mixing or light.

a PDMS axis to form a cylinder (Figure 6.3c–e). The reflection peaks of γ-CD–PDMS complexes are similar to those of the γ-CD/PDMS complex and different from those of the γ-CD [16].

^{1}H NMR spectra of the complex between γ-CD and PDMS at various conditions are shown in Figure 6.4. The mole ratio of PDMS to γ-CD was calculated in the complex by comparing the integral of the CD(1H) peak and that of the methyl group on PDMS; two monomer units were found to bind to a γ-CD molecule. The mole ratios of the complexes are 1.5, 3, and 1 (monomer unit), which is similar to those obtained from the conversion of the complex. The length of the two monomer units corresponds to the depth of the γ-CD cavity. However, the mole ratios of the complexes (monomer unit/γ-CD) are 1.5, 3, and 1 for (monomer

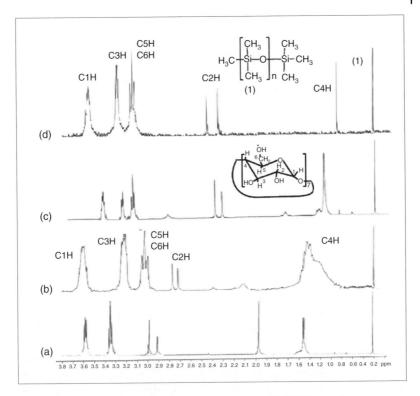

Figure 6.4 ^{1}H NMR spectra of the complex between γ-CD/PDMS complexes at various conditions: (a) PDMS–γ-CD, 7 days at room temperature; (b) PDMS–γ-CD under sonic energy at room temperature; (c) PDMS–γ-CD, 7 days at room temperature without mixing or light; and (d) PDMS–Br, 7 days at room temperature.

unit/γ-CD) under sonic energy, without light or mixing, and PDMS–Br at room temperature, respectively [16].

Thermal stability and structural properties of the formed inclusion complexes were characterized by DSC, which is a useful tool to determine the melting and crystallization temperatures and provide both quantitative and qualitative information about the physiochemical state of the guest inside the CD complexes. DSC thermograms of cyclodextin and the inclusion complex of PDMS/γ-CD in the temperature range from −100 °C to +350 °C are shown in Figure 6.5. The DSC plot of γ-CD showed that two endothermic peaks were observed in the temperature range between 100 and 110 °C due to the loss of water and near 320 °C due to the γ-CD fusion [16].

Glass transition temperature (T_g) of PDMS cannot be observed in Figure 6.5 because its value has been reported to be about −120 °C

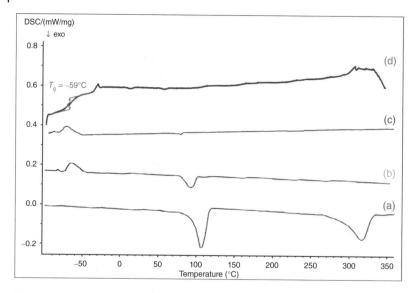

Figure 6.5 DSC spectra of (a) γ-CD and the complex between PDMS/γ-CD, (b) 7 days at room temperature without mixing or light, (c) under sonic energy at room temperature, and (d) 7 days at room temperature.

[18–20]. The DSC thermograms for the PDMS/γ-CD systems show sifting the persistence exothermic peak of PDMS (at −120 °C) in all products. These results are characteristic of the microphase-separated morphology of the inclusion complex. On the basis of the DSC results, one can conclude that the inclusion complex of PDMS/γ-CD has successfully been synthesized [16].

6.3 Supramolecular Pentablock Copolymer Containing Via ATRP of Styrene and Vinyl Acetate Based on PDMS/CD Inclusion Complexes as Macroinitiator

Bromoalkyl-terminated poly(dimethylsiloxane)/cyclodextrins macroinitiator (Br–PDMS/γ-CD) was synthesized and used as a macroinitiator for synthesized organic–inorganic pentablock copolymers containing styrene (St) and vinyl acetate (VAc) monomers at 60 °C via atom transfer radical polymerization (ATRP) using CuCl/N,N,N'',N''',N'''-pentamethyldiethylenetriamine (PMDETA) as a catalyst system. For macroinitiator preparation, Br–PDMS–Br was reacted with γ-CD in different conditions to obtain inclusion complex between CD and

Br–PDMS–Br, and then characterized through hydrogen nuclear magnetic resonance (^{1}H NMR) and differential scanning calorimetry (DSC). The resulting Br–PDMS–Br/γ-CD inclusion complexes were taken as macroinitiators for ATRP of St and VAc. The well-defined poly (styrene)-*b*-poly(vinyl acetate)-*b*-poly(dimethylsiloxane/γ-cyclodextrin)-*b*-poly(vinyl acetate)-*b*-poly(styrene) (PSt-*b*-PVAc-*b*-PDMS/γ-CD-*b*-PVAc-*b*-PSt) pentablock copolymer was characterized by ^{1}H NMR, gel permeation chromatograph (GPC), and DSC. The obtained pentablock copolymers consisting of Br–PDMS–Br/γ-CD inclusion complex as central blocks (inorganic block) and PVAc and PSt as terminal blocks were synthesized and can undergo a temperature-induced reversible transition upon heating of the copolymer complex from white complex at 22 °C to green complex in 55 °C, which is characterized with XRD and ^{1}H NMR. The (Br–PDMS–Br/γ-CD)-based pentablock copolymers were synthesized by copper(I)-mediated ATRP of St and PVAc according to the following typical procedure [16, 21–25]. A required amount of CuCl was introduced into the three-neck round-bottomed flask equipped with a magnetic stirrer. The flask was sealed with a rubber septum and was cycled between vacuum and nitrogen three times. The "freeze–pump–thaw" cycle was carried out three times to remove oxygen from the flask containing the reaction mixture. The flask was sealed under vacuum and then immersed in a preheated oil bath at a desired temperature (i.e., 60 °C). Depending on the monomer used, the reaction mixture was removed at the given reaction time from the oil bath and diluted with THF, filtered, and dried under vacuum to a constant weight. The dried copolymer was dissolved in THF and passed through a short column of neutral alumina to remove the remaining copper catalyst [25]. The samples were then dried under vacuum at 50 °C up to a constant weight and used in ^{1}H NMR, DSC, and GPC analyses. The conversion of the monomers St and PDMS was determined gravimetrically [25–27] (Scheme 6.1).

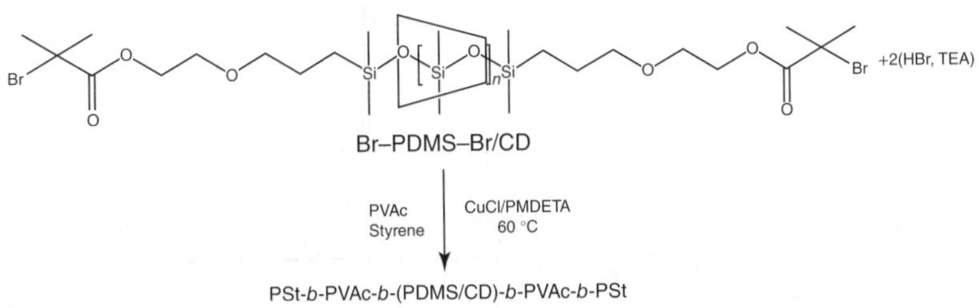

Scheme 6.1 ATRP route of PVAc and PSt initiated with (Br–PDMS–Br/CD) macroinitiators.

6.3.1 Complex Formation of γ-CD with Br–PDMS–Br

Br–PDMS–Br was added to aqueous solutions of γ-CD, and the mixture was mixed at room temperature for 7 days. The heterogeneous solution became turbid, and the complexes were formed. This is the first observation in which γ-CD forms a complex with inorganic polymers at room temperature without sonic energy. γ-CD forms a complex as soon as Br–PDMS–Br is added, and the reaction yield becomes higher after 6 days at room temperature (67%) than under sonic energy (61%) and without light and mixing (47%) (Figure 6.6).

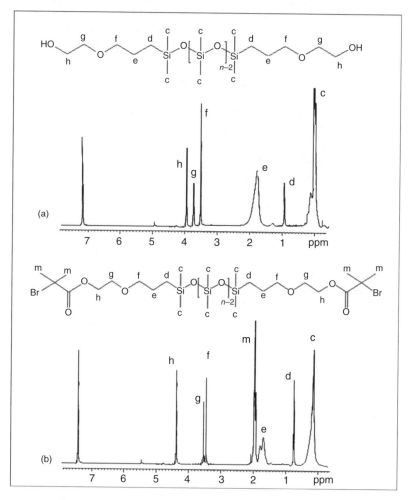

Figure 6.6 ^{1}H NMR spectra of HO–PDMS–OH (a) and Br–PDMS–Br macroinitiator (b) in the $CDCl_3$ solvent.

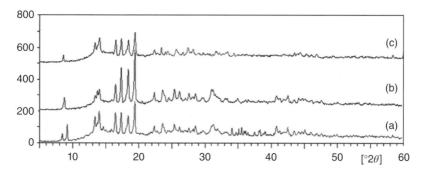

Figure 6.7 Powder X-ray diffraction patterns (solid state) of γ-CD/Br–PDMS–Br inclusion complexes at various conditions: (a) Br–PDMS–Br/γ-CD under sonic energy at room temperature; (b) Br–PDMS–Br/γ-CD, 7 days at room temperature; and (c) Br–PDMS–Br/γ-CD without mixing or light, 7 days at room temperature.

The inclusion complexes were isolated by centrifugation, then washed, and dried. The inclusion complexes were thermally stable. The complexes were insoluble in water, even under boiling conditions. The X-ray diffraction pattern of the complex between γ-CD and Br–PDMS–Br shows that the complexes are crystalline. The complexes are soluble in dimethylsulfoxide, dimethylformamide, and pyridine. The X-ray diffraction studies (powder) show that all of the complexes are crystalline, although linear Br–PDMS–Br is a liquid at room temperature. Figure 6.7 shows the powder X-ray diffraction patterns of γ-CD/Br–PDMS–Br complexes under various conditions. The γ-CD/Br–PDMS–Br were found to have a cage-type structure at room temperature; however, the γ-CD/Br–PDMS–Br complexes under sonic energy or at room temperature without light and mixing adopt a head-to-head channel-type structure in which γ-CD molecules are stacked along a Br–PDMS–Br axis to form a cylinder [16, 17, 25–28].

6.3.2 Characterization of Polyrotaxane-based Pentablock Copolymers

First-order kinetic polymerization plots of St and PVAc initiated by Br–PDMS–Br/γ-CD macroinitiator are shown in Figure 6.8. The linear fit indicated that the concentration of propagating radical species is constant and radical termination reactions are not significant over the timescale of the reaction. It was found that the molecular weights increase almost linearly with conversion, indicating that the number of chains was constant, and the chain transfer reactions were rather negligible [25–28]. Another important feature observed for controlled radical polymerization was that the molecular weight distribution

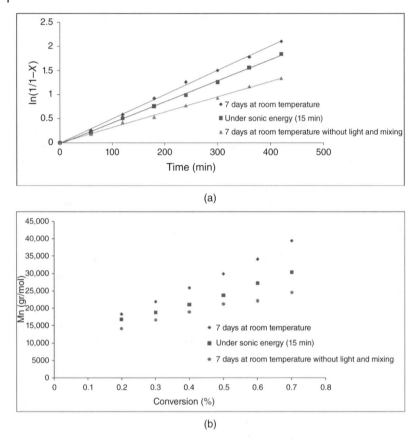

Figure 6.8 (a) Time dependence of $\ln[M]_0/[M]$ (M: monomer) and (b) dependence of the pentablock copolymer M_n on the conversion for the ATRP of St and PVAc with Br–PDMS–Br/γ-CD macroinitiator, which is synthesized from various conditions at 60 °C.

decreases with the progress of polymerization, indicating that nearly all the chains start to grow simultaneously [19–23]. Thus, from Figure 6.8, one may conclude that the polymerization was controlled and had narrow molecular weight distribution with a good agreement between theoretical and experimental results of molecular weight (Table 6.3).

GPC is an effective analytical technique to discover the structure of polyrotaxanes. If polyrotaxanes are not efficiently end-capped, dethreading of CDs will produce multiple GPC peaks. The GPC curves of polyrotaxanes-based pentablock copolymers (PSt-*b*-PVAc-*b*-PDMS/γ-CD-*b*-PVAc-*b*-PSt) are depicted in Table 6.3. M_n of PSt-*b*-PVAc-*b*-PDMS-*b*-PVAc-*b*-PSt (27,800) and M_n γ-CD (1270), the number of CDs capped in polyrotaxane can be calculated. By measuring Mn of

Table 6.3 GPC and ^{1}H NMR results of PSt-*b*-PVAc-*b*-[γ-CD/PDMS]-*b*-PVAc-*b*-PSt initiated with γ-CD/Br–PDMS–Br synthesized from various conditions.

Reaction	Reaction	$X^{a)}$(%)	$\overline{M}_{n,\,^1H\,NMR}$ (g/mol)	$\overline{M}_{n,\,GPC}$ (g/mol)	PDI
A	7 days at room temperature without light and mixing	40	41,900	44,100	1.53
B	Under sonic energy (15 min)	56	46,300	45,400	1.37
C	7 days at room temperature	72	48,500	49,500	1.24

a) Final conversion measured by gravimetric method.

PSt-b-PVAc-b-PDMS-b-PVAc-b-PSt and Mn γ-CD, the number of CDs capped in polyrotaxane can be calculated. When Br–PDMS–Br/γ-CD was formed by mixing at room temperature and used as macroinitiator, the yield of copolymer was greater than that of Br–PDMS/γ-CD formed under sonic energy and 13 γ-CDs capped in polyrotaxane. When Br–PDMS–Br/γ-CD under sonic energy at room temperature was used as macroinitiator, 11 γ-CDs capped in polyrotaxane. It is implied that the number of entrapped γ-CD can be modified in this ATRP process [25, 26].

^{1}H NMR spectra of the PSt-*b*-PVAc-*b*-PDMS/γ-CD-*b*-PVAc-*b*-PSt pentablock copolymer are illustrated in Figure 6.9. All signals of the ^{1}H NMR spectra were assigned to their corresponding monomers, and it can be declared that the synthesis of pentablock copolymers has proceeded successfully [25–28]. The ^{1}H NMR of pentablock showed a signal at 0.0–0.2 ppm associated with CH_3—Si methyl protons of PDMS in addition to some other signals at 2.1–2.3 ppm and 6.6–7.1 ppm corresponding to the $OCOCH_3$ group from PVAc segment and PSt segment, respectively. Therefore, it is possible to synthesize the polyrotaxane-based pentablock copolymers via ATRP of monomers such as vinyl acetate and styrene at the presence of bis(haloalkyl)-terminated PDMS macroinitiator. Signals observed at about 4.5 ppm were attributed to the OH of CDs segment [25–28].

It was concluded from Figure 6.9 that when Br–PDMS–Br/γ-CD (used as macroinitiator) was formed by mixing at room temperature for 7 days, the yield of copolymer was greater than that of Br–PDMS–Br/γ-CD formed under either sonic energy or 7 days at room temperature without light and mixing. This is explained by the higher content of γ-CD in the complex that is able to trap more PVAc and PSt units. The well-defined microstructure can be used in macromolecular design of the pentablock copolymer, assuming that one can increase the mole ratio of the monomer or use different molecular weights of the VAc or PDMS/CD macroinitiator in the structure [27, 28].

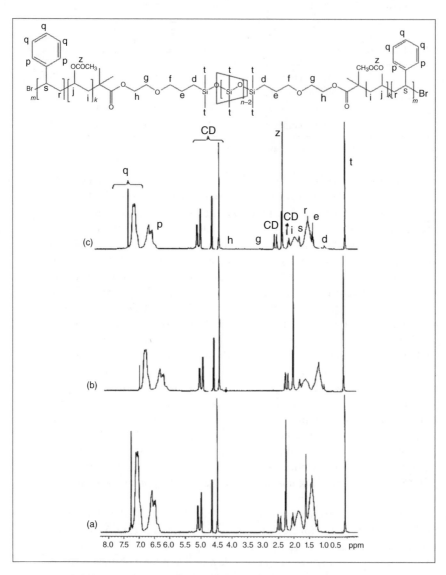

Figure 6.9 ^{1}H NMR of PSt-*b*-PVAc-*b*-PDMS/γ-CD-*b*-PVAc-*b*-PSt initiated with inclusion complex of γ-CD/Br–PDMS–Br synthesized from (a) 7 days at room temperature, (b) under sonic energy (15 min), and (c) 7 days at room temperature without light and mixing.

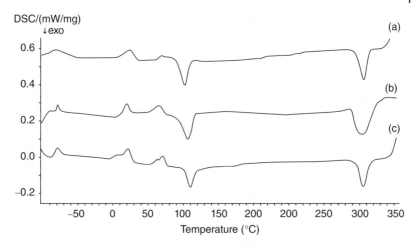

Figure 6.10 DSC thermograms for the (a) PSt-*b*-PVAc-*b*-(PDMS/γ-CD)-*b*-PVAc-*b*-PSt initiated with inclusion complex of γ-CDs/Br–PDMS–Br synthesized at (a) 7 days at room temperature, (b) under sonic energy (15 min), and (c) 7 days at room temperature without light and mixing.

The ratios between the sequence lengths of PVAc and PSt to PDMS/γ-CD mixed 7 days in room temperature were reported as 2.11 and 2.19, respectively, while that of PVAc and PSt to PDMS/γ-CD under sonic energy were 2.06 and 2.03, respectively, and finally that of PVAc and PSt to PDMS/γ-CD 7 days in room temperature without mixing were 1.95 and 2.02, respectively. Thereby, pentablock copolymers of PSt-*b*-PVAc-*b*-PDMS/γ-CD-*b*-PVAc-*b*-PSt were synthesized. Therefore, it is possible to synthesize the polyrotaxane-based pentablock copolymers via ATRP of monomers such as vinyl acetate and styrene using haloalkyl-terminated PDMS/CD macroinitiator [25, 26]. DSC thermograms of PVAc-based pentablock copolymers at the temperature range −100 to +350 °C are depicted in Figure 6.10. Glass transition temperature (T_g) of PDMS has been reported to be about −120 °C [27, 28]. PVAc and PSt segments in the corresponding block copolymers exhibited T_g values of 32 and 80 °C, respectively, which can be observed in Figure 6.10 [25, 26].

The DSC plot of γ-CD revealed two endothermic peaks, one at temperature range 100–110 °C due to the loss of water and another one near 330 °C, which corresponds to the γ-CD fusion, though they are not observed in Figure 6.10. These results were characteristic of microphase-separated morphology of the pentablock copolymers [12, 25, 26]. It can be judged that PVAc-based pentablock copolymers have been synthesized successfully taking into account the DSC results.

The thermoresponsiveness of the pentablock copolymers was investigated with ^{1}H NMR in CDCl$_3$. The complex formation of pentablock copolymer with cyclodextrin studied by ^{1}H NMR in thermal cycle of 22—55 °C and 22 °C is shown in Figure 6.11. The interaction of cyclodextrin with aromatic end group of the block copolymer, indicated in peaks around 6.0–8.0 ppm, shows temperature dependency. These peaks shift to 7.0–8.0 ppm during the temperature cycle from 55 to 22 °C. This shift affects further the observed peaks in the region around 3.0–5.0 ppm and 1.0–3.0 ppm. This indicates temperature-induced effect on the complex in a reversible process that dislocates cyclodextrin from the end group of the block copolymer. The appearance of sharp peaks related to s, p, and q protons at ≈1.5 ppm and 7.0–7.5 ppm indicated in this process confirms this effect, as a temperature-induced reversible transition or sol–gel transition of the inclusion complex, where the sharp peaks for cyclodextrin in Figure 6.5 from 4.5–5.0 ppm shift to 2.0–2.5 ppm region in Figure 6.11.

Figure 6.11 shows the ^{1}H NMR spectra obtained for the PSt-*b*-PVAc-*b*-PDMS-*b*-PVAc-*b*-PSt at 22 °C (Figure 6.11a), temperature increased to 55 °C (Figure 6.11b), and temperature decreased from 55 to 22 °C (Figure 6.11c). The integral of the peaks associated with protons adjacent to the styrene and vinyl acetate (6.5–7 ppm and 1.2–2.1 ppm, respectively) was observed to decrease with increasing temperature from 22 to 55 °C relative to the integral of the protons in the PMDS (Figure 6.11b). This reflects the mobility of the γ-cyclodextrin in block copolymer structure; upon decreasing the temperature to 22 °C, the integral of the peaks associated with protons adjacent to the styrene and vinyl acetate, the copolymers, undergo increase to reach initial integral (Figure 6.11c) [25–28].

Figure 6.12 shows the XRD spectra obtained for the PSt-*b*-PVAc-*b*-PDMS-*b*-PVAc-*b*-PSt at 22 °C (Figure 6.12a), temperature increased to 55 °C (Figure 6.12b), and temperature decreased from 55 to 22 °C (Figure 6.12c). Characteristic peaks of PDMS, PVAc, and PSt occur at $2\theta = 8, 13, 18, 2\theta = 13$, and at $2\theta = 6, 14, 12, 11, 8$–$17, 25$–$29$, respectively. At 22 °C, crystallinity percent of PVAc and PSt are 13% and 29%, respectively (Figure 6.12a). By increasing the temperature from 22 to 55 °C, crystallinity percent of PVAc and PSt become 9% and 43%, respectively (Figure 6.12b), which indicated migration of CD from PVAc to PSt. By decreasing the temperature from 55 to 22°C, crystallinity percent of PVAc and PSt become 10% and 38%, respectively (Figure 6.12c), which indicated migration of CD from PSt to PVAc [25–28].

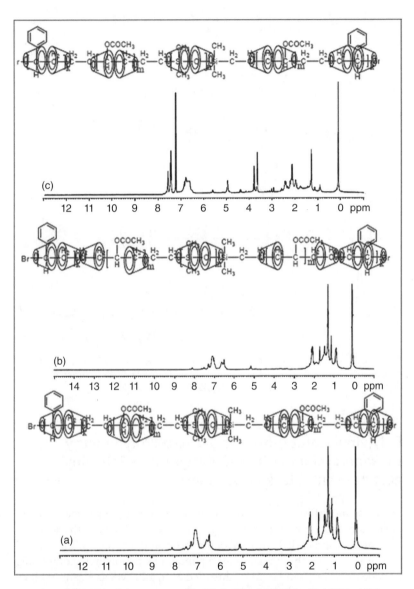

Figure 6.11 ^{1}H NMR of thermoreversible PSt-*b*-PVAc-*b*-PDMS/γ-CD-*b*-PVAc-*b*-PSt block copolymers initiated with inclusion complex of γ-CD/PDMS (synthesized 7 days at room temperature) at (a) 22 °C, (b) 55 °C, and (c) 22 °C.

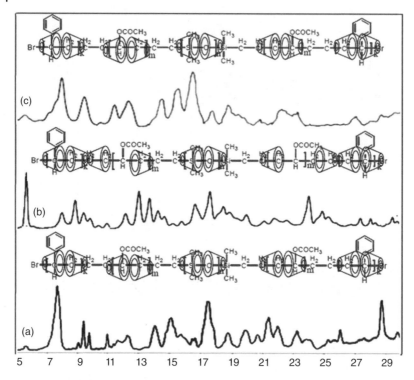

Figure 6.12 XRD of thermoreversible PSt-*b*-PVAc-*b*-PDMS/γ-CD-*b*-PVAc-*b*-PSt block copolymers initiated with inclusion complex of γ-CD/Br–PDMS–Br (synthesized 7 days at room temperature) at (a) 22 °C, (b) 55 °C, and (c) 22 °C.

6.4 Synthesis and Characterization of Poly(vinylacetate)-*b*-Polystyrene-*b*-(Polydimethyl siloxane/cyclodextrin)-*b*-Polystyrene-*b*-Poly(vinyl acetate) Pentablock Copolymers

Polyrotaxanes (PRs) are one of the most thoroughly investigated types of supramolecular polymers having many cyclic molecules threaded on a linear axis and ended with bulky moieties on both sides of their linear axis. A series of polyrotaxane-based pentablock copolymers comprising cyclodextrins (γ, β, and α-CDs) threaded onto a PDMS as a central block and PVAc outer blocks as end-stoppers are prepared via ATRP of PVAc and PSt in this work. The presence of styrene in the block copolymer can enhance rigidity of the chain, and it can thus be used in the macromolecular design of silicone-based copolymers [15, 17, 32–34]. The structure of the resultant copolymers was characterized in detail by ^{1}H NMR,

GPC, and DSC techniques. The results of the current research uncovered that pentablock copolymers containing γ-CD/PDMS−Br, which were synthesized at room temperature for 7 days, have lower PDI than other conditions (under sonic energy, at room temperature for 7 days, and kept out of light). The results of this study also demonstrated that pentablock copolymers containing γ-CD/PDMS−Br have lower PDI than β-CD/PDMS−Br and α-CD/PDMS−Br inclusion complexes (PDI = 1.24, 1.31, and 1.5, respectively). Temperature-dependent complexations with CDs have been extensively studied earlier due to their ability to produce supramolecular structures of desired architectures [18–20]. Generally speaking, the rate of polymer complexation with CD increases significantly with raising the temperatures. Rheological effects of temperature-dependent complexations were also studied [21–23]. Thermosensitive micellization phenomena for a supramolecular polymeric host–guest system consisting of PVAc and PSt with a α, β, γ-cyclodextrin (α, β, γ-CD) core (as the host polymer) and bis(boromoalkyl)-terminated poly(dimethylsiloxane) (PDMS) (as the guest polymer) in an aqueous solution showed that interesting host–guest system exhibited a dual thermoresponsive behavior due to two existing kinds of thermoresponsive segments, PDMS and cyclodextrin. In other words, this polymeric host–guest system was able to form CABAC-type supramolecular pentablock copolymers via inclusion complexation in aqueous solution. This pentablock copolymer underwent a reversible temperature-induced transition to aggregate under correct conditions. The critical micelle temperature (CMT) for this thermoresponsive host–guest system was dependent on the composition and concentration of the components. The size of the resultant noncovalently connected micelle could be tuned simply not only via adjusting temperature and concentration of the components but also by modifying host/guest ratio and the length of the PDMS blocks in the guest polymer [18–20, 23, 24]. This pentablock copolymer revealed a thermosensitive micelle formation behavior in the solid state. As temperature increased, the PDMS/CD chains were dehydrated, while the color of the powder was changed from white to green through hydrophobic–hydrophobic interactions. Further increase of the temperature caused the phase transition related to PDMS to occur on the corona of noncovalently connected micelles. Therefore, the micelles were destabilized, which resulted in micelle aggregation and precipitation. The supramolecular system presented here is rather different from the reported systems. This work has formed dual thermoresponsive pentablock copolymers in order to propose a novel supramolecular approach in designing and constructing the noncovalently connected polymeric micelles with modifiable properties.

6.4.1 Preparation of PDMS Macroinitiator (Br–PDMS–Br)

First, a solution of HO–PDMS–OH in the anhydrous THF was prepared and placed in the three-neck round-bottomed flask. According to its molecular weight and data reported in the literature [14], the microstructure of HO–PDMS–OH can be shown as HO–R–PDMS$_n$–R–OH (Scheme 6.2 with $n \approx 0$). Triethylamine was added to the solution followed by slow addition of 2-bromo-2-methylpropionyl bromides at 0 °C with stirring. The solution was left overnight at room temperature (Scheme 6.2). The triethylammonium bromide salt was removed by filtration of the solution, and the solvent was then removed under vacuum. The resulting oil (yellow in color) was redissolved in dichloromethane and washed two times with saturated sodium bicarbonate solution [25, 26]. The organic layer was isolated, dried over anhydrous magnesium sulfate, and filtered. ^{1}H NMR of the functionalized PDMS was used to verify the quantitative modification of the end groups to the terminal bromine atoms. Then, the final product was obtained as oil after removing volatiles under vacuum. GPC analysis showed that $M_n = 7400$, with MWD = 1.93 and maximum yield 82% determined gravimetrically. ^{1}H NMR (CDCl$_3$): $\delta = 0.0$–0.3 ppm (protons of methyl groups of —Si(CH$_3$)$_2$O) and $\delta = 3.2$ ppm (methylene group(—CH$_2$—CH$_2$—) next to the bromide) [25, 26].

GPC is an effective analytical technique to discover the structure of polyrotaxanes. If polyrotaxanes are not efficiently end-capped, dethreading of CDs will produce multiple GPC peaks. The GPC results of polyrotaxanes-based pentablock copolymers (PVAc-*b*-PSt-*b*-[γ-CD/PDMS]-*b*-PSt-*b*-PVAc) are depicted in Table 6.4. The molecular weight was increased during 7 days at room temperature. It is implied that the number of entrapped γ-CD can be modified in this ATRP process [25, 26].

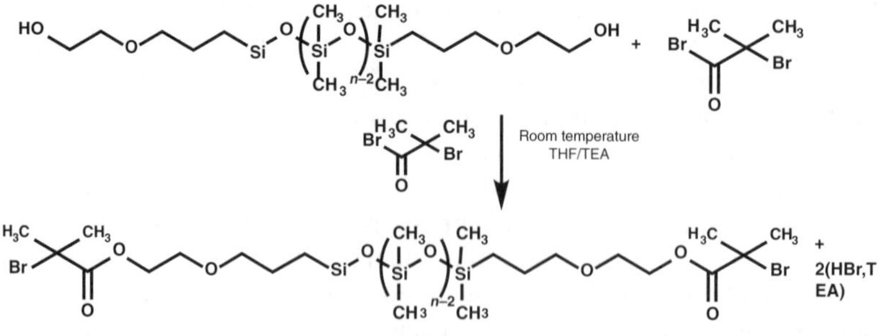

Scheme 6.2 Reaction scheme for synthesis of bis(2-bromoisobutyrate)-terminated PDMS macroinitiator from bis(hydroxyalkyl)-terminated PDMS.

Table 6.4 Summary of the results obtained from GPC and ^{1}H NMR analyses of PSt-b-PVAc-b-[γ-CD/PDMS]-b-PVAc-b-PSt initiated with γ-CD/PDMS synthesized from various conditions [VAc:St:(PDMS/CDs) = 1:2:0.5].

Exp. No.	Reaction	$X^{a)}$(%)	$M_{n, ^1H\,NMR}$ (g/mol)	$M_{n, GPC}$ (g/mol)	PDI
1	7 days at room temperature without light and mixing	40	24,900	24,100	1.53
2	Under sonic energy (15 min)	56	26,300	25,400	1.37
3	7 days at room temperature	72	28,500	29,500	1.24

a) Final conversion measured by gravimetric method.

^{1}H NMR spectra of the PVAc-b-PSt-b-PDMS/γ-CD-b-PSt-b-PVAc pentablock copolymer are illustrated in Figure 6.13. All signals of the ^{1}H NMR spectra were assigned to their corresponding monomers, and it can be declared that the syntheses of pentablock copolymers have proceeded successfully [17–20, 27]. The ^{1}H NMR of pentablock showed a signal at 0.0–0.2 ppm associated with CH_3—Si methyl protons of PDMS in addition to some other signals at 2.1–2.3 ppm and 6.6–7.1 ppm corresponding to the $OCOCH_3$ group from PVAc segment and PSt segment, respectively.

Therefore, it is possible to synthesize the polyrotaxane-based pentablock copolymers via ATRP of monomers such as vinyl acetate and styrene in the presence of bis(haloalkyl)-terminated PDMS macroinitiator. Signals observed at about 4.5 ppm were attributed to the OH of CDs segment [25–28]. It was concluded from Figure 6.13 that when Br–PDMS/γ-CD (used as macroinitiator) was formed by mixing at room temperature for 7 days, the yield of copolymer was greater than that of Br–PDMS/γ-CD formed under either sonic energy or 7 days at room temperature without light and mixing. This is explained by higher content of γ-CD in the complex that is able to trap more PVAc and PSt units.

6.4.2 Polymerization of St and PVAc Initiated by PDMS–CDs Macroinitiator

First-order kinetic polymerization plots of St and PVAc initiated by Br–PDMS/CD macroinitiator are shown in Figure 6.14. The linear fit indicated that the concentration of propagating radical species is constant, and the radical termination reactions are not significant over the timescale of the reaction. It was found that the molecular weights increase almost linearly with conversion, indicating that the number

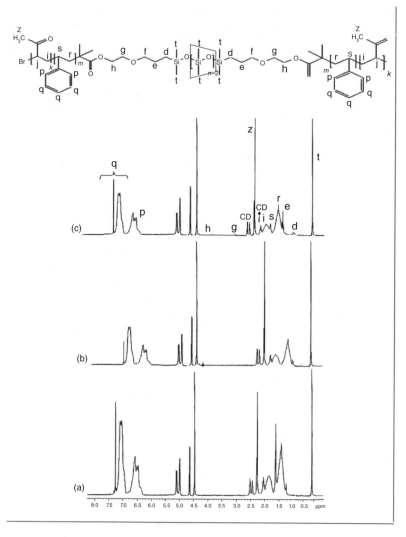

Figure 6.13 ¹H NMR of PVAc-*b*-PSt-*b*-[γ-CD/PDMS]-*b*-PSt-*b*-PVAc initiated with inclusion complex of γ-CD/PDMS synthesized from (a) 7 days at room temperature, (b) under sonic energy, and (c) 7 days at room temperature without light and mixing [VAc:St:(PDMS/CDs) = 1:2:0.5].

of chains was constant and the chain transfer reactions were rather negligible [18–20].

Another important feature observed for controlled radical polymerization was that the molecular weight distribution decreases with the progress of polymerization, indicating that nearly all the chains start to

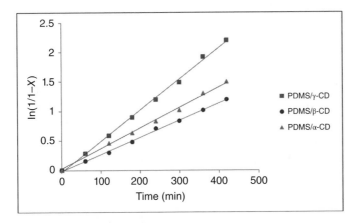

Figure 6.14 First-order kinetic plot for the ATRP of St and PVAc initiated by Br–PDMS/CDs at 60 °C for 6 h.

Table 6.5 Summary of the results obtained from GPC and ^{1}H NMR analyses of PSt-b-PVAc-b-(PDMS/CDs)-b-PVAc-b-PSt initiated with PVAc macroinitiator from bulk telomerization in the presence of Co(acac)$_2$ at 60 °C for 6 h [VAc:St:(PDMS/CDs) = 1:2:0.5].

Exp. No.	Reaction	$X^{a)}$(%)	$M_{n,}{}^1_{H\,NMR}$ (g/mol)	$M_{n,\,GPC}$ (g/mol)	PDI
1	PSt-b-PVAc-b-(PDMS/α-CD)-b-PVAc-b-PSt	48	21,400	23,000	1.5
2	PSt-b-PVAc-b-(PDMS/β-CD)-b-PVAc-b-PSt	59	29,700	31,000	1.31
3	PSt-b-PVAc-b-(PDMS/γ-CD)-b-PVAc-b-PSt	72	28,500	29,500	1.24

a) Final conversion measured by gravimetric method.

grow simultaneously [17–20]. Thus, from Table 6.5, one may conclude that the polymerization was controlled and had narrow molecular weight distribution with a good agreement between theoretical and experimental results of molecular weight. GPC results demonstrated that PDI is raised by increasing PDMS and/or its yield value [21–23]. The growing polymer peak has shifted to higher molecular weights continuously with monomer conversion. These results verify that the number of chains was constant [25–28].

^{1}H NMR spectra of the PVAc-b-PSt-b-PDMS/(α, β, γ-CD)-b-PSt-b-PVAc pentablock copolymer are illustrated in Figure 6.15. All signals of the ^{1}H NMR spectra were assigned to their corresponding monomers,

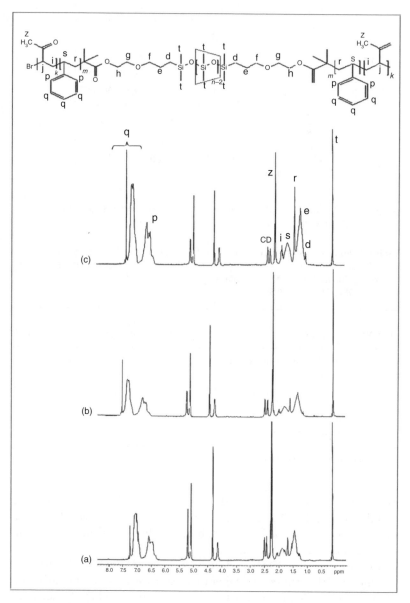

Figure 6.15 ^{1}H NMR thermograms for the PVAc-*b*-PSt-*b*-[CD/PDMS]-*b*-PSt-*b*-PVAc: (a) γ-CD/PDMS, (b) β-CD/PDMS, and (c) α-CD/PDMS initiated with inclusion complex of CD/PDMS at 60 °C for 6 h [VAc:St:(PDMS/CDs) = 1:2:0.5].

and it can be declared that the syntheses of pentablock copolymers have been successful [25–28]. The ^{1}H NMR of pentablock showed one signal at 0.0–0.2 ppm associated with CH_3—Si methyl protons of PDMS along with other signals at 2.1–2.3 ppm and 6.6–7.1 ppm related to $OCOCH_3$ group from PVAc and PSt segments, respectively. Therefore, it is possible to synthesize the polyrotaxane-based pentablock copolymers via ATRP of monomers such as vinyl acetate and styrene in the presence of haloalkyl-terminated PDMS/CD macroinitiator. Signals of about 4.5 ppm were also assigned to the OH of CDs segment [25–28].

From Figure 6.14 it was concluded that when Br–PDMS/γ-CD was formed as macroinitiator by mixing at room temperature for 7 days, the yield value of copolymer was greater than that of Br–PDMS/γ-CD developed under sonic energy or 7 days at room temperature without light and mixing. This may be related to the higher content of γ-CD in the complex that can trap more PVAc and PSt units [25, 26].

6.4.3 Microstructural Studies of the Pentablock Copolymers

A well-defined microstructure of vinyl acetate identifies the special peak of the PVAc with total intensity of

$$\sum I_{VAc} = (I_k + I_g + I_f)_{mainchain} + (I_i + I_h)_{end\ chains}$$

as a major peak of the PVAc. Dimethylsiloxane macroinitiator in Figure 6.1 indicates major ^{1}H NMR peak related to the brominated poly(dimethylsiloxane) (BPDMS) with total intensity characterized as:

$$\sum PDMS/CD = (I_d)_{DMS} + (I_e + I_g + I_f + I_m)_{BPDMS} + I_{CD}$$

The intensity of peaks for PSt segments is shown in Figure 6.1.

$$\sum PSt = I_p + I_g + I_s + I_r$$

The intensities are used to estimate the number of each monomer in pentablock copolymer. The number of styrene units is calculated as:

$$N_{St} = \frac{\sum PSt}{\sum PSt + \sum PDMS/CD + \sum PVAc}$$

The number of PDMS unit is:

$$N_{PDMS/CD} = \frac{\sum PDMS/CD}{\sum PSt + \sum PDMS/CD + \sum PVAc}$$

And the number of PVAc unit in the chain is:

$$N_{PVAc} = \frac{\sum PVAc}{\sum PSt + \sum PDMS/CD + \sum PVAc}$$

The well-defined microstructure can be used in macromolecular design of the pentablock copolymer, assuming that one can increase the mole ratio of the monomer or use different molecular weights of the VAc or PDMS/CD macroinitiator in the structure [25, 26].

The ratios between the sequence lengths of PVAc and PSt to PDMS/α-CD were reported as 2.08 and 2.14, respectively, while that of PVAc and PSt to PDMS/β-CD were 2.11 and 2.21, respectively, and finally that of PVAc and PSt to PDMS/γ-CD were 2.18 and 2.22, respectively. Thereby, pentablock copolymers of PVAc-b-PSt-PDMS/CD-b-PSt-b-PVAc were synthesized. Therefore, it is possible to synthesize the polyrotaxane-based pentablock copolymers via ATRP of monomers such as vinyl acetate and styrene using haloalkyl-terminated PDMS/CD macroinitiator [25, 26].

DSC thermograms of PVAc-based pentablock copolymers at the temperature range −100 to +250 °C are depicted in Figure 6.16. Glass transition temperature (T_g) of PDMS cannot be addressed in Figure 6.16, since its value has been reported to be about −120 °C. PVAc and PSt segments in the corresponding block copolymers exhibited T_g values

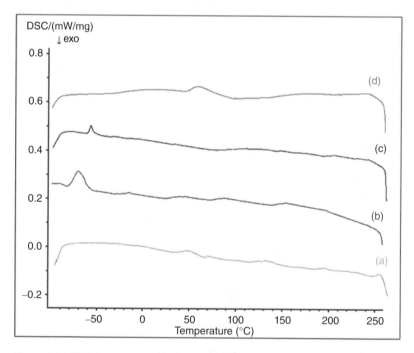

Figure 6.16 DSC thermograms for the (a) PSt-b-PVAc-b-PDMS-b-PVAc-b-PSt, (b) PSt-b-PVAc-b-(PDMS/α-CD)-b-PVAc-b-PSt, (c) PSt-b-PVAc-b-(PDMS/β-CD)-b-PVAc-b-PSt, and (d) PSt-b-PVAc-b-(PDMS/α-CD)-b-PVAc-b-PSt initiated with inclusion complex of CDs/PDMS at 60 °C for 6 h [VAc:St:(PDMS/CDs) = 1:2:0.5].

of 32 and 80 °C, respectively, which cannot be observed in Figure 6.16 [25–28]. The DSC plot of γ-CD revealed two endothermic peaks, one at temperature range 100–110 °C due to the loss of water and another one near 330 °C that corresponds to the γ-CD fusion, though they are not observed in Figure 6.16. These results were characteristic of microphase-separated morphology of the triblock copolymers. It can be judged that PVAc-based pentablock copolymers have been synthesized successfully taking into account the DSC results [25–28].

6.5 Conclusion

Polyrotaxane-based pentablock copolymers were successfully synthesized via ATRP of St and PVAc using CuCl/PMDETA as a catalyst system at 60 °C. Copolymers are of major interest because of their possibility to form amphiphilic pentablock copolymers of PVA-*b*-PSt-*b*-PDMS-*b*-PSt-*b*-PVA by hydrolysis of the PVAc blocks. ^{1}H NMR technique was utilized to analyze microstructure of the pentablock copolymers as well as the number of PSt, PDMS/CD, and PVAc units in the block copolymer. There was a very good agreement between the number-average molecular weight calculated from ^{1}H NMR spectra and the one calculated theoretically. The narrow dispersity indices from GPC of the synthesized pentablock copolymers produced a lower PDI block copolymer due to the living/controlled characteristic of the reaction. The obtained results showed that ATRP of St and PVAc initiated with inclusion complex of PDMS/CDs at the presence of PVAc from bulk telomerization using Co(acac)$_2$ occurs with well-controlled compositions and molecular weights as well as narrow polydispersity indices. Furthermore, DSC results uncovered that PDMS-based pentablock copolymers have been synthesized successfully.

References

1 (1) Yang, C., Yang, J., Ni, X. and Li, J. (2009) Novel supramolecular block copolymer: a polyrotaxane consisting of many threaded α- and γ-cyclodextrins with an ABA triblock architecture. *Macromolecules*, **42**, 3856–3859.

2 Sauvage, J.P. and Dietrich, B. (1999) Undirectional rotary motion in a molecular system. *Nature*, **401**, 150–152.

3 Semlyen, J.A. (2000) *Cyclic Polymers*, 2nd edn, Kluwer Academic Publisher, New York.

4 Bender, M.L. and Komiyama, M. (1987) Importance of apolar binding in complex formation of cyclodextrins with adamantanecarboxylate. *J. Am. Chem. Soc.*, **100**, 2259–2267.
5 Wenz, G., Han, B.H. and Muller, A. (2006) Cyclodextrin rotaxanes and polyrotaxanes. *Chem. Rev.*, **106**, 782–817.
6 Harada, A. and Kamachi, M. (1990) Complex formation between poly(ethylene glycol) and α-cyclodextrin. *Macromolecules*, **23**, 2821–2823.
7 (a) Jones, R.G. (2000) Kluwer Academic Publishers. *Dordrecht: The Netherlands*, 79; (b) Duo, Q., Wang, C., Cheng, C. et al. (2006) PDMS-modified polyurethane films with low water contact angle hysteresis. *Macromol. Chem. Phys.*, **207**, 2170–2179.
8 Braunecker, W.A. and Matyjaszewski, K. (2007) Controlled/living radical polymerization: features, developments, and perspectives. *Prog. Polym. Sci.*, **32**, 93–146.
9 Matyjaszewski, K. (2005) Macromolecular engineering: from rational design through precise macromolecular synthesis and processing to targeted macroscopic material properties. *Prog. Polym. Sci.*, **30**, 858–75.
10 Debuigne, A., Caille, J.R. and Jerome, R. (2005) Synthesis of end functionalized poly(vinyl acetate) by cobalt mediated radical polymerization. *Macromolecules*, **38**, 5452–5458.
11 Okumura, H., Kawaguchi, Y. and Harada, A. (2001) Preparation and characterization of inclusion complexes of poly(dimethylsiloxane)s with cyclodextrins. *Macromolecules*, **34**, 6338–6343.
12 Okumura, H., Kawaguchi, Y. and Harada, A. (2003) Preparation and characterization of the inclusion complexes of poly(dimethylsilane)s with cyclodextrins. *Macromolecules*, **36**, 6422–6429.
13 Storsberg, A. and Helmut, R. (2000) Cyclodextrins in polymer synthesis: polymerization of methyl methacrylate under atom-transfer conditions (ATRP) in aqueous solution. *Macromol. Rapid Commun.*, **21**, 3250–3253.
14 Tamer, U., Mariana, R. and Alan, E.T. (2004) Polymerization of styrene in cyclodextrin channels: can confined free-radical polymerization yield stereoregular polystyrene? *Macromol. Rapid Commun.*, **25**, 1382–1386.
15 Semsarzadeh, M.A. and Amiri, S. (2012) Preparation and characterization of inclusion complexes of poly(dimethylsiloxane)s with γ-cyclodextrin without utilizing sonic energy. *Silicon*, **4**, 151–156.
16 Ivaylo, D., Barbara, T., Andrzej, D. and Christo, B.T. (2007) Thermosensitive water-soluble copolymers with doubly responsive reversibly interacting entities. *Prog. Polym. Sci.*, **32**, 1275–1343.

17 Yu-Cai, W., Ling-Yan, T., Yang, L. and Jun, W. (2009) Thermoresponsive block copolymers of poly(ethylene glycol) and polyphosphoester: thermo-induced self-assembly, biocompatibility, and hydrolytic degradation. *Biomacromolecules*, **10**, 66–73.

18 Giancarlo, M., Marco, D. and Vittorio, C. (2008) ATRP synthesis and association properties of thermoresponsive anionic block copolymers. *J. Polym. Sci.: A: Polym. Chem.*, **46**, 4830–4842.

19 Neeraj, K., Majeti, N.V.R. and Domba, A.J. (2001) Biodegradable block copolymers. *Adv. Drug Deliv. Rev.*, **53**, 23–44.

20 Dai, X.H., Dong, C.M., Fa, H.B. et al. (2006) Supramolecular polypseudorotaxanes composed of star-shaped porphyrin-cored poly(ε-caprolactone) and α-cyclodextrin. *Biomacromolecules*, **7**, 3527–3533.

21 Fujita, H., Ooya, T. and Yui, N. (1999a) Synthesis and characterization of a polyrotaxane consisting of β-cyclodextrins and a poly(ethylene glycol)–poly(propylene glycol) triblock copolymer. *Macromol. Chem. Phys.*, **200**, 706–713.

22 Bromberg, L.E. and Ron, E.S. (1998) Temperature-responsive gels and thermogelling polymer matrices for protein and peptide delivery. *Adv. Drug Deliv. Rev.*, **31**, 197–221.

23 Choi, H.S., Kontani, K., Huh, K.M. et al. (2002) Rapid induction of thermoreversible hydrogel formation based on poly(propylene glycol)-grafted dextran inclusion complexes. *Macromol. Biosci.*, **2**, 298–303.

24 Semsarzadeh, M.A. and Amiri, S. (2013) Synthesis and characterization of PVAc-b-PDMS-b-PVAc triblock copolymers by atom transfer radical polymerization initiated by PDMS macroinitiator. *J. Inorg. Organomet. Polym.*, **23**, 553–559.

25 Semsarzadeh, M.A. and Amiri, S. (2013) Silicone macroinitiator in atom transfer radical polymerization of styrene and vinyl acetate: synthesis and characterization of pentablock copolymers. *J. Inorg. Organomet. Polym. Mater.*, **23**, 432–438.

26 Semsarzadeh, M.A. and Amiri, S. (2013) Preparation and properties of polyrotaxane from α-cyclodextrin and poly (ethylene glycol) with poly (vinyl alcohol). *Bull. Mater. Sci.*, **36**, 989–996.

27 Semsarzadeh, M.A. and Abdollahi, M. (2012) Atom transfer radical polymerization of styrene and methyl (meth)acrylates initiated with poly(dimethylsiloxane) macroinitiator: synthesis and characterization of triblock copolymers. *J. Appl. Polym. Sci.*, **123**, 2423–2430.

28 Trefonas, P., Djurovich, P.I., Zhang, X.H. et al. (1983) Organosilane high polymers: synthesis of formable homopolymers. *J. Polym. Sci. Polym. Lett. Ed.*, **21**, 819–823.

29 Huh, K.M., Ooya, T., Lee, W.K. et al. (2001) Supramolecular-structured hydrogels showing a reversible phase transition by inclusion complexation between poly(ethylene glycol) grafted dextran and α-cyclodextrin. *Macromolecules*, **34**, 8657–8662.
30 Silver Stain, R.M., Bassler, G.C. and Morrill, T.C. (1981) *Spectrometric Identification of Organic Compounds*, 5th edn, Wiley, New York.
31 Joachim, S., Markus, H., Axel, H.E.M. and Helmut, R. (2000) Cyclodextrins in polymer synthesis: polymerization of methyl methacrylate under atom-transfer conditions (ATRP) in aqueous solution. *Macromol. Rapid Commun.*, **21**, 1342–1346.
32 Storsberg, J., Hartenstein, M., Muller, A.H.E. and Ritter, H. (2000) Cyclodextrins in polymer synthesis: free radical polymerization of cyclodextrin host–guest complexes of methyl methacrylate or styrene from homogenous aqueous solution Joachim. *Macromol. Rapid Commun.*, **21**, 1342–1346.
33 Hsin-Fang, L., Hwo-Shuenn, S., U-Ser, J. et al. (2005) Preparation and supramolecular self-assembly of a polypeptide-*block*-polypseudorotaxane. *Macromolecules*, **38**, 6551–6558.
34 (a) Harada, A. and Kamachi, M. (1990) Complex-formation between cyclodextrin and poly (propylene glycol). *J. Chem. Soc., Chem. Commun.*, **19**, 1322–1323; (b) Harada, A. and Kamachi, M. (1992) The molecular necklace-a rotaxane containing many threaded alpha-cyclodextrins. *Nature*, **356**, 325–327; (c) Harada, A. and Kamachi, M. (1993a) Preparation and properties of inclusion complexes of poly (ethylene glycol) with alphacyclodextrin. *Macromolecules*, **26**, 5698–5703; (d) Harada, A. and Kamachi, M. (1993b) Synthesis of a tubular polymer from threaded cyclodextrins. *Nature*, **364**, 516–518; (e) Harada, A. and Kamachi, M. (1994b) Preparation and characterization of a polyrotaxane consisting of monodisperse poly (ethylene glycol) and alpha-cyclodextrins. *J. Am. Chem. Soc.*, **116**, 3192–3196; (f) Harada, A. and Kamachi, M. (1995) Preparation and characterization of inclusion complexes of poly (propylene glycol) with cyclodextrins. *Macromolecules*, **28**, 8406–8411.

7

Cyclodextrin Applications

Cyclodextrins (CDs) are a family of enzymatically modified starches consisting of α-1,4-linked glucose monomers with a hydrophilic outer surface and a lipophilic central cavity. CDs and their derivatives can be used for the formation of inclusion complexes with a wide range of monomers because of their specific molecular structure and shape, and the formed inclusion complexes facilitate the alteration of physical, chemical, and biological properties of guest molecules.

The formation of inclusion complex increases the aqueous solubility of poorly soluble components, increases their bioavailability and stability, reduces irritation, prevents incompatibility, and masks undesirable odor and taste. CDs are widely used in various industries such as foods, flavors, cosmetics, packaging and textiles, drug delivery, gene delivery, and separation science. The formation of inclusion complex between CD and guest molecules protects compounds against light, heat, and oxygen or increases shelf life and makes controlled release of compounds into the environment.

7.1 Cyclodextrin Industrial Applications

The industrial applications of CDs are as follows:
- Control of solubility (to increase or decrease the solubility)
- Stabilization of volatile compounds
- Masking the effects or odor of the guest
- Enhancing the complexation efficiency
- Reduction of volatility
- Directing chemical reactions
- Control of fluorescence and light absorption
- Compatibility of cyclodextrins with other materials

Cyclodextrins: Properties and Industrial Applications, First Edition. Sahar Amiri and Sanam Amiri.
© 2017 John Wiley & Sons Ltd. Published 2017 by John Wiley & Sons Ltd.

- Encapsulation of a wide range of guest molecules and changing their properties
- Controlled release of guest molecules in drug delivery systems

7.1.1 Pharmaceutical Applications of Cyclodextrins

CDs can form inclusion complex with drugs for numerous purposes such as stabilization of volatile drugs, improvement of bioavailability of drugs and water solubility of poorly soluble drugs, encapsulation of drug and target drug delivery, and controlled drug release [1–3]. They are able to form inclusion complex with many drugs including hydrophobic or hydrophilic drugs in aqueous solutions and entrap drug in their cavity or surface by taking up the drug molecule or some lipophilic moiety of the molecule [4, 5]. Formation of inclusion complex is based on hydrogen bonding or van der Waals interaction and no covalent bonds are formed or broken during complexation, so interaction between CD and drug is reversible and drug molecules in the complex are in rapid equilibrium with free molecules in the solution [5, 6].

In drug delivery applications, the required amount of drug must be delivered to target site in the necessary period of time without undesirable effects on the other cells in the body both efficiently and precisely. For these reasons and to achieve target delivery, CDs are used for the formation of inclusion complex with drugs that encapsulate drug, mask the undesirable properties of drug molecules, and deliver drug with a controlled rate to target site [7].

Major challenge in drug delivery is poor bioavailability of drugs, which is related to poor solubility and low permeability of drugs. CDs improved aqueous solubility of limited water soluble or poorly water-soluble drugs, increased their bioavailability and stability, stabilized them, prevented drug–drug or drug–additive interactions, and delivered drug to target sites. CDs formed inclusion complex with hydrophobic drugs and encapsulated drugs in their cavity, improved their aqueous solubility, permeability, dissolution rate, and presystemic metabolism [8].

Limited bioavailability of drugs causes various problems when released from the formulation in dissolved form. Formation of inclusion complex improves dissolution and absorption rates and solubility [9].

CD complexation also improves chemical, physical, and thermal stability of drugs. Some drugs degrade upon exposure to oxygen, water, radiation or heat, chemical reactions or light, but inclusion complex formation entraps drug in CD cavity, makes diffusion of drug from CD difficult, and protects guest from reacting with active agents, thus improving stability. Complexation with CDs reduces local concentration of the free drug in the body, thus reducing irritancy of drugs that irritate the stomach, skin, or eye, and so these kinds of drugs can be used [10, 11].

Inclusion complexation of drugs with CDs masks unpleasant odor and taste of drugs, which is done by entrapping functional groups of drugs in CD cavity. Moreover, some drugs are oils/liquids at room temperature, and it is very difficult to handle them and convert them to stable solid dosage forms. Encapsulation of oil/liquid drugs leads to the formation of microcrystalline or amorphous powders [12].

7.1.2 Inclusion Complex Formation Advantages with Drugs

7.1.2.1 Water Solubility

Improved aqueous solubility of drugs is the most important and impressive property which is expected by the formation of inclusion complex between CDs and drugs. Formation of inclusion complex of drug with CDs alters guest compound solubility, that is, may increase or decrease the solubility.

In the case of hydrophobic drugs, drug is incorporated into the CD cavity, and hydrophilic surface of CD interacts with the solvent and increases the solubility of drug. In the case of volatile drugs, formation of inclusion complex stabilizes drugs and may increase their solubility [13, 14].

7.1.2.2 Drug Bioavailability

Nearly 40% of drugs are water insoluble or water soluble and show low bioavailability and hence low efficacy, safety, and patient compliance. For this reason, cyclodextrins are the best candidate for complex formation with drugs which entrap insoluble drugs in their cavity, improve their water solubility, dissolution, and permeability. By improving the bioavailability of drugs, drug permeability at the surface of the biological barrier, skin, mucosa, or the eye cornea can be improved. CDs are able to remove cholesterol, increase membrane fluidity, and induce membrane invagination through a loss of bending resistance [15].

The important parameter is that if solubility of drug increases so much, its availability may decrease [16, 17]. In the case of water-soluble drugs, cyclodextrins increase the drug permeability by direct action on mucosal membranes and enhance drug absorption and bioavailability [18].

In addition to improving the bioavailability of drugs, inclusion complex formation improves drug irritation which in turn improves drug life time and contact time at the absorption site in delivery systems and improved of drugs [19, 20].

7.1.2.3 Drug Safety

Some drugs may irritate and affect the body and so small dosage of them can be used, which means their usage is limited. Formation of inclusion complex between drug and CD increases applicability and drug solubility, thus resulting in the decrease of dose usage and reduction of drug toxicity. Moreover, by encapsulation of drug in CD cavity, direct contact of drug with biological membranes is limited which may decrease their side effects and local irritation with no drastic loss of therapeutic benefits [21, 22].

7.1.2.4 Drug Stability

Unstable drug usage is limited because of their short shelf life and reactions with active agents such as dehydration, hydrolysis, oxidation, and photodecomposition. For overcoming these problems, drugs can be reacted with CDs, encapsulated in CD cavity, interaction between drug and active agents and bioconversion at the absorption site can be inhibited, and drug stability can thus be improved.

Encapsulation process may cover the drugs and protect them against various degradation processes which depend on the nature and effect of the included functional group on drug stability and vehicle [23]. For example, photostability and shelf life of trimeprazine and promethazine are increased by inclusion complex formation between CDs and drugs [24–26].

Moreover, formation of inclusion complex postponed drug hydrolysis compared to free drugs and thus protected it [27–29].

7.1.2.5 Mask Unpleasant Odor, Taste, and Side Effects of Drugs

By improving drug water solubility, dissolution rate and absorption rate of drugs are improved.Moreover, the contact time between the drug and the body is decreased, drug irritation and amount of contact with taste receptors are limited, and drug safety which is very important in the preparation of oral solutions is improved. Complexation has been used to mask the unpleasant bitter taste of a number of drugs such as acetaminophen [30].

7.1.2.6 Drug Stability

As previously mentioned, the formation of inclusion complex has improved the stability of drug against hydrolysis, hydrolytic dehalogenation, oxidation, decarboxylation, and isomerization, in both liquid and solid states. CD kind and position of the guest molecule inside the CD cavity affected stabilization or destabilization process. By incorporation of drug into the CD cavity, interaction of the CD hydroxyls (or derivative functional groups) with a hydrolytically prone site is inhibited and hence stability improved [31–36].

Big guest molecules also form inclusion complex with CD and their chemical and physical stability are increased. The CDs will typically interact with functional groups present on exposed surfaces of the macromolecules and often form multiple complexes (several CDs per molecule) [13].

7.1.2.7 Drug Solubility and Dissolution

CDs and their derivatives find enormous applications in improved water solubility and stability of poorly water-soluble drugs by the formation of inclusion complex. CDs can act as nanocarriers for controlled drug release and drug delivery to target site [21, 37].

By improving drug solubility, crystallinity on complexation or solid dispersion is decreased and drug solubility and dissolution rate are increased in both solid state and liquid medium [13, 38, 39].

Water solubility of antibiotics is highly improved by complex formation with CDs [14].

7.1.2.8 Reduction in Volatility

Several drugs, spices, flavors, and essential oils such as lemon oil are volatile substances at room temperature and cannot be used as ordinary applications. Their compounds form inclusion complex with CDs and become stable at room temperature [15, 16].

7.1.3 CD-based Carriers

Encapsulation of drugs with CDs protect drugs from active agents, enzymatic degradation, hydrolysis, and oxidation and release the drug in a sustained pattern at specific and controlled rate, so that high concentrations of drug is not present at target site. Furthermore, CD-based nanocapsules easily permeate through biological membranes and reach the target site, so drug bioavailability and solubility of water-insoluble drugs are improved [18].

7.2 Drug Delivery Systems Based on Cyclodextrins

It is known that CDs have various applications in drug delivery systems which are able to form complex with drugs, act as functional carrier in pharmaceutical formulations for efficient and precise delivery of required dosage of drugs to a target site for a necessary time. CDs can be used in the design of some novel delivery systems such as liposomes, microspheres, osmotic pump, peptide and protein delivery, and specific site-targeting nanoparticles. These systems are more attractive in colon-specific drug delivery, brain drug delivery, or brain targeting with controlled release [39]. CD-based drug delivery systems can be used for oral, transdermal, or ocular drug delivery.

7.2.1 Oral Drug Delivery System

Oral drug delivery systems are the most common route for designing new drug delivery system due to limit effects influencing gastrointestinal absorption such as pH, food, stomach emptying, and drug released is either dissolution controlled, diffusion controlled, osmotically controlled, density controlled, or pH controlled [10]. Oral delivery systems can be used for drugs in the form of suspensions, oily drops, gels, ointments, and solid inserts, and mask their unwanted side effects such as eye irritation and blurred vision. Formation of complex between drug and CD may stabilize the liquid drug in the form of powder and improve the solubility of poorly water-soluble drugs so that drug permeability through biological membranes is improved and drugs can be permeated from an aqueous medium to the lipophilic absorption surface in the gastrointestinal tract. CD acts as a carrier by which hydrophobic drug is incorporated into CD cavity, solubility of drug is improved, and the drug is delivered to the surface of the ocular barrier. CD can moderate drug concentration in target site and prevent a rapid increase in the systemic drug concentration and systemic and hepatic first-pass metabolism. Chemically modified hydrophobic and hydrophilic CDs may serve as novel slow-release carriers for water-soluble drugs, including proteins and peptide drugs, and enhance drug absorption [40–43].

7.2.2 Rectal Drug Delivery System

In the case of drugs which have bitter and nauseating taste, drug delivery remains a challenge especially for children or infants and elderly. Rectal drugs delivery systems are a good choice to overcome these problems and deliver drugs to the unconscious patients, children, and infants. The dissolution media volume, low rectal bioavailability of some drugs, limited absorbing surface and drug degradation by microorganisms present in the rectum, and disordered release of drugs limited rectal drug delivery systems. One of the methods to improve efficiency of rectal drug delivery systems is complexation of drug with CDs, which improves water solubility of drugs and bioavailability of drugs through biological membrane, and masks bitter and undesirable taste. Inclusion complex of CD and drug led to rectal drug delivery systems, which prevent irritation of the rectal mucosa, inhibit the reverse diffusion of drugs into the vehicle, and also prevent their auto-oxidation [44–46].

In addition, using CDs in rectal drug delivery systems improved hydrophilic drug release from vehicles and the dissolution rates with a controlled rate to target site such as rectal epithelial cells [11].

7.2.3 Nasal Drug Delivery System

Some drugs which have low oral bioavailability and incitement gastrointestinal cannot be used directly. Bioavailability is due to low water solubility of drug through biological membrane. In the case of hydrophilic drugs, low nasal bioavailability leads to poor transport properties across the nasal mucosa. Lipophilic drugs and large hydrophilic drugs such as peptides and proteins show insufficient nasal absorption due to their poor water solubility.

Complexation of CDs with nasal drugs leads to higher solubility of drug, improved drug delivery through biological barriers, and enhanced nasal absorption. CD-based nasal drug delivery did not show local, systemic, ciliostatic, irritating, and allergenic effects, but increased drug aqueous solubility and permeation through nasal epithelium [46, 47].

7.2.4 Transdermal Drug Delivery System

Drugs which must be absorbed for both local and systemic action without undesirable intestinal or gastrointestinal irritation and also higher drug dosage can be delivered via transdermal drug delivery system. In this system, drug is released in blood circulation over a long desired period of time, but this did not have any harmful effect to the body cell. The common thing between this method and other drug delivery systems is solubility, absorption, and permeation characteristics.

The transdermal drug transport is greatly limited by stratum corneum permeation characteristics and drugs must maintain sufficient stability until delivered to target site. Formation of inclusion complex between drug and CDs stabilized drug in transdermal drug delivery for safe therapeutic effects and didn't interact or irritate the skin or change the pH of the skin [48–52].

7.2.5 Ocular Drug Delivery System

The ocular drug delivery is dependent on drug's physicochemical properties and anatomical barriers (including different layers of cornea, sclera, and retina including blood aqueous and blood–retinal barriers, choroidal and conjunctival blood flow, lymphatic clearance, and tear dilution) of the eye, and eye drop is the most attractive form of this type. Because eye drop uses drugs in these systems, concentration of drug and colloidal dosage forms must be low, which cannot overcome static and dynamic barriers, so ocular drug delivery system's usage is restricted [53–55]. CDs can form inclusion complex with drugs to improve water solubility and stability of drugs, which prevent side effects of drugs such as irritation and discomfort without changing the physical and chemical properties of drugs. Moreover, the formed complex must be nontoxic, inert in nature, and improve the permeability of the drug through the corneal mucosa without irritating the ocular surface [30, 55–57].

7.2.6 Liposomal Drug Delivery

Liposomal-based drug delivery systems can be used for drug targeting but they cannot improve the solubility of drugs, so these systems reacted with CDs and show two properties. Hydrophilic drugs are incorporated in liposomes in the aqueous phase and hydrophobic drugs in the lipid bilayers, and preserve drugs in route to their target site and release at a specific rate [58, 59].

After formation of water-soluble complexes, CD improves solubility of insoluble drugs in aqueous media, increases drug-to-lipid mass ratio and insoluble drugs encapsulation and makes controlled drug delivery with reduced drug toxicity [17, 60].

Inclusion complexation highly increases the chemical stability of volatile drugs in multilamellar liposomes containing riboflavin/ γ-CD complex. Indomethacin/HP-β-CD inclusion complex improves the stability of hydrolyzable drugs. Liposomal entrapment in the presence of CD is higher than free drug, and also improves drug solubility and stability [20, 61].

Most of the drugs have poor aqueous solubility and this problem makes the formulation of efficient drugs difficult. Complexation of drugs with cyclodextrin will mask the undesirable properties and defects of drug, alter drug properties and cause better medical effect and preserve required time without adverse effects on other cells. Cyclodextrin-based carriers are necessary for safe and efficient drug delivery. Preparation of CD-based microsphere was first studied by Loftsson [49].

CD-based microspheres show stabilizing effect due to increased hydrophilicity of proteins caused by shielding of their hydrophobic residues in CD cavity which decreased aggregation and denaturation by keeping them away from methylene chloride/water interface and limited the interactions. One of the most promising agents for stabilizing lysozyme and bovine serum albumin (BSA) during primary emulsification of poly(D,L-lactide-co-glycolide) (PLGA) microsphere preparation is HP-β-CD. Chitosan microspheres based on CD release nifedipine in a slow rate and improve drug-loading efficiency. Uncomplex drug released out of the microspheres is primarily due to lower drug availability rather than CD/drug complex. Moreover, chitosan forms inclusion complex with CD and forms a hydrophilic layer around the lipophilic drug that further decreases the drug matrix permeability [62, 63].

Another technique used for the formation of CD-based microspheres consisting of β-CD/poly acrylic acid (PAA) is water/oil solvent evaporation technique which achieves high encapsulation efficiency [64].

7.2.7 Osmotic Pump Tablet

The most important application of osmotic pump systems is for oral formulations. These systems can be used for designing new systems for oral drug delivery with controlled drug release at a specific rate based on the principle of osmosis so that patient compliance is improved and higher dosage of drug may reach the target site. These tablets are rigid with a semipermeable outer membrane and one or more active drug through in. By using the tablet, the semipermeable membrane via osmosis absorbs water, then osmotic pressure or hydrostatic pressure pushes the active drug through the opening in the tablet and exposes the drug to an aqueous environment. Osmotic pump tablets are suitable for delivery of drugs with moderate to high water solubility such as oxybutynin chloride, nifedipine, and glipizide, and their efficiency is affected by drug solubility and the osmotic pressure of the core [65–68].

As previously mentioned, 60% of drugs have low water solubility and they cannot be used in these systems. To overcome this problem, these systems can be complexed with CDs. Testestrone, prednisolone, chlorpromazine, indometacin, and naproxen are poorly water-soluble drugs, which can form inclusion complex with CDs and CD derivatives and their solubility can be increased. CDs can act as solubilizing agent and osmotic agent and enhance the solubility of drugs in tablet formulations [69–71].

7.2.8 Peptide and Protein Delivery

In the past decades peptide and protein were attractive for therapeutic applications but degradation, chemical and enzymatic instability, poor solubility and absorption through biological membranes, immunogenicity, and target delivery problems limited their use, and are currently delivered intravenously or subcutaneously [32–34].

Peptides can be designed to target a broad range of molecules, giving them almost limitless possibilities in fields such as oncology, immunology, infectious disease, and endocrinology. For overcoming this problem, oral peptide delivery must be improved, which is done by using absorption enhancers, enzyme inhibitors, carrier systems, and stability enhancers [72–74]. By improving the absorption of peptides and proteins, these compounds can be used for transdermal delivery which limit gastrointestinal tract and improve patient compliance. CDs are used as carriers for the delivery of proteins, peptides, and oligonucleotide drugs which increase their solubility, bio adaptability, and bioavailability and improve peptides permeation through the blood–brain barrier (BBB) [74–77].

One of the methods that can be used for the modification of chemical and physical properties of peptides and proteins is the formation of inclusion complex with CDs. Peptides and proteins are highly hydrophilic and can form complex with CDs. But as their size is so bulky, formation of inclusion complexes is decreased due to topological constraints of their backbone. Complex between CDs and peptides and proteins is locally which led to incorporation of hydrophobic segments into CD cavity and formation of inclusion complex. This complexation limited intermolecular association in peptides and proteins structure and modified chemical and biological properties, protected peptides and proteins against enzymatic and chemical degradation, improved absorption of peptides and proteins, and increased permeation through physical and metabolic barriers.

Modified peptides and proteins via CDs can be used for various target delivery such as nasal absorption enhancers have been demonstrated for luteinizing hormone-releasing hormone agonists, insulin, adreno corticotropic hormone analogue, calcitonin, granulocyte colony-stimulating factor, insulin-like growth factor-I, and so on [78–80].

7.3 Cyclodextrin-based Targeting Systems

Efficacy and applicability of some kind of drugs such as anticancer and antitumor drugs are affected by toxicities and drug resistance and cannot reach the target site successfully. Nanoscale-based targeting drug delivery systems deliver required amount of drug to target organs, tissues, or cells for required time without side effects to other cells. Among the nanoscale carriers, CDs are good candidate for target drug delivery which the hydroxyl functions are orientated to the cone exterior with the primary hydroxyl groups of the sugar residues at the narrow edge of the cone and the secondary hydroxyl groups at the wider edge [81–83].

Formation of inclusion complex of drug and CD is based on hydrogen bonding which free drug and CD molecules show equilibrium in water and free drug delivered to target site. By complex formation of drugs and CDs, functional nanostructures obtained with great potential and versatility for defined drug delivery purposes to target sites are divided into nanospheres, micelles, polymer–drug therapeutics, dendrimers, liposomes, nano core–shells, and nanoparticles [83–85].

7.3.1 Nanoparticles

Using nanoparticles as carriers in drug delivery reduces toxicity and side effects of drugs. Nanoparticles consist of macromolecular substances which can be used in the form of solid or colloidal particles. In these systems drug is dissolved in preferred solvent and entrapped or encapsulated into a nanomatrix system. Nanoparticles can be prepared using various methods and are formed as nanospheres or nanocapsules that can be used for the encapsulation or drug delivery systems [86, 87].

Nanoparticles are able to improve solubility, efficacy, and bioavailability of drugs with poor solubility in water and act as containers for delivering drug to target site without any side effects to other cells. The size of nanoparticles helps them to permeate through various cellular and biological barriers such as albumin, gelatin, and phospholipids for liposomes, and thus can be used for drug delivery. Nanoparticles can be formed as nanocapsules, nanospheres, and nanosponges. Nanoparticles are able to entrap very low drug concentration, so their efficiency is limited. CD-based nanoparticles can be used for designing new systems for target drug delivery which improve drug solubility. Formation of inclusion between drug and CDs improves loading capacity of nanoparticles [27, 28, 88].

The most important parameters for designing nanoparticles for drug delivery systems are safety of the nanoparticle formulations and drug-loading capacity of polymeric nanoparticles. In recent years, CDs are used for synthesis of nanoparticles for target drug delivery due to increased solubility and stability of active ingredients. By formation of inclusion complex between active ingredients and CDs, active sites are increased in the polymerization medium and large amounts of CD can be attached to the nanoparticles and so the drug concentration is increased [89–92].

7.3.2 Nanocapsules

In nanocapsules, drug is surrounded by a protective polymer coating (shell) and a core which consists of one or more active materials in which the drug is physically and uniformly dispersed [93].

The protective coating of nanocapsules easily undergoes oxidation which is so important for drug delivery process, that is, to permeate barriers, reach the target site, and maintain drug release. Drug selectivity and improved bioavailability are obtained while drug toxicity is reduced. Nanocapsules can be used widely in pharmaceutical, biochemical, electrical, optical, or magnetic applications [94, 95].

7.3.3 Microsphere

In the presence of a high percentage of highly soluble hydrophilic excipients, complexation may not improve the drug dissolution rate from microspheres. Nifedipine release from chitosan microspheres was slowed down on complexation with HP-β-CD in spite of the improved drug-loading efficiency. Since it is highly unlikely for CD molecules to diffuse out of the microspheres, even with a low stability constant, the complex must first release the free drug that can permeate out of the microspheres. Hence, the observed slow nifedipine release from the microspheres was reported to be due to lesser drug availability from the complex and also due to formation of hydrophilic chitosan/CD matrix layer around the lipophilic drug that further decreases the drug matrix permeability [21–23, 97]. Sustained hydrocortisone release with no enhancement of its dissolution rate was observed from chitosan microspheres containing its HP-β-CD complex. The sustained hydrocortisone release was reported to be due to the formation of a layer adjacent to the interface by the slowly dissolving drug during the dissolution process that makes the microsphere surface increasingly hydrophobic [97].

Encapsulation of active agents in CDs has various properties such as:

- increase of skin occlusion;
- increase of skin hydration and elasticity;
- enhancement of skin permeation and drug targeting;
- improvement of benefit/risk ratio;
- enhancement of UV blocking activity; and
- enhancement of chemical stability of chemically labile compounds.

7.3.4 Nanosponges

Nanosponges consist of microscopic particles with few nanometer wide cavities and encapsulated both lipophilic and hydrophilic active agents as core, led to the improvement of solubility of poorly water-soluble molecules and protect degradable molecules. Nanosponges are water soluble but do not break up chemically in water and are able to mask unpleasant flavors, improve dissolution rate, solubility, and stability of drugs, and deliver drug to target site. Only small molecules can be incorporated into nanosponges. Nanosponges have nanoporous structure so that they can entrap hydrophobic drugs and be formulated as oral, parenteral, topical, or inhalation dosage forms and can be used as carriers for gases such as oxygen and carbon dioxide, and in biomedical applications, selectively soak up biomarkers for the diagnosis and can harvest rare cancer marker from blood [98–100].

7.4 CDs in the Food Industry

The use of CDs in the food industry has increased significantly in the last years. These compounds are highly recommended for applications in food processing and as food additives. Thus, volatile compounds can be encapsulated in CDs in order to limit aroma degradation or loss during processing and storage. Moreover, the use of CDs–flavor inclusion complexes allows using minor amounts of flavors. On the other hand, CDs can also be used for deodorizing and removing undesirable components such as off flavors or bitter components present in the food in its natural form, or for improving the nutritional characteristics of many dairy foods such as milk, mayonnaise, lard, or cream through their complexation with cholesterol molecules [101–103].

In addition, CDs can play an important role as food color modulators due to the ability of these compounds to form complexes with the polyphenol-oxidase enzyme that is responsible for catalyzing browning reactions [102, 103]. Another interesting application of CDs is their incorporation in the food packaging materials as antiseptic or conserving agents. Food ingredients, essential oils, and other bioactive components that include lipids, vitamins, peptides, fatty acids, antioxidants, minerals, and living cells such as probiotics must be protected against undesirable reactions such as evaporation, chemical reactions, or migration in food. These compounds are sensitive to various parameters such as variation of pH, mechanical stress, transport conditions, and digestive enzymes in the stomach. Therefore, for the best efficiency, preventing undesirable interactions with other components in food products, masking unpleasant feelings during eating, and reaching specific site via a controlled delivery system, these compounds may be encapsulated in appropriate carriers such as natural polymers such as alginate, agarose, carrageenan, chitosan, pectin, and other polysaccharides [103–107].

By encapsulation of probiotics, their stability and bioavailability are increased and by addition to food formula texture and taste of the food are not changed. Food flavors usually consist of volatile actives that may undergo evaporation and degradation. By encapsulation of these compounds with CDs, a protective shell formed surround the flavors and protect the flavors against evaporation, chemical reactions such as flavors–flavors interactions, light-induced reactions, oxidation, or migration in food products. One of the effective methods for improving bioavailability, water solubility, and stability of active agents in food industries is complexation with polysaccharides such as cyclodextrins [103–108].

As discussed in Chapter 5, encapsulation process consists of incorporation of an active agent within protective shell (capsule) with a diameter of a few nanometers to a few millimeters. This method is applicable to improve delivery of bioactive molecules and living cells into foods. Protective shell must be biodegradable and form a barrier between the internal phase and its surroundings. Various methods are used for encapsulation of food ingredients such as spray drying, spray-chilling, freeze-drying, melt extrusion, and melt injection. A new method that is applicable for the encapsulation of food ingredients is the formation of complex with polysaccharides such as cyclodextrins, proteins, and lipids. Encapsulation of proteins and lipids show poor water solubility, so inclusion complex formation between CD and food ingredients is attractive and it improves water solubility, masks bad tasting or smelling, stabilizes food ingredients, and increases their bioavailability and life time [108–111].

Encapsulation of active agents in the CD cavity has various benefits such as:

- easier handling;
- stabilization of active agent in food processing systems, storage (temperature, oxygen, and light), in the gastrointestinal tract (pH, enzymes, and presence of other nutrients), and in final usage;
- stabilization of volatile actives and protection of them in the presence of oxygen or water;
- improved safety (e.g., reduced flammability of volatiles like aroma, no concentrated volatile oil handling);
- improved stability in final products and during processing;
- masking of undesirable taste and odor;
- controlled release of active agent;
- increased complexity of production process and/or supply chain;
- masking of unpleasant bitter taste of polyphenols at high antioxidant activities;
- improved effectiveness, stability, bioactivity, and bioavailability of polyphenols;
- improved delivery of bioactive molecules (e.g., antioxidants, minerals, vitamins, phytosterols, lutein, fatty acids, and lycopene), and living cells (e.g., probiotics) into foods; and
- providing adequate concentration and uniform dispersion of an active agent.

By using CDs, it is possible that solids, liquids, or gaseous materials are entrapped in small capsules in which the shell membrane protects the core part against motive factors, to prevent undesirable interactions with food matrix, and to deliver at a controlled rate over prolonged periods under specific conditions. Complexation of these materials with CDs also provides a protective barrier around bioactive materials and increase stability and bioavailability of food ingredients [109–112].

In addition to encapsulation, CDs can be used for the removal of cholesterol from animal products, providing more safe products with better nutritional characteristics. Using CDs, 41% and 100% of cholesterol can be removed from milk and mayonnaise, respectively [15, 113].

Various reported is based on using CD for removal cholesterol from butter [115, 116], lard [117], cheese [114, 116, 118–120], egg yolks [121–123], and cream [124].

CDs can also be used in food packaging materials, which reduce residual organic volatile contaminants in packaging materials and improve barrier properties of the packaging materials, thus improving food quality and safety (Wood [126]) especially in meat, fish paste, or in frozen seafood products, or release of α-tocopherol from antioxidative active packaging [125, 126].

CDs may provide an antifungal volatile during storage of frozen and greenhouse fruits, for reducing post-harvest diseases [127].

7.5 Cyclodextrins in Skin Delivery and Cosmetic

Cyclodextrins are used in skin delivery, cosmetic, and toiletry products such as creams, lotions, shampoos, toothpastes, skin creams, softeners, toothpaste, room fresheners, detergents, and perfumes in formulations because most of active ingredients in creams are poorly water soluble and this problem reduces their efficiency and permeability through biological barriers [132, 133].

Formation of inclusion complex between CDs and active ingredients increases physical and chemical stability, reduces skin irritation, controls odor, masks unwanted body odor, stabilizes emulsions and suspensions, limits interactions between various formulation ingredients and makes controlled release of various active ingredients into skin. Encapsulation of essential oils, drug, or cream in CD cavity as active ingredients is done and this method is used to create long-lasting fragrances [128–131]. The significant parameter in complexation with CDs is physicochemical stability problems, which are caused owing to the interaction of CDs with formulation ingredients. In pomade formulation, oil/water emulsions are desirable due to their pleasant touch, but their water solubility is low and they are thermodynamically unstable liquid–liquid dispersed systems. Moreover, in oil/water emulsions, phase separation may occur which is prevented by addition of emulsifier. By formation of inclusion complex, stability of emulsions is increased and they diffuse from the hydrophilic phase to the lipophilic phase. CDs also interacted with emulsifying agents and stabilized them without any potential irritant [132–134].

Vegetable oils also can be used in skin delivery, which are composed primarily of triglycerides. Fatty acid chain of triglyceride is amphiphilic and formed partial inclusion complex with CDs. It is important to consider a potential competition and/or interaction between excipient (polymer, cosolvent, emulsifying agent, and preservative) and CDs [135–137].

Clotrimazole is a poorly water-soluble antifungal medicine that prevents the growth of fungus and is used for treating skin infections such as athlete's foot, jock itch, ringworm, and other fungal skin infections (candidiasis). To improve the solubility, oral bioavailability, and permeability of clotrimazole, inclusion complex of CDs and clotrimazole was formed [138].

Inclusion complex formation between CDs and propylene glycol and poly(ethylene glycol) also improved cream skin formulations and increased the solubility of drugs. After complex formation and dissolved in water, propylene glycol molecules displace piroxicam from inclusion with HP-β-CD cavity and then reduce complex stability whereas interactions between α-CD and poly(ethylene glycol) result in the formation of crystalline inclusion complexes [139, 140].

In cosmetic or dermatological creams, bacterial amount must be the least and must be applied for a long time period in multiple doses. CDs can be used as preservatives and they form inclusion complex with active agents and change antimicrobial activity. For example, *p*-hydroxybenzoic esters (methyl-, ethyl-, propyl-, and butylparabens) form inclusion complex with α- and β-CDs and their derivatives [141–144], and the antimicrobial activity of parabens against *Candida albicans* (yeast) is reduced. The antimicrobial activity depends only on the fraction of preservative that is included in the CD molecule [143, 144].

Thimerosal, phenylmercury acetate, bronopol, phenolic substances, and aliphatic aryl–substituted alcohols can form inclusion complex with CDs [144–146].

CDs can also be used to reduce body odors in tissues and underarm shields, convert liquid ingredient to a solid form, in lipsticks for flavor and color protection, to improve thermal stability of body oils, to mask odors in dishwashing and laundry detergents, to enhance triclosan availability in silica-based toothpastes, and to improve shelf life and performance of self-tanning emulsions or creams [147–150].

7.6 Agricultural Applications

Cyclodextrins can form inclusion complex with a wide variety range of hydrophobic and hydrophilic guest molecules, encapsulate guest molecules and alter their physical, chemical, and biological properties [151, 152].

Agricultural chemicals such as herbicides, insecticides, fungicides, repellents, pheromones, and growth regulators are hydrophobic compounds, so their solubility in water and aqueous medium is low. Cyclodextrins can form inclusion complex with these compounds and improve their solubility, stability, and prevent oxidation of these compounds against UV light. Complexation with CDs delay and control germination of seeds and plant growth and improve plant growth yielding a 20–45% larger harvest [152–155].

Environmental disadvantages of hydrophobic organic pesticides are: they are absorbed by soil, removal of residues from the soil is difficult, and the efficiency of pesticides is limited. Pesticides such as diuron, isoproturon, and 2,4-dichlorophenoxyacetic acid (2,4-D) react with CDs to increase water solubility and bioavailability and to decrease toxicological and contamination effects [156, 157].

Pesticides that are adsorbed on the soil can be used by using CDs and this guarantees soil health [158, 159].

Inclusion complex formation of CDs with some herbicides delays photodegradation and volatility of them and prevents their entry into the groundwater while maintaining effective pest control [160].

IC formation of controlled release profile of herbicides, insecticides such as aldicarb and sulprofos from complex prevent leaching while maintaining effective pest control [161, 162].

IC formation of volatile agricultural chemicals stabilizes compounds, improves physical stability, and masks unpleasant smell. The complex formulations exist as dry solid thereby producing a higher energy barrier to volatilization and making controlled release of active agents from the complex [163, 164].

7.7 Self-healing Coating

Major challenges in nanostructured inorganic–organic hybrid coatings for a long-term protection of aluminum alloys and surface protection against atmospheric corrosion are based on using an encapsulated corrosion inhibitor which acts as a barrier to protect the underlying substrate and leads to a long-term corrosion protection with slow release of inhibitors [165–167].

These coatings are intelligent coatings, which can be repaired in situ after they had been damaged. These coatings which are prepared by incorporation of an encapsulated healing agent prevent corrosion. Encapsulated corrosion inhibitors are released slowly from the host molecule and led to a protective surface comparable to that of the original undamaged surface [165, 167, 168].

To design and create such hybrid coatings with desired environmental protection properties, sol–gel technology is the method of choice because of its simplicity. Sol–gel technology offers various ways to prepare functional hybrid coatings with unique chemically tailored properties [166, 167, 170].

To enhance the corrosion protection properties of sol–gel derived coatings, nanoparticles are incorporated in the hybrid sol systems, which has improved mechanical properties and reduced cracking potential which is strongly affected by the concentration and size of the nanoparticles [171, 172].

Addition of corrosion inhibitors into the hybrid sol can also improve the self-repairing properties of hybrid sol–gel coating, which can repress corrosion initiated at the coating defects. Cylodextrins are good candidate for incorporation of organic corrosion inhibitors via inclusion complexes into the coating material. CDs are described as truncated cone-shaped structures with a hydrophilic exterior surface and a hydrophobic interior cavity, so various molecules can enter CD cavity as guests, forming inclusion complexes with these hosts [173–175].

Cyclodextrins are known as effective complexing agents, which have an ability to form inclusion complexes with various organic guest molecules that fit the size of the cyclodextrin cavity. Organic aromatic and heterocyclic compounds are usually predominant candidates for the inclusion complexation reaction [173–176]. The encapsulation of organic corrosion inhibitors with cyclodextrins in the form of inclusion complexes are more bulky, stable, and expected to be more easily trapped within the cross-linked nanoporous coating materials making the inhibitor more difficult to leach out and thus prolonging the inhibition effect of the doping agent. Inclusion complexes of inhibitor and CD are effective delivery systems of organic inhibitors in active corrosion protection applications [177, 178]. When a defect is formed in the sol–gel film, inhibitor is released slowly and results in an effective long-term self-healing effect that protects the metallic substrates from corrosion [179, 180].

The encapsulation of organic corrosion inhibitors in the form of their inclusion complexes with CD has several advantages over the encapsulation of these inhibitors in their free molecular forms, which encourages us to use encapsulated corrosion inhibitors as effective delivery of organic inhibitors in active corrosion protection applications. Long-term delivery of corrosion inhibitor and healing of a damaged coating is due to slow release of organic corrosion inhibitor from the cyclodextrin cavity [181–184].

Amiri et al. designed self-repairing anticorrosion coating by using the encapsulated corrosion inhibitors (2-mercaptobenzimisazole (MBI) and 2-mercaptobenzothiazole (MBT)) and α-, β-, and γ-cyclodextrins as smart corrosion inhibitor nanocontainers in different conditions (at room temperature and under sonic energy) [181–184]. Inclusion complex formation of MBI or MBT with CDs led to encapsulated corrosion inhibitors which became active in corrosive electrolytes, and slowly diffused out of the host material, ensuring continuous delivery of the inhibitors to corrosion sites and long-term corrosion protection.

The encapsulated corrosion inhibitor particles were used as reservoirs for repairing agent and chemical initiator, which became active in corrosive electrolytes, and slowly diffused out of the host material, ensuring continuous delivery of the inhibitors to corrosion sites and long-term corrosion protection [181–184].

References

1 Loftsson, T. and Duchene, D. (2007) Cyclodextrins and their pharmaceutical applications. *Int. J. Pharm.*, **329**, 1–11.
2 Hakkarainen, B., Fujita, K., Immel, S. *et al.* (2005) ^{1}H NMR studies on the hydrogen-bonding network in mono-altro-β-cyclodextrin and its complex with adamantane-1-carboxylic acid. *Carbohydr. Res.*, **340**, 1539–1545.
3 Irie, T. and Uekama, K. (1997) Pharmaceutical applications of cyclodextrins-toxicological issues and safety evaluation. *J. Pharm. Sci.*, **86**, 147–162.
4 Gould, S. and Scott, R.C. (2005) 2-Hydroxypropyl-β-cyclodextrin (HP-β-CD): a toxicology review. *Food Chem. Toxicol.*, **43**, 1451–1459.
5 Szente, L., Szejtli, J. and Kis, G.L. (1998) Spontaneous opalescence of aqueous Γ-cyclodextrin solutions: complex formation or self-aggregation. *J. Pharm. Sci.*, **87**, 778–781.
6 Liu, L. and Guo, Q.X. (2002) The driving forces in the inclusion complexation of cyclodextrins. *J. Incl. Phenom. Macrocycl. Chem.*, **42**, 1–14.
7 Aggarwal, S., Singh, P.N. and Mishra, B. (2002) Studies o−n solubility and hypoglycemic activity of gliclazide beta-cyclodextrin-hydroxypropyl methylcellulose complexes. *Pharmazie*, **57**, 191–193.
8 Archontaki, H.A., Vertzoni, M.V. and Athanassioum Malaki, M.H. (2002) Study on the inclusion complexes of bromazepam with beta- and beta-hydroxypropyl-cyclodextrins. *J. Pharm. Biomed. Anal.*, **28**, 761–769.

9 Arima, H., Yunomae, K., Miyake, K. et al. (2001) Comparative studies of the enhancing effects of cyclodextrins o–n the solubility and oral bioavailability of tacrolimus in rats. *J. Pharm. Sci.*, **90**, 690–701.
10 Arima, H., Miyaji, T., Irie, T. et al. (2001) Enhancing effect of hydroxypropyl b-cyclodextrin on cutaneous penetration and activation of ethyl-4-biphenyl acetate in hairless mouse skin. *Eur. J. Pharm. Sci.*, **6**, 53–59.
11 Arias, M.J., Mayano, J.R., Munoz, P. et al. (2000) Study of omeprazole-g-cyclodextrin complexation in the solid-state, *Drug Dev. Ind. Pharm.*, **26**, 253–259.
12 Asai, K., Morishita, M., Katsuta, H. et al. (2002) The effects of water-soluble cyclodextrins on the histological integrity of the rat nasal mucosa. *Int. J. Pharm.*, **246**, 25–35.
13 Choi, H.G., Lee, B.J., Han, J.H. et al. (2001) Terfenadine-beta-cyclodextrin inclusion complex with antihistaminic activity enhancement. *Drug Dev. Ind. Pharm.*, **27**, 857–862.
14 Chutimaworapan, S., Ritthidej, G.C., Yonemochi, E. et al. (2001) Effect of water-soluble carriers o–n dissolution characteristics of nifedipine solid dispersions. *Drug Dev. Ind. Pharm.*, **26**, 1141–1150.
15 Dalmora, M.E., Dalmora, S.L. and Oliveira, A.G. (2001) Inclusion complex of piroxicam with beta-cyclodextrin and incorporation in cationic microemulsion: in vitro drug release and in vivo topical anti-inflammatory effect. *Int. J. Pharm.*, **222**, 45–55.
16 Doliwa, A., Santoyo, S. and Ygartua, P. (2001) Transdermal Iontophoresis and skin retention of piroxicam from gels containing piroxicam: hydroxypropyl-beta-cyclodextrin complexes. *Drug Dev. Ind. Pharm.*, **27**, 751–758.
17 Duchene, D., Ponchel, G. and Wouessidjewe, D. (1999) Cyclodextrins in targeting: application to nanoparticles. *Adv. Drug. Deliv. Rev.*, **36**, 29–40.
18 Emara, L.H., Badr, R.M. and Elbary, A.A. (2002) Improving the dissolution and bioavailability of nifedipine using solid dispersions and solubilizers. *Drug Dev. Ind. Pharm.*, **28**, 795–807.
19 Evrard, B., Chiap, P., DeTullio, P. et al. (2002) Oral bioavailability in sheep of albendazole from a suspension and from a solution containing hydroxyl propyl beta cyclodextrin. *J. Control. Release*, **85**, 45–50.
20 Fatouros, D.G., Hatzidimitriou, K. and Antimisiaris, S.G. (2001) Liposomes encapsulating prednisolone and prednisolone–cyclodextrin complexes: comparison of membrane integrity and drug release. *Eur. J. Pharm. Sci.*, **13**, 287–296.

21 Fathy, M. and Sheha, M. (2000) In-vitro and in-vivo evaluation of an Amylobarbitone-HPb-CD complex prepared by freeze drying method. *Pharmazie*, **55**, 513–517.
22 Fernandes, C.M., Teresa, V.M. and Veiga, F.J. (2002) Physicochemical characterization and in vitro dissolution behavior of nicardipine-cyclodextrins inclusion compounds. *Eur. J. Pharm. Sci.*, **15**, 79–88.
23 Ficarra, R., Tommasini, S., Raneri, D. et al. (2002) Study of flavonoids/beta-cyclodextrins inclusion complexes by NMR, FT-IR, DSC, X-ray investigation. *J. Pharm. Biomed. Anal.*, **29**, 1005–1014.
24 Fresta, M., Fontana, G., Bucolo, C. et al. (2004) Ocular tolerability and in vivo bioavailability of poly (ethylene glycol) (PEG)-coated polyethyl-2-cyanoacrylate nanosphere-encapsulated acyclovir. *J. Pharm. Sci.*, **90**, 288–297.
25 Garcia-Rodriguez, J.J., Torrado, J. and Bolas, F. (2005) Improving bioavailability and anthelmintic activity of albendazole by preparing albendazole-cyclodextrin complexes. *Parasite*, **8**, 188–190.
26 Ghorab, M.K. and Adeyeye, M.C. (2001) Elucidation of solution state complexation in wet-granulated oven-dried ibuprofen and beta-cyclodextrin: FT-IR and ^{1}H-NMR studies. *Pharm. Dev. Technol.*, **6**, 315–324.
27 Granero, G., Bertorello, M.M. and Longhi, M. (2006) Solubilization of a naphthoquinone derivative by hydroxypropyl-beta-cyclodextrin (HP-beta-CD) and polyvinylpyrrolidone (PVP-K30).The influence of PVP-K30 and pH o–n solubilizing effect of HP-beta-CD. *Boll. Chim. Farm.*, **141**, 63–66.
28 Gudmundsdottir, H., Sigurjonsdottir, J.F., Masson, M. et al. (2006) Intranasal administration of midazolam in a cyclodextrin based formulation: bioavailability and clinical evaluation in humans. *Pharmazie*, **56**, 963–969.
29 Gudmundsdottir, E., Stefansson, E., Bjarnadottir, G. et al. (2003) Methazolamide 1% in cyclodextrin solution lowers IOP in human ocular hypertension. *Invest. Ophthalmol. Vis. Sci.*, **41**, 3552–3554.
30 Brewster, M.E. and Loftsson, T. (2002) The use of chemically modified cyclodextrins in the development of formulations for chemical delivery systems. *Pharmazie*, **57**, 94–101.
31 Cappello, B., Carmignani, C., Iervolino, M. et al. (2001) Solubilization of tropicamide by hydroxypropyl-beta-cyclodextrin and water-soluble polymers: in vitro/in vivo studies. *Int. J. Pharm.*, **213**, 75–81.
32 Ceschel, G.C., Mora, P.C., Borgia, S.L. et al. (2002) Skin permeation study of dehydroepiandrosterone (DHEA) compared with its alpha-cyclodextrin complex form. *J. Pharm. Sci.*, **91**, 2399–2407.

33 Charoenchaitrakool, M., Dehghani, F. and Foster, N.R. (2003) Utilization of supercritical carbondioxide for complex formation of ibuprofen and methyl-beta-cyclodextrin. *Int. J. Pharm.*, **239**, 103–112.
34 Chang, S.L. and Banga, A.K. (1999) Transdermal iontophoretic delivery of hydrocortisone from cyclodextrin solutions. *J. Pharm. Pharmacol.*, **50**, 635–640.
35 Chowdhary, K.P.R. and Hymavathi, R. (2001) Enhancement of dissolution rate of Meloxicam. *Indian J. Pharm. Sci.*, **2**, 150–154.
36 Chowdhary, K.P.R. and Nalluri, B.N. (2000) Nimesulide and b-cyclodextrin inclusion complexes: physico-chemical characterization and dissolution rate studies. *Drug Dev. Ind. Pharm.*, **26**, 1217–1219.
37 Leuner, C. and Dressman, J. (2005) Improving drug solubility for oral delivery using solid dispersions. *Eur. J. Pharm. Biopharm.*, **50**, 47–60.
38 Kinoshita, M., Kazuhiko, B., Atushi, N. *et al.* (2002) Improvement of solubility and oral bioavailability of a poorly water soluble drug. *J. Pharm. Sci.*, **91**, 362–370.
39 Kamada, M., Hirayama, F., Udo, K. *et al.* (2002) Cyclodextrin conjugate-based controlled release system: repeated-and prolonged-releases of ketoprofen after oral administration in rats. *J. Control. Release*, **82**, 407–416.
40 Baboota, S. and Agarwal, S.P. (2003) Inclusion complexes of meloxicam with b-cyclodextrins. *Indian J. Pharm. Sci.*, **64**, 408–411.
41 Bayomi, M., Abanumay, K. and Ai-Angary, A. (2002) Effect of inclusion complexation with cyclodextrins o–n photostability of nifedipine in solid state. *Int. J. Pharm.*, **243**, 107–117.
42 Loftsson, T. and Stefansson, E. (1997) Effect of cyclodextrins on topical drug delivery to the eye. *Drug Dev. Ind. Pharm.*, **23**, 473–481.
43 Duchene, D. and Wouessidjewe, D. (1996) Pharmaceutical and medical applications of cyclodextrins, in *Polysaccharides in Medical Applications* (ed. S. Dumitriu), Marcel Dekker, New York, pp. 575–602.
44 Arima, H., Kondo, T., Irie, T. and Uekama, K. (1992) Enhanced rectal absorption and reduced local irritation of the anti-inflammatory drug ethyl 4-biphenylylacetate in rats by complexation with water-soluble cyclodextrin derivatives and formulation as oleaginous suppository. *J. Pharm. Sci.*, **81**, 1119–1125.
45 Kondo, T., Irie, T. and Uekama, K. (2006) Combination effects of alpha-cyclodextrin and xanthan gum o–n rectal absorption and metabolism of morphine from hollow-type suppositories in rabbits. *Biol. Pharm. Bull.*, **19**, 280–286.

46 Beraldo, H., Sinisterra, R.D., Teixeira, L.R. et al. (2002) An effective anticonvulsant prepared following a host–guest strategy that uses hydroxypropyl-beta-cyclodextrin and benzaldehyde semicarbazone. *Biochem. Biophys. Res. Commun.*, **296**, 241–246.

47 Becirevic, L.M. and Filipovic-Grcic, J. (2000) Effect of hydroxypropyl-beta-cyclodextrin o-n hydrocortisone dissolution from films intended for ocular drug delivery. *Pharmazie*, **55**, 518–520.

48 Bibby, D.C., Davies, N.M. and Tucker, I.G. (2000) Mechanisms by which cyclodextrins modify drug release from polymeric drug delivery systems. *Int. J. Pharm.*, **197**, 1–11.

49 Loftsson, T. and Masson, M. (2001) Cyclodextrins in topical drug formulations: theory and practice. *Int. J. Pharm.*, **225**, 15–30.

50 Lopez, R.F.V., Collett, J.H. and Bentley, M.V.L.B. (2000) Influence of cyclodextrin complexation on the in vitro permeation and skin metabolism of dexamethasone. *Int. J. Pharm.*, **200**, 127–132.

51 Uekama, K., Adachi, H., Irie, T. et al. (1992) Improved transdermal delivery of prostaglandin E1 through hairless mouse skin: combined use of carboxy methyl-ethyl-beta-cyclodextin and penetration enhancers. *J. Pharm. Pharmacol.*, **44**, 119–121.

52 Yang, W., Chow, K.T., Lang, B. et al. (2010) In vitro characterization and pharmacokinetics in mice following pulmonary delivery of itraconazole as cyclodextrin solubilized solution. *Eur. J. Pharm. Sci.*, **39**, 336–347.

53 Barar, J., Javadzadeh, A.R. and Omidi, Y. (2008) Ocular novel drug delivery: impacts of membranes and barriers. *Expert Opin. Drug Deliv.*, **5**, 567–581.

54 Sunkara, G.K.U. (2003) Membrane transport processes in the eye, in *Ophthalmic Drug Delivery Systems* (ed. A.K. Mitra), Marcel Dekker, Inc, New York, pp. 13–58.

55 Geroski, D.H. and Edelhauser, H.F. (2001) Transscleral drug delivery for posterior segment disease. *Adv. Drug Deliv. Rev.*, **52**, 37–48.

56 Pitkanen, L., Ranta, V.P., Moilanen, H. and Urtti, A. (2005) Permeability of retinal pigment epithelium: effects of permeant molecular weight and lipophilicity. *Invest. Ophthalmol. Vis. Sci.*, **46**, 641–646.

57 Urtti, A. (2006) Challenges and obstacles of ocular pharmacokinetics and drug delivery. *Adv. Drug Deliv. Rev.*, **58**, 1131–1135.

58 Sukegawa, T., Furuike, T., Niikura, K. et al. (2002) Erythrocyte-like liposomes prepared by means of amphiphilic cyclodextrin sulfates. *Chem. Commun.*, **5**, 430–431.

59 Jicsinszky, L., Petrikovics, I., Petro, M., Horvath, G., Szejtli, J., and Way, J.L. (2007) Improved drug delivery by conjugation with

cyclodextrins. Proceedings of 14th European Carbohydrate Symposium, September 2–7, Lubeck, Germany.
60 Lim, H.J., Cho, E.C., Shim, J. and Kim, D.H. (2008) Polymer-associated liposomes as a novel delivery system for cyclodextrin bound drugs. *J. Colloid Interface Sci.*, **2**, 460–468.
61 Skalko, B.N., Pavelic, Z. and Becirevic, L.M. (2000) Liposomes containing drug and cyclodextrin prepared by the one-step spray-drying method. *Drug Dev. Ind. Pharm.*, **26**, 1279–1284.
62 Kang, F., Jiang, G., Hinderliter, A. *et al.* (2002) Lysozyme stability in primary emulsion for PLGA microsphere preparation: effect of recovery methods and stabilizing excipients. *Pharm. Res.*, **19**, 629–633.
63 Grcic, F.J., Voinovich, D., Moneghini, M. *et al.* (2000) Chitosan microspheres with hydrocortisone and hydrocortisone–hydroxypropyl-b-cyclodextrin inclusion complex. *Eur. J. Pharm. Sci.*, **9**, 373–379.
64 Bibby, D.C., Davies, N.M. and Tucker, I.G. (1999) Investigations into the structure and composition of beta-cyclodextrin/polyacrylic acid microspheres. *Int. J. Pharm.*, **180**, 161–168.
65 Kang, F. and Singh, J. (2003) Conformational stability of a model protein (bovine serum albumin) during primary emulsification process of PLGA microspheres synthesis. *Int. J. Pharm.*, **260**, 149–156.
66 McClelland, G.A., Sulton, S.C., Engle, K. and Zentner, G.M. (1991) The solubility–modulated osmotic pump: in vitro/in vivo release of diltiazem hydrochloride. *Pharm. Res.*, **8**, 88–92.
67 Kharmnna, S.C. (1991) Therapeutic system for sparingly soluble active ingredients. US patent 4: 992: 278.
68 Kuczynski, A.L., Ayer, A.D. and Wong, P.S. (1997) Dosage form for administering oral hypoglycemic glipizide, US Patent 5: 545: 413.
69 Geerke, J.H. (1997) Method and apparatus for forming dispenser delivery ports, US Patent 5: 658: 474.
70 Okimoto, K., Rajewski, R.A. and Stella, V.J. (1999) Release of testosterone from an osmotic pump tablets utilizing (SBE) 7M-β-CD as both a solubilizing and an osmotic pump agent. *J. Control. Release*, **58**, 29–38.
71 Mehramizi, A., Monfared, A.E., Pourfarzib, M. *et al.* (2007) Influence of β-cyclodextrin complexation on lovastatin release from osmotic pump tablets (OPT). *DARU*, **15**, 71–78.
72 Craik, D.J., Fairlie, D.P., Liras, S. and Price, D. (2013) The future of peptide-based drugs. *Chem. Biol. Drug Des.*, **81**, 136–147.

73 Aungst, B., Saitoh, H., Burcham, D. *et al.* (1996) Enhancement of the intestinal absorption of peptides and nonpeptides. *J. Control. Release*, **41**, 19–31.

74 Borchardt, T., Jeffrey, A., Siahaan, T.J. *et al.* (1997) Improvement of oral peptide bioavailability: peptidomimetics and prodrug strategies. *Adv. Drug Deliv. Rev.*, **27**, 235–256.

75 Torchilin, V. (2009) Intracellular delivery of protein and peptide therapeutics. *Drug Discov. Today Technol.*, **5**, 95–103.

76 Maher, S. and Brayden, D. (2012) Overcoming poor permeability: translating permeation enhancers for oral peptide delivery. *Drug Discov. Today Technol.*, **9**, 113–119.

77 Hamman, J.H., Enslin, G.M. and Kotze, A.F. (2005) Oral delivery of peptide drugs: barriers and developments. *BioDrugs*, **19**, 165–177.

78 Irie, T. and Uekama, K. (1999) Cyclodextrins in peptide and protein delivery. *Adv. Drug Deliv. Rev.*, **36**, 101–123.

79 Sayani, A. and Chien, Y.W. (1996) Systemic delivery of peptides and proteins across absorptive mucosae. *Crit. Rev. Ther. Drug Carrier Syst.*, **13**, 85–184.

80 Soares, A., Francisca, C., Rui, A. and Francisco, V. (2007) Oral administration of peptides and proteins: nanoparticles and cyclodextrins as biocompatible delivery systems. *Nanomedicine*, **2**, 183–202.

81 Hattori, K. (2007) Synthesis and evaluation of novel targeting cyclodextrins as drug carrier. Program of 4th Asian Cyclodextrin Conference, May 18.

82 Wu, W.M., Wu, J. and Bodor, N. (2002) Effect of 2-hydroxypropyl-betacyclodextrin on the solubility, stability, and pharmacological activity of the chemical delivery system of TRH analogs. *Pharmazie*, **57**, 130–134.

83 Begley, D.J. (1996) The blood brain barrier: principles for targeting peptides and drugs to the central nervous system. *J. Pharm. Pharmacol.*, **48**, 136–140.

84 Sailor, M.J. and Park, J.H. (2012) Hybrid nanoparticles for detection and treatment of cancer. *Adv. Mater.*, **24**, 3779–3802.

85 Chaudhuri, P., Paraskar, A., Soni, S. *et al.* (2009) Fullerenol-cytotoxic conjugates for cancer chemotherapy. *ACS Nano.*, **3**, 2505–2514.

86 Barratt, G.M. (2000) Therapeutic applications of colloidal drug carriers. *Pharmaceut. Sci. Technol. Today*, **3**, 163–171.

87 Couvreur, P. (1995) Controlled drug delivery with nanoparticles: current possibilities and future trends. *Eur. J. Pharm. Biopharm.*, **41**, 2–13.

88 Hedges, A.R. (2005) Industrial applications of cyclodextrins. *Chem. Rev.*, **98**, 2035–2044.

89 Eguchi, M., Da, Y.Z., Ogawa, Y. et al. (2006) Effects of conditions for preparing nanoparticles composed of aminoethylcarbamoyl-β-cyclodextrin and ethylene glycol diglycidyl ether on trap efficiency of a guest molecule. *Int. J. Pharm.*, **311**, 215–222.
90 Gao, H., Wang, Y.N., Fen, Y.G. and Ma, J.B. (2007) Conjugation of poly (DL-lactide-co-glycolide) on amino cyclodextrins and their properties as protein delivery system. *J. Biomed. Mater. Rest. A*, **80A**, 111–122.
91 Gao, H., Wang, Y.N., Fen, Y.G. and Ma, J.B. (2006) Conjugates of poly (DL-lactic acid) with ethylenediamino or diethylenetriamino bridged bis(β-cyclodextrins) and their nanoparticles as protein delivery systems. *J. Control. Release*, **112**, 301–311.
92 Bilensoy, E., Dogan, A.L., Sen, M. and Hincal, A.A. (2007) Complexation behaviour of antiestrogen drug tamoxifen citrate with natural and modified cyclodextrins. *J. Incl. Phenom. Macrocycl. Chem.*, **57**, 651–655.
93 Andrieu, V., Fessi, H., Dubrasquet, M. et al. (1989) Pharmacokinetic evaluation of indomethacin nanocapsules. *Drug Des. Deliv.*, **4**, 295–302.
94 Shen, Y., Jin, E., Zhang, B. et al. (2010) Prodrugs forming high drug loading multifunctional nanocapsules for intracellular cancer drug delivery. *J. Am. Chem. Soc.*, **132**, 4259–4265.
95 Jager, A., Stefani, V., Guterres, S.S. and Pohlmann, A.R. (2007) Physico-chemical characterization of nanocapsule polymeric wall using fluorescent benzazole probes. *Int. J. Pharm.*, **338**, 297–305.
96 Felton, L.A., Wiley, C.J. and Godwin, D.A. (2002) Influence of hydroxypropyl-beta-cyclodextrin o−n the transdermal permeation and skin accumulation of oxybenzone. *Drug Dev. Ind. Pharm.*, **28**, 1117–1124.
97 Cavalli, R., Akhter, A.K., Bisazza, A. et al. (2010) Nanosponge formulations as oxygen delivery systems. *Int. J. Pharm.*, **402**, 254–257.
98 Longo, C., Gambara, G., Espina, V. et al. (2011) A novel biomarker harvesting nanotechnology identifies Bak as a candidate melanoma biomarker in serum. *Exp. Dermatol.*, **20**, 29–34.
99 Amber, V., Shailendra, S. and Swarnalatha, S. (2008) Cyclodextrin based novel drug delivery systems. *J. Incl. Phenom. Macrocycl. Chem.*, **62**, 23–42.
100 Wandrey, C., Bartkowiak, A. and Harding, S.E. (2009) Materials for encapsulation, in *Encapsulation Technologies for Food Active Ingredients and Food Processing* (eds N.J. Zuidam and V.A. Nedovic), Springer, Dordrecht, The Netherlands, pp. 31–100.

101 Fang, Z. and Bhandari, B. (2010) Encapsulation of polyphenols – a review. *Trends Food Sci. Technol.*, **21**, 510–523.
102 Vos, P., Faas, M.M., Spasojevic, M. and Sikkema, J. (2010) Review: encapsulation for preservation of functionality and targeted delivery of bioactive food components. *Int. Dairy J.*, **20**, 292–302.
103 Dewettinck, K. and Huyghebaert, A. (1999) Fluidized bed coating in food technology. *Trend Food Sci. Technol.*, **10**, 163–168.
104 Kim, K.I., Yoon, Y.H. and Baek, Y.J. (1996) Effects of rehydration media and immobilization in Ca-alginate on the survival of *Lactobacillus casei* and *Bifidobacterium bifidum*. *Korean J. Dairy Sci.*, **18**, 193–198.
105 Augustin, M.A., Sanguansri, L., Margetts, C. and Young, B. (2001) Microencapsulation of food ingredients. *Food Australia*, **53**, 220–223.
106 Buffo, R.A. and Reineccius, G.A. (2001) Comparison among assorted drying processes for the encapsulation of flavors. *Perfumer Flavorist*, **26**, 58–67.
107 Madene, A., Jacquot, M., Scher, J. and Desobry, S. (2006) Aroma encapsulation and controlled release – a review. *Int. J. Food Sci. Technol.*, **41**, 1–21.
108 Roos, K.B. (2003) Effect of texture and microstructure on flavour retention and release. *Int. Dairy J.*, **13**, 593–605.
109 Astray, G., Gonzalez-Barreiro, C., Mejuto, J.C. *et al.* (2009) A review on the use of cyclodextrins in foods. *Food Hydrocolloids*, **23**, 1631–1640.
110 Hedges, R.A. (1998) Industrial applications of cyclodextrins. *Chem. Rev.*, **98**, 2035–2044.
111 Martín, D.V.E.M. (2004) Cyclodextrins and their uses: a review. *Process Biochem.*, **39**, 1033–1046.
112 Bhandari, B., Darcy, B. and Young, G. (2001) Flavour retention during high temperature short time extrusion cooking process: a review. *Int. J. Food Sci. Technol.*, **36**, 453–461.
113 Kwak, H.S., Kim, S.H., Kim, J.H. *et al.* (2004) Immobilized b-cyclodextrin as a simple and recyclable method for cholesterol removal in milk. *Arch. Pharm. Res.*, **27**, 873–877.
114 Jung, T.H., Kim, J.J., Yu, S.H. *et al.* (2005) Properties of cholesterol-reduced butter and effect of gamma linolenic acid added butter on blood cholesterol. *Asian-Australas. J. Animal Sci.*, **18**, 1646–1654.
115 Kim, J.J., Jung, T.H., Ahn, J. and Kwak, H.S. (2006) Properties of cholesterol-reduced butter made with b-cyclodextrin and added evening primrose oil and phytosterols. *J. Dairy Sci.*, **89**, 4503–4510.
116 Bae, H.Y., Kim, S.Y. and Kwak, H.S. (2008) Comparison of cholesterol-reduced Camembert cheese using cross-linked

β-cyclodextrin to regular Camembert cheese during storage. *Milchwissenschaft*, **63**, 153–156.

117 Kim, S.H., Kim, H.Y. and Kwak, H.S. (2007) Cholesterol removal from lard with crosslinked β-cyclodextrin. *Asian-Australas. J. Animal Sci.*, **20**, 1468–1472.

118 Kim, H.Y., Bae, H.Y. and Kwak, H.S. (2008) Development of cholesterol-reduced Blue cheese made by crosslinked β-cyclodextrin. *Milchwissenschaft*, **63**, 53–56.

119 Han, E.M., Kim, S.H., Ahn, J. and Kwak, H.S. (2008) Comparison of cholesterol reduced cream cheese manufactured using crosslinked β-cyclodextrin to regular cream cheese. *Asian-Australas. J. Animal Sci.*, **21**, 131–137.

120 Kwak, H.S., Jung, C.S., Shim, S.Y. and Ahn, J. (2002) Removal of cholesterol from Cheddar cheese by β-cyclodextrin. *J. Agric. Food Chem.*, **50**, 7293–7298.

121 Awad, A.C., Bennink, M.R. and Smith, D.M. (1997) Composition and functional properties of cholesterol reduced egg yolk. *Poultry Sci.*, **76**, 649–653.

122 Mine, Y. and Bergougnoux, M. (1998) Adsorption properties of cholesterol-reduced egg yolk low-density lipoprotein at oil-in-water interfaces. *J. Agric. Food Chem.*, **46**, 2153–2158.

123 Smith, D.M., Awad, A.C., Bennink, M.R. and Gill, J.L. (1995) Cholesterol reduction in liquid egg yolk using β-cyclodextrin. *J. Food Sci.*, **60**, 691–694.

124 Wood, W.E. (2001) Improved aroma barrier properties in food packaging with cyclodextrins. TAPPI – Polymers, Laminations and Coatings Conference, pp. 367–377.

125 Shim, S.Y., Ahn, J. and Kwak, H.S. (2003) Functional properties of cholesterol removed whipping cream treated by β-cyclodextrin. *J. Dairy Sci.*, **86**, 2767–2772.

126 Hara, H. and Hashimoto, H. (2002). Japan Kokai, JP 2002029901.

127 Cal, K. and Centkowska, K. (2008) Use of cyclodextrins in topical formulations: practical aspects. *Eur. J. Pharm. Biopharm.*, **68**, 467–478.

128 Prasad, N., Strauss, D., and Reichart, G. (1999) Cyclodextrins inclusion for food, cosmetics and pharmaceuticals, European Patent 1,084,625.

129 Almenar, E., Auras, R., Rubino, M. and Harte, B. (2007) A new technique to prevent the main post harvest diseases in berries during storage: inclusion complexes β-cyclodextrin-hexanal. *Int. J. Food Microbiol.*, **118**, 164–172.

130 Masson, M., Loftsson, T., Masson, G. and Stefansson, E. (1999) Cyclodextrins as permeation enhancers: some theoretical evaluations and in vitro testing. *J. Control. Release*, **59**, 107–118.
131 Matsuda, H. and Arima, H. (1999) Cyclodextrins in transdermal and rectal delivery. *Adv. Drug Deliv. Rev.*, **36**, 81–99.
132 Challa, R., Ahuja, A., Ali, J. and Khar, R.K. (2005) Cyclodextrins in drug delivery: an updated review. *AAPS PharmSciTech*, **6**, 329–357.
133 Tatsuya, S. (1999) Stabilisation of fragrance in bathing preparations, Japanese Patent 11,209,787.
134 Duchene, D., Wouessijewe, D. and Poelman, M.C. (1991) in *New Trends in Cyclodextrins and Derivatives* (ed. D. Duchene), Editions de Sante, Paris, pp. 448–481.
135 Shimada, K., Ohe, Y., Ohguni, T. *et al.* (1991) Emulsifying properties of a-cyclodextrins, b-cyclodextrins and c-cyclodextrins. *J. Jpn. Soc. Food Sci. Technol.*, **38**, 16–20.
136 Shimada, K., Kawano, K., Ishii, J. and Nakamura, T. (1992) Structure of inclusion complexes of cyclodextrins with triglyceride at vegetable oil–water interface. *J. Food Sci.*, **57**, 655–656.
137 Duchene, D., Bochot, A., Yu, S.C. *et al.* (2003) Cyclodextrins and emulsions. *Int. J. Pharm.*, **266**, 85–90.
138 Prabagar, B., Yoo, B.K., Woo, J.S. *et al.* (2007) Enhanced bioavailability of poorly water-soluble clotrimazole by inclusion with beta-cyclodextrin. *Arch. Pharm. Res.*, **30**, 249–254.
139 Doliwa, A., Santoyo, S. and Ygartua, P. (2001) Influence of piroxicam: hydroxypropyl-*b*-cyclodextrin complexation on the in vitro permeation and skin retention of piroxicam. *Skin Pharmacol. Appl. Skin Physiol.*, **14**, 97–107.
140 Harada, A. (1996) Preparation and structures of supramolecules between cyclodextrins and polymers. *Coord. Chem. Rev.*, **148**, 115–133.
141 Uekama, K., Ikeda, Y., Hirayama, F. *et al.* (1980) Inclusion complexation of *p*-hydroxybenzoic acid esters with a and b-cyclodextrins: dissolution behaviour and antimicrobial activities. *Yakugaku Zasshi*, **100**, 994–1003.
142 Chan, L.W., Kurup, T.R.R., Muthaiah, A. and Thenmozhiyal, J.C. (2000) Interaction of *p*-hydroxybenzoic esters with b-cyclodextrin. *Int. J. Pharm.*, **195**, 71–79.
143 Lehner, S.J., Muller, B.W. and Seydel, J.K. (1993) Interactions between *p*-hydroxybenzoic acid-esters and hydroxypropyl b-cyclodextrin and their antimicrobial effect against *Candida albicans*. *Int. J. Pharm.*, **93**, 201–208.

144 Lehner, S.J., Muller, B.W. and Seydel, J.K. (1994) Effect of hydroxypropyl-*b*-cyclodextrin on the antimicrobial action of preservatives. *J. Pharm. Pharmacol.*, **46**, 186–191.
145 Tanaka, M., Iwata, Y., Kouzuki, Y. et al. (1995) Effect of 2-hydroxypropyl-b-cyclodextrin on percutaneous absorption of methyl paraben. *J. Pharm. Pharmacol.*, **47**, 897–900.
146 Loftsson, T., Stefansdottir, O., Frioriksdottir, H. and Guomundsson, O. (1992) Interactions between preservatives and 2-hydroxypropyl-*b*-cyclodextrin. *Drug Dev. Ind. Pharm.*, **18**, 1477–1484.
147 Foley, P.R., Kaiser, C.E., Sadler, J.D., Burckhardt, E.E., and Liu, Z. (2000) Detergent composition with cyclodextrin perfume complexes to mask malodours, PCT Int Appl WO 01 23,516.
148 Angell, W.F. and France, PA. (2001) Detergent composition having granular cyclodextrin, PCT Int Appl WO 01 18,163.
149 Loftsson, T., Leeves, N., Bjornsdottir, B. et al. (1999) Effect of cyclodextrins and polymers on triclosan availability and substantivity in toothpastes in vivo. *J. Pharm. Sci.*, **88**, 1254–1258.
150 Scalia, S., Villani, S. and Casolari, A. (1999) Inclusion complexation of the sunscreen agent 2-ethylhexyl-*p*-dimethylaminobenzoate with hydroxypropyl-beta-cyclodextrin: effect on photostability. *J. Pharm. Pharmacol.*, **51**, 1367–1374.
151 Szetjli, J. (1998) Introduction and general overview of cyclodextrin chemistry. *Chem. Rev.*, **98**, 1743–1753.
152 Szejtli, J. (1985) Cyclodextrins in pesticides. *Starch/Staerke*, **37**, 382–386.
153 Szente, L. and Szejtli, J. (1996) Cyclodextrin, in *Comprehensive Supramolecular Chemistry*, vol. 3, Elsevier Science Ltd, New York, pp. 483–502.
154 Szejtli, J. and Osa, T. (1996) *Chemistry, Physical and Biological Properties of Cyclodextrins*, Elsevier Science, Oxford, pp. 503–514.
155 Martin del Valle, E.M. (2004) Cyclodextrins and their uses: a review. *Process Biochem.*, **39**, 1033–1046.
156 Saikosin, R., Limpaseni, T. and Pongsawasdi, P. (2002) Formation of inclusion complexes between cyclodextrins and carbaryl and characterization of the complexes. *J. Incl. Phenom. Macrocycl. Chem.*, **44**, 191–196.
157 Dupuy, N., Marquis, S., Vanhove, G. et al. (2004) Characterization of aqueous and solid inclusion complexes of diuron and isoproturon with β-cyclodextrin. *Appl. Spectrosc.*, **58**, 711–718.
158 Pérez-Martínez, J.I., Morillo, E. and Gines, J.M. (1999) β-CD effect on 2,4-D soil adsorption. *Chemosphere*, **39**, 2047–2056.

159 Morillo, E., Perez-Martínez, J.I. and Gines, J.M. (2001) Leaching of 2,4-D from a soil in the presence of β-cyclodextrin: laboratory columns experiments. *Chemosphere*, **44**, 1065–1069.
160 Villaverde, J., Perez-Martínez, J.I., Maqueda, C. et al. (2005) Inclusion complexes of α- and γ-cyclodextrins and the herbicide norflurazon: I. Preparation and characterisation. II. Enhanced solubilisation and removal from soils. *Chemosphere*, **60**, 656–664.
161 Dailey, O.D., Dowler, C.C. and Glaze, N.C. (1990) in *Pesticide Formulations and Application Systems* (eds L.E. Bode, J.L. Hazen and D.G. Chasin), American Society for Testing and Materials, Philadelphia, PA, pp. 105–116.
162 Dailey, O.D. (1991) in *Biotechnology of Amylodextrin Oligosaccharides*, ACS Symposium Series 458 (ed. R.B. Friedman), American Chemical Society, Washington, DC.
163 Bergamasco, R.C., Zanin, G.M. and De Moraes, F.F. (2005) Sulfluramid volatility reduction by β-cyclodextrin. *J. Agric. Food Chem.*, **53**, 1139–1143.
164 Szente, L. and Szejtli, J. (1981) Cyclodextrin complex of a volatile insecticide (DDVP). *Acta Chim. Acad. Sci. Hung.*, **107**, 195–202.
165 Wu, Y., Meure, S. and Solomon, D. (2008) Self-healing polymeric materials: a review of recent developments. *Progr. Polym. Sci.*, **33**, 479–522.
166 Wool, R.P. (2008) Self-healing materials: a review. *Soft Matter*, **4**, 400–418.
167 Syrett, J., Becer, C. and Haddleton, D. (2010) Self-healing and self-mendable polymers. *Polym. Chem.*, **1**, 978–987.
168 Blaiszik, B.J., Kramer, S.L.B., Olugebefola, S.C. et al. (2010) Self-healing polymers and composites. *Annu. Rev. Mater. Res.*, **40**, 179–211.
169 Ghosh, S. (2009) *Self-healing Materials: Fundamentals, Design Strategies, and Applications*, Wiley-VCH, Weinheim.
170 Zandi-zand, R., Ershad-langroudi, A. and Rahimi, A. (2005) Silica based organic–inorganic hybrid nanocomposite coatings for corrosion protection. *Progr. Org. Coat.*, **53**, 286–291.
171 Khramov, A.N., Voevodin, N.N., Balbyshev, V.N. and Donley, M.S. (2004) Hybrid organo-ceramic corrosion protection coatings with encapsulated organic corrosion inhibitors. *Thin Solid Films*, **447–448**, 549–557.
172 Semsarzadeh, M.A. and Amiri, S. (2012) Preparation and characterization of inclusion complexes of poly(dimethylsiloxane)s with gamma-cyclodextrin without sonic energy. *Silicon*, **4**, 151–156.

173 Semsarzadeh, M.A. and Amiri, S. (2013) Preparation and properties of polyrotaxane from α-cyclodextrin and poly (ethylene glycol) with poly (vinyl alcohol). *Bull. Mater. Sci.*, **36**, 989–996.

174 Semsarzadeh, M.A. and Amiri, S. (2013) Polyrotaxane macroinitiator in atom transfer radical polymerization of styrene and vinyl acetate: synthesis and characterization of PVAc-*b*-PSt-*b*-(PDMS/cyclodextrin)-*b*-PSt-*b*-Pvac pentablock copolymers. *J. Incl. Phenom. Macrocycl. Chem.*, **77**, 489–499.

175 Amiri, S. and Rahimi, A. (2016) Hybrid nanocomposite coating by sol–gel method: a review. *Iranian Polym. J.*, **25**, 559–577.

176 Zheludkevich, M.L., Shchukin, D.G., Yasakau, K.A. et al. (2007) Anti-corrosion coatings with self-healing effect based on nanocontainers impregnated with corrosion inhibitor. *Chem. Mater.*, **19**, 402–411.

177 Khramov, A.N., Voevodin, N.N., Balbyshev, V.N. and Mantzc, R. (2005) Sol–gel-derived corrosion-protective coatings with controllable release of incorporated organic corrosion inhibitors. *Thin Solid Films*, **483**, 191–196.

178 Amiri, S. and Rahimi, A. (2014) Preparation of supramolecular corrosion inhibitor nanocontainers for self-protective hybrid nanocomposite coatings. *J. Polym. Res.*, **21**, 556. doi: 10.1007/s10965-014-0566-5

179 Amiri, S. and Rahimi, A. (2014) Self-healing hybrid nanocomposite coatings containing encapsulated organic corrosion inhibitors nanocontainers. *J. Polym. Res.*, **22**, 624. doi: 10.1007/s10965-014-0624-z

180 Amiri, S. and Rahimi, A. (2015) Synthesis and characterization of supramolecular corrosion inhibitor nanocontainers for anticorrosion hybrid nanocomposite coatings. *J. Polym. Res.*, **22**, 66. doi: 10.1007/s10965-015-0699-1

181 Amiri, S. and Rahimi, A. (2015) Anti-corrosion hybrid nanocomposite coatings with encapsulated organic corrosion inhibitors. *J. Coat. Technol. Res.*, **12**, 587–593.

182 Amiri, S. and Rahimi, A. (2016) Self-healing anticorrosion coating containing 2-mercaptobenzothiazole and 2-mercaptobenzimidazole nanocapsules. *J. Polym. Res.*, **23**, 83. doi: 10.1007/s11998-014-9652-1

183 Mohanraj, V.J. and Chen, Y. (2006) Nanoparticles-a review. *Trop. J. Pharm. Res.*, **5**, 561–573.

184 Wang, D. and Bierwagen, G.P. (2009) Sol–gel coatings on metals for corrosion protection. *Progr. Org. Coat.*, **64**, 327–338.

Index

a

Agricultural   27, 213, 284, 285
Amphiphilic block copolymers   80–83, 92, 98
Anthocyanidins   207, 208, 211
Atom transfer radical polymerization   77, 78, 124, 239, 246

b

Block copolymers   25, 55, 75–84, 89, 97, 99, 124, 239–241, 246, 247, 249–265

c

Catechins   207, 211
Catenanes   5, 29, 49, 50, 52, 53
Centrifugal extrusion   213, 215
Coacervation   155, 213, 217
Coprecipitation   14, 17, 18, 135
Corrosion inhibitors   143, 149, 150, 159, 160, 166, 167–172, 178, 184, 285–287
Cosmetics   2, 3, 22, 28, 33, 48, 269, 296
Cyclohexaglucan   3
Cyclomaltohexaose   3
Cyclomaltooctaose   3
Cyclomaltose   1

d

Dealloying corrosion   147, 148
Drug delivery   2, 9, 23, 25, 38, 48, 65, 75, 80, 84–92, 96, 103, 109, 111, 113, 121, 122, 128–140, 216, 218, 219, 222, 249, 269, 273–279
Dry mixing   12, 15

e

Electroresponsive   100–102
Encapsulation   3, 21–24, 28, 29, 33, 82, 84, 99, 100, 108, 155, 167, 168, 171, 178, 182, 187, 211–227, 233–238, 271–275, 278, 279, 281–286, 295
Erosion corrosion   143, 149
Extrusion   18, 86, 213, 215, 282, 295
Extrusion–spheronization technique   215

f

Flavones   202, 206–208
Flavanoids   202, 206
Fluid bed coating   28, 213, 214
Freeze-drying   221, 282

g

Galvanic corrosion 144, 145
Gene delivery 26, 33, 38, 76, 92, 103, 124, 132, 219, 269
Glucopyranose 1, 3, 4, 41, 170, 222, 239

h

Host–guest complex 6, 8, 29, 47, 123
Hydrogen bonding 7, 8, 11, 17, 25, 29, 42–45, 49–52, 55, 62, 64, 76, 87, 90, 97, 107, 115, 118, 120, 152–154, 167, 170, 172, 182, 218, 223, 270, 278

i

In situ polymerization 60, 213, 220, 221
Interfacial polymerization 155, 213, 220–222
Intergranular corrosion 143, 147
Isoflavonoids 208, 211

l

Light-sensitive 21, 96, 97
Lignans 202, 203, 208
Liposomal drug delivery 276

m

Macroinitiator 239, 250–264
Micelles 47, 52, 80–84, 86, 88, 96–100, 108, 109, 121, 218, 219, 241, 257, 278
Microencapsulation 28, 155, 157, 192, 196, 218, 220, 221, 233–238, 295
Microsphere 180, 214, 217, 221, 273, 276, 277, 280
Molecular tubes 62–65

n

Nanocapsules 166, 168, 172, 180, 185, 186–189, 273, 279
Nanoparticles 80, 113, 121–123, 166–169, 180, 216, 217, 239, 273, 279, 286
Nanosponges 279, 280
Nasal drug delivery 275
Nitroxide-mediated polymerization 77, 78

o

Ocular drug delivery 219, 273, 275
Oligosaccharides 1, 49, 76, 168, 170, 171, 222, 239
Oral drug delivery 274, 277
Osmotic pump 273, 277

p

Paste complexation 12, 14, 15, 18
Peptide and protein delivery 273, 277
Pharmaceuticals 23, 28, 35, 90, 218, 296
Photoresponsive 97–100, 109, 110, 113–117, 121, 122
pH-responsive 86, 90, 92, 93, 96, 100, 104–106
Physicochemical methods 213, 216
Pitting corrosion 145, 146, 184
Poly(ethylene glycol) 56, 60, 64, 95, 100, 284
Polyethylene oxide 59, 95
Polyisobutylene 59
Poly(propyleneoxide) 59
Polypseudorotaxanes 54, 57, 58, 60, 62, 64, 73, 74, 105, 110, 116

Polyrotaxane 5, 26, 50–62, 64, 73, 74, 105, 108–111, 239, 240, 249, 250, 253, 256, 258, 259, 263–265

r

Rectal drug delivery 274
Reverse addition–fragmentation chain transfer polymerization 77

s

Schardinger dextrins 1
Self-assembling 54
Self-healing coatings 151, 152, 180
Slurry complex formation 14, 18
Smart polymers 75, 81, 82, 103, 120
Sol–gel 89, 113, 116, 160–169, 172, 179, 180, 254, 286
Spray-cooling/chilling 216
Spray drying 19, 28, 213, 214, 227, 234, 235, 238
Stilbenes 202, 203, 208
Stimuli-responsive 20, 75, 76, 82, 83, 86, 87, 89, 90, 96, 103, 104, 114, 120, 121, 124, 154

Stress corrosion cracking 147, 148, 149
Supramolecular 3, 5, 24, 25, 27, 29, 33, 36–38, 41–73, 75, 76, 82–90, 104, 107, 113, 119, 120, 153, 167, 182, 218, 239–241, 256, 257
Supercritical fluids 213, 215

t

Tannins 202, 208
Targeting systems 278
Toxicity 1–6, 24, 28, 205, 209, 272, 276, 279
Transdermal drug delivery 275

u

Ultrasonication 217, 242
Uniform corrosion 143, 144

v

Vesicles 52, 80, 83–86, 91–96, 99, 100, 156, 219